Sustainable Development of Coal Based Energy: Technology, Environment, Humanities, Economy, and Education

Sustainable Development of Coal Based Energy: Technology, Environment, Humanities, Economy, and Education

Stratum Movement and Mining Subsidence

Collection Editors

Cun Zhang
Fangtian Wang
Shiqi Liu
Erhu Bai

MDPI
Basel • Beijing • Wuhan • Barcelona • Belgrade • Novi Sad • Cluj • Manchester

Collection Editors
Cun Zhang
School of Energy and Mining Engineering
China University of Mining and Technology (Beijing)
Beijing
China

Fangtian Wang
School of Mines
China University of Mining and Technology
Xuzhou
China

Shiqi Liu
School of Energy and Mining Engineering
China University of Mining & Technology (Beijing)
Beijing
China

Erhu Bai
School of Energy Science and Engineering
Henan Polytechnic University
Jiaozuo
China

Editorial Office
MDPI AG
Grosspeteranlage 5
4052 Basel, Switzerland

This is a reprint of the Topical Collection, published open access by the journal *Sustainability* (ISSN 2071-1050), freely accessible at: https://www.mdpi.com/journal/sustainability/topical_collections/mining_tech_envir_energy.

For citation purposes, cite each article independently as indicated on the article page online and as indicated below:

Lastname, A.A.; Lastname, B.B. Article Title. *Journal Name* **Year**, *Volume Number*, Page Range.

Volume I
ISBN 978-3-7258-6492-8 (Hbk)
ISBN 978-3-7258-6493-5 (PDF)
https://doi.org/10.3390/books978-3-7258-6493-5

Set
ISBN 978-3-7258-6828-5 (Hbk)
ISBN 978-3-7258-6829-2 (PDF)

Contents

About the Editors

Cun Zhang

Cun Zhang is a Professor affiliated with the School of Energy and Mining Engineering at China University of Mining and Technology, Beijing. His research interests primarily include mine pressure and strata control, mine water resource utilization, coal and gas co-mining, and the reuse of abandoned mines. He has presided over more than 10 national and provincial-level vertical projects. His notable achievements include being a recipient of the Excellent Doctoral Dissertation Award from the Chinese Society of Rock Mechanics and Engineering, being accepted to the Young Talent Promotion Project of the China Association for Science and Technology, and being elected as a Fellow of the International Association of Advanced Materials (IAAM Fellow). Furthermore, he has been continuously listed in the Stanford University's Top 2% of the World's Most-Cited Scientists since 2022. In terms of academic contributions, he has received one first prize award for provincial-level teaching and five first prize awards for provincial-level scientific and technological progress, among other awards. He has produced over 200 publications, including more than 50 JCR Q1 papers as first or corresponding author, and 8 ESI highly cited papers, and has published 4 monographs and holds more than 10 invention patents as the first inventor. In addition to this, he serves as an Associate Editor for the SCI *Journal of Geophysics and Engineering* and is an Editorial Board Member for *Rock Mechanics Bulletin* and *Sustainability*.

Fangtian Wang

Fangtian Wang is a Professor at the School of Mines, China University of Mining and Technology. He has led more than 10 national and provincial-level research projects. Among his key honors, he is a recipient of the Jiangsu Outstanding Youth Fund and has been recognized as a young academic leader in the national key discipline of mining engineering. He has also been awarded the Sun Yueqi Youth Science and Technology Award and was selected as a young and middle-aged academic leader under the Jiangsu Provincial "Qinglan Project". In addition, he serves as a member of the International Society for Mine Surveying and on the Editorial Board of the *Journal of Mining and Strata Control Engineering*. His research and teaching activities are centered on mining rock mechanics and the intelligent, green mining of coal-related resources. To date, he has published over 60 high-quality papers, 8 of which have been recognized as ESI Highly Cited Papers or selected for the F5000 program. He has authored 3 books, secured 31 invention patents, and registered 13 software copyrights. His academic contributions have earned him several prestigious awards, including three first prizes for scientific and technological progress at the provincial and ministerial level. Furthermore, under his mentorship, students have achieved top honors such as the Gold Award in the China International College Students' Innovation Competition and the Grand Prize in the National "Challenge Cup" Competition.

Shiqi Liu

Shiqi Liu, born in Hengshui, Hebei Province, is currently an Associate Professor and Doctoral Supervisor at the School of Energy and Mining, China University of Mining and Technology, Beijing. Dr. Liu earned his Bachelor's, Master's, and Ph.D. degrees in Mining Engineering from China University of Mining and Technology, Beijing, between 2006 and 2016, after which he began his teaching career at the university. His research focuses on mine water utilization and prevention, shaft damage control, and comprehensive utilization of abandoned mines. In the past three years, he has led over 20 research projects, published 3 industry standards, authored 2 monographs, and produced more than 40 academic papers. His achievements include 3 provincial/ministerial first-class and 7 second-class scientific awards, 11 authorized national invention patents, and 5 registered software copyrights. Dr. Liu serves as a Youth Editorial Board Member of *Coal Science and Technology* and Guest Associate Editor of *Sustainability* (SCI). He has been recognized with honors such as the "National Huang Danian-style Teacher Team", "Senior Expert of Shanxi Coal Society Thousand-Person Think Tank", and "Outstanding Young Teacher Teaching Award".

Erhu Bai

Erhu Bai, Ph.D., born in July 1989, is a Professor currently serving as Deputy Director of the Intelligent Mining Department at Henan Polytechnic University's School of Energy Science and Engineering, and holds the position of Youth Editorial Board Member for the *Journal of Mining and Safety Engineering*. His research focuses on green coal mining; "three-under" mining (extraction under structures, water bodies, and railways); and mining damage and protection. He has led two National Natural Science Foundation projects, one sub-project of an NSFC joint key program, one national major project sub-task, and six provincial/ministerial-level projects, while participating in four NSFC joint key projects and four general projects. His honors include 16 provincial/ministerial scientific awards (with 5 government awards, 1 ranked first), the National Coal Industry Teaching Achievement Special Prize, and Henan Higher Education Teaching Achievement First Prize. He has published nearly 30 SCI/EI papers as first/corresponding author, as well as 6 monographs and textbooks, with 1 textbook winning the Henan Textbook Construction First Prize and National Coal Industry Outstanding Textbook Special Prize. His course "Mining Damage and Environmental Protection" is recognized as a National First-Class Undergraduate Course. He holds over 10 invention patents, a Silver Medal at the Geneva International Exhibition of Inventions, and has contributed to 1 national energy industry standard. Additional accolades include Zhongyuan Young Talents, Henan Higher Education Institutions Innovation Talent, and Green Mining Youth Science and Technology Award.

Preface

Large-scale coal mining-induced overburden movement and fracture lead to surface subsidence and a series of geological hazards in mining areas. These issues not only damage surface structures and impact residential living environments but also cause groundwater depletion, grassland degradation, vegetation deterioration, and slope instability, severely disrupting the surface ecosystem. This has become a critical challenge for green mine construction and the sustainable development of the ecological environment in mining regions.

This reprint focuses on the theme of "Strata movement and mining Subsidence", compiling 11 representative innovative research that systematically explore the laws of overburden failure and surface subsidence under various mining and geological conditions. The studies cover complex scenarios such as high-relief terrain, steeply inclined coal seams, complicated mining conditions, mountainous areas, and underwater extraction. They not only deepen the theoretical understanding of the extent of overburden failure and differential surface damage but also provide theoretical support and design methods for engineering practice, forming an innovation-driven cycle from field observation to theoretical exploration and practical application.

By integrating theoretical analysis, physical simulation, laboratory testing, and field monitoring, the research clarifies the differential behavior of overburden displacement and the degree of failure, analyzes the height of water-conducting fracture zones based on micro seismic data and overburden failure extent, reveals the zonal response mechanisms of surface subsidence under different geological conditions, and develops a new subsidence prediction model incorporating time effects. These advancements promote the development of the field toward handling more complex conditions and achieving more accurate predictions.

Aimed at researchers, engineers, and policymakers in mining engineering, mine surveying, mining subsidence, and ecological restoration—particularly professionals engaged in mine safety and surface hazard prevention—this Reprint offers cutting-edge insights. By presenting the latest theoretical understanding and innovative practices regarding overburden and surface subsidence mechanisms, it seeks to enhance global technical capabilities in mining damage mitigation and protection, contributing wisdom toward green energy development and high-quality ecological sustainability in mining areas.

Cun Zhang, Fangtian Wang, Shiqi Liu, and Erhu Bai
Collection Editors

sustainability

MDPI

Article

A Novel Analytical Model of Mining Subsidence Considering Time Effect Based on the Probability Integral Theory

Peng Li [1,2,3], Yingchun Li [2,4,*], Qiang Li [5], Zhiguo Cao [2,3] and Lujun Wang [2,3]

1 School of Energy and Mining Engineering, China University of Mining and Technology, Beijing 100083, China
2 State Key Laboratory of Water Resource Protection and Utilization in Coal Mining, Shenhua Shendong Coal Group Corporation Limited, Shenmu 719315, China
3 National Institute of Clean and Low Carbon Energy, Beijing 102211, China
4 State Key Laboratory of Coastal and Offshore Engineering, Dalian University of Technology, Dalian 116024, China
5 Shaanxi Yanchang Petroleum Group, Xi'an 710075, China
* Correspondence: yingchun_li@dlut.edu.cn

Abstract: Surface subsidence caused by underground coal mining has received wide attention due to its impact on the ecological environment. To obtain first-hand data on mining subsidence, we arranged line measurement stations in the mining area of the Weiqiang coalfield and implemented surface movement observation at its first mining working face and obtained the dynamic subsidence curve. The subsidence curves reflected the initial, active and declining stages of subsidence at each measurement point at a basic level, and were consistent with the general rule of surface subsidence. To accurately analyze the surface movement caused by coal mining, based on the probability integral method and considering the actual advancement progress of the working face, a new three-dimensional dynamic prediction model, namely the strip unit mining model, was proposed. The core assumption of this model was that the longwall mining is regarded as the result of superposition of many units of mining. Considering the influence of different mining times and periods after mining, the subsidence time function of two-time factors was introduced into the strip unit mining model, and the final formulation of the dynamic subsidence prediction of the three-dimensional surface is derived by considering the time factor. Based on the analysis of the measured data of the Weiqiang coalfield, the prediction parameters of surface movement were obtained, and the surface movement and deformation are predicted via the probability integral method. Finally, the measured surface subsidence was compared with the theoretically-predicted one. The good match showed that the parameters of the probability integral method, determined according to the measured data, meet the requirements of geological conditions and mining settings in the mining area, and the predicted subsidence curves agreed well with the field measurements, demonstrating the effectiveness of the newly-developed strip unit mining model.

Keywords: surface movement; prediction model; probability integral method; strip unit; time factor

1. Introduction

The rock is a very complex medium and when the useful minerals are extracted from the rock formations, the original stress equilibrium of the rock around the mining area is disrupted and the stress is redistributed to reach a new equilibrium. In this process, continuous movement, deformation and discontinuous damage (cracking, falling, etc.) of the rock layer and the surface are produced, which is called "mining subsidence" [1]. When the rock mass is affected by various mining methods, the mining subsidence is a very complicated process in terms of both time and space [2].

The surface subsidence caused by underground mining is a global problem. The surface subsidence caused by underground mining attracted the attention of people in the mining field and related fields in the mid-19th century, and gradually formed the current mining subsidence prediction theory [3]. In 1838, the Belgian engineer Gonot proposed the first theory of mining subsidence prediction, "perpendicular theory". After the improvement of Gonott and French engineer Tauritz, it was developed into "normal theory", which laid the foundation for the in-depth study of mining subsidence prediction theory [4]. During the period from 1923 to 1940, Schmitz et al. [5] successively proposed and developed the geometric mining subsidence prediction theory of mining influence distribution. In the 1940s, the Soviet scholar Avershingave established the subsidence basin profile equation and mathematical plasticity theory, and then Savustovich [6,7] proposed the elastic foundation beam theory for rock subsidence. In the 1950s, the Polish scholar Litvenishin [6,8] proposed the stochastic medium theory for mining subsidence prediction. In the early 1960s, Liu and Liao [9,10] solved the problem of the surface moving plane prediction based on the stochastic medium theory. From the 1980s to the end of the last century, the theoretical and practical research on mining subsidence prediction in China has undergone rapid development. Chinese scholars have successfully solved the problem of surface movement prediction space, the movement prediction problem in the overlying rock and the movement prediction problem of open-pit mining [3,5], and developed them into a prediction system for special surface terrain problems through the probability integral method. The integral method is a prediction system for special surface terrain problems. The probabilistic integration method has become one of the more mature and most widely used mining subsidence prediction methods in China [11]. Coal seam mining subsidence is a time-space four-dimensional problem; however, most scholars have treated it as a simplified plane or three-dimensional problem. Although many valuable results have been achieved [12,13], there are still some problems that have not yet been solved. Therefore, it is necessary to establish a surface movement model to satisfy the needs of engineering practice. Additionally, the establishment of a three-dimensional dynamic prediction model favors more straightforward and accurate prediction of the surface movement caused by coal seam mining.

Existing dynamic predictions of subsidence are usually implemented in the form of a sinking velocity integral [14], or by attaching a time function to the final state prediction method. Typical time functions include the Knothe time function [15], the modified Knothe time function [16], the normalized time function [17], the Sroka-Schober time function [18] and the normal distribution time function [19]. Dynamic prediction can better evaluate the deformation process and damage degree of surface structures during the mining process. Recently, extensive studies have also been conducted on the surface subsidence prediction. Wang and Deng [20] proposed a model based on the Richards function model to describe the biological growth process, and proposed a model based on the Richards function model to predict mining subsidence. The logistic function model posed by Zhang et al. [21] analyzed the probability-integral function model, where the downhole advance distance is expressed in terms of advance velocity (v) and advance time (t), and divides the time into three parts by considering two factors, namely whether full exploitation is reached and the start-up distance, to achieve the probability-integral method of dynamic prediction. Cheng et al. [22] proposed a stepwise prediction model based on the influence function method, which involved a dynamic iterative calculation process to derive the movement and deformation of each formation and modified the influence function method. Zhang and Cui [23] discussed the shortcomings of this time function and proposed corresponding improvement methods to address the existing problems, and then constructed a new segmented Knothe time function model with a wider range of application and higher prediction accuracy. Alternatively, subsidence prediction via numerical simulations using the finite element method, boundary element method, finite different method and distinct element method have been conducted [24–30]. However, these approaches usually cannot capture the time-dependent variation of the subsidence curves after progressive mining.

This paper establishes a three-dimensional dynamic prediction model of coal seam mining. The model is based on the mining subsidence prediction theory of probability integral method. The influence of different mining times and post-mining experience times of strip units were considered in the process of establishing the model. Firstly, the surface movement caused by mining is studied using the random medium theory. In addition, considering the actual implementation of mining in the working face, the establishment of a strip unit mining model along the entire working face will be closer to the actual situation. It can be regarded as the result of the superposition of many units of body mining, the subsidence time function is introduced into the strip unit mining model [15], and the strip unit mining model with time factor is established. At the same time, based on the observation on the surface movement of the first working face of the Weiqiang coalfield, the measured surface subsidence curve caused by mining and the theoretically predicted curve are compared in order to analyze the selected theoretically predicted model and parameters, and the fitting situation of the actual measurement. The fitting results show that the parameters of the probability integration method, determined according to the measured data, meet the requirements under the geological mining conditions of the mining area, and the predicted results are in positive agreement with the measured results.

2. Surface Movement Observation in Weiqiang Coalfield

The Weiqiang coalfield is located in the north of the Yuheng mining area and the south area of the Jurassic coalfield in northern Shaanxi. As a result of the loess covering the surface of the mining area being cut, by crisscrossing the loess gullies to form a complex surface shape, the mining damage in the area is much more serious than that under standard conditions. So far, due to special conditions, such as topography, the mining area has not carried out a relatively complete observation and study of the movement law of the mining surface rock layers. The movement law of the surface caused by mining and its strata movement parameters are the basis for studying and solving the "three-under" coal mining and environmental protection problems in the mining area, and are also the key technology to ensure the safety of surface construction facilities [10]. Therefore, carrying out special research on the observation of the law of surface movement and deformation caused by mining in these areas has important theoretical research significance and application value for the rational development and utilization of coal seams, the protection of surface structures in the mining area, the protection of the ecological environment and the sustainable and coordinated development of the economy of the mining area [31].

2.1. Overview of the First Working Face 1301

2.1.1. Location of Working Face

The 1301 working face is located in the north of the Heng Shan County planning reserve area (wellfield boundary), west of the Baomao Expressway approach road and 600 m east of Majialiang Village. The workings are located in the southern part of the main road, in the west wing of the first pan area, and are the first workings of the mine. It is the first working face of the mine. It is located south of the shaft boundary, west of the unmined area, and east of the coal pillar protection boundary.

The surface elevation of the 1301 working face is 1037.7~1140.3 m, the underground elevation is 789~794.92 m, the recovery area is 634,200 m^2, and the daily advance speed of the working face is about 8.8 m/d in the early stage of mining.

2.1.2. Overview of the Coal Seam

The average thickness of the coal seam at the 1301 working face is 3.2 m, and the dip angle of the coal seam is <1°; the depth of the coal seam is 291–325 m, and the 3-1 coal seam is long-flame coal with a simple structure and stable seam.

2.1.3. Overview of the Roof and Floor

The friable immediate roof of the coal seam is gray silty mudstone, 0.1~0.2 m thick, and the thickness is not constant. It is easy to delaminate and collapse with coal seam during mining, and the rock hardness is relatively soft. The immediate roof is dark gray medium-thick layered silty mudstone, 1.14~7.55 m thick, with argillaceous siltstone interbedded.

The main roof is siltstone, medium-grained sandstone, siltstone mudstone and fine-grained sandstone, 17.43–38.4 m thick.

The floor is siltstone and fine-grained sandstone, 1.62–4.77 m thick. The main floor is light grey medium-thick laminated siltstone with undulating laminations, with obvious contact with the underlying layers, and the layers and thickness are relatively stable.

2.1.4. Coal Mining Technique

The 1301 working face was retrieved using the longwall mining and the roof is managed using the full caving method. The length of the working face in the strike and dip are 2114 m and 300 m, respectively (Figure 1).

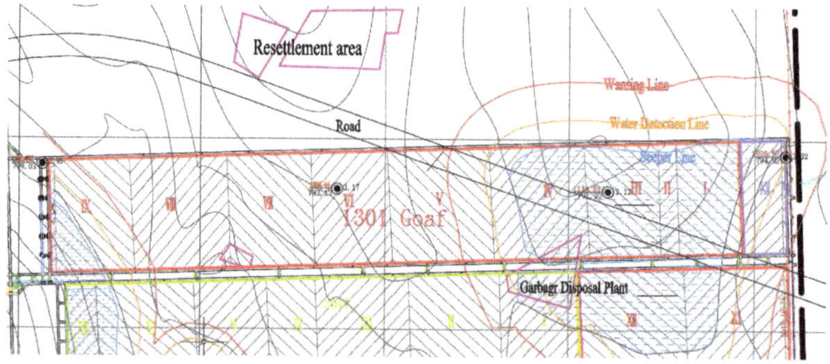

Figure 1. Diagram of the first working face 1301.

2.2. Layout of Mobile Surface Observation Stations

The comprehensive analysis of the observation area shows that the length of the mining influence radius is 161 m. Considering the randomness of the mining influence and the fact that the sliding range of the ditch slope increases, the mining influence radius is 175 m. In order to enable the observation line to effectively control the impact of mining on the half-basin, the observation station plans to arrange two observation lines:

1. Arranging a strike observation line (Z line) along the strike direction at the center of the first mining face. The strike observation line is perpendicular to the working face open-off cut, and the length of LZ should try to control the half-basin range.
2. An observation line (A line) is arranged along the inclination, and the observation line is perpendicular to the strike observation line (Z line) and intersects with the strike observation line. According to requirements, the dip observation line shall be able to control the full basin along the inclination.

Due to the influence of the terrain, the lengths of the two survey lines were adjusted accordingly when laying the survey lines: the Z line was 500 m long, and the closest distance to the open-off cut was 475 m; the A line was 800 m long, and the distance to the open-off cut was 800 m.

According to the buried depth of the working face of the Weiqiang coalfield, the distance between measuring points is 25 m. According to the actual shape of the surface, the final survey line points are arranged as follows: the A Line has a total of 33 survey points, and 29 survey points are set up in the early stage, numbered A1~A29, then add 4 measuring points, numbered JA1~JA4; the Z line has a total of 20 measuring points, Z1~Z20. The layout of the actual measuring points is shown in Figure 2.

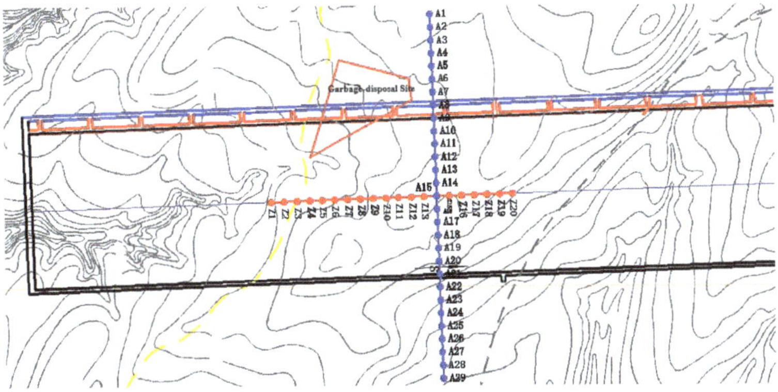

Figure 2. Layout of actual measuring points of surface movement observation station.

2.3. *Implementation of Surface Mobile Observation*

The internal collation method of the continuous measurement results of the observatory is the same as the conventional method. Finally, the plane coordinates and elevation of the control point of the observatory need to be obtained. In the process of connecting and measuring the working face of the Weiqiang coalfield, RTK measuring technology [32] was used to measure the coordinates and the elevation of control points.

To accurately determine the spatial position of the working survey point before the start of the surface movement, two comprehensive observations were independently carried out after the connection measurement and before the start of the surface movement. The content of the comprehensive observation includes the determination of the plane position and the elevation of each measuring point, as well as the distance between each measuring point, the appropriate increase in level measurement work between the first and last comprehensive observations, and the time interval of repeated level measurement, depending on the surface subsidence rate.

By November 2016, a total of 2 comprehensive observations and 27 daily observations had been carried out in the observation area of the first mining face of the Weiqiang coalfield, and a total of 1508 sets of observation data had been obtained.

2.4. *Calculation of Observed Results*

In order to ensure the correctness of the observation results, the field observation results should be checked again before the collation of the internal results, and then the calculation and adjustment of various correction data should be carried out.

The internal calculation of observation data is primarily used to calculate the elevation of each measuring point and the horizontal distance between two adjacent measuring points, followed by calculating the movement deformation value and subsidence speed value of each measuring point. Attached below is the internal calculation data of some measuring points, as shown in Table 1.

Table 1. Coordinates of some measuring points.

Date	27 January 2016	1 April 2016	4 April 2016	8 April 2016
z1	1114.947	1114.695	1114.542	1114.140
z2	1118.375	1118.276	1118.190	1117.863
z3	1120.593	1120.520	1120.482	1120.292
z4	1122.302	1122.256	1122.248	1122.169
z5	1122.115	1122.083	1122.083	1122.032
z6	1124.702	1124.687	1124.681	1124.650

After that, the subsidence calculation of each measuring point is carried out, and the subsidence of n points in m times of observation is:

$$w_n = H_n^0 - H_n^m \tag{1}$$

where w_n is the subsidence value of n point, mm; H_n^0 and H_n^m are the elevation of point n at the first and m observations, respectively, mm.

As shown in Figure 3, the trend subsidence curve illustrates that the subsidence value of the middle measurement point is relatively smaller than that of the two end measurement points after critical mining, primarily because the slope of the middle measurement point is small, the slip amount caused by the slope is small, and therefore the subsidence of the measuring point in the middle gentle area is slightly smaller.

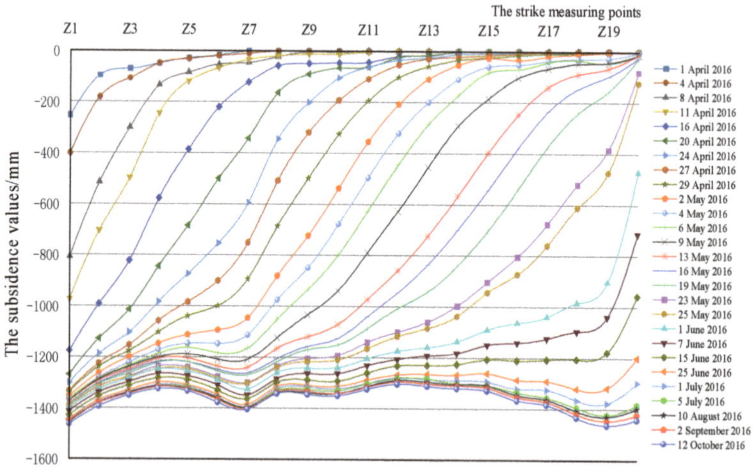

Figure 3. Dynamic subsidence curve of strike observation line.

It can be seen from the subsidence curve of the dip observation line in Figure 4 that the change of the surface subsidence basin is relatively gentle near the edge, while the change of its value increases near the goaf directly above, which indicates that the surface subsidence and slope degree of the goaf boundary is small. Affected by the larger buried depth, the subsidence curve has good continuity and a smooth transition, which is also a feature of surface subsidence in general deep-buried coal mines. From the comparative analysis of the relationship between the subsidence value and the time change in the above figure, it can be seen that the amount of surface subsidence was not obvious in the previous observations. For the strike observation line, the subsidence of each measuring point is mainly divided into the subsidence initiation stage, the subsidence active stage, and the subsidence decline stage. Among all the survey points so far, the last major subsidence occurred at the Z19 survey point, and the maximum subsidence value was 1.460 m. From the last few subsidence curves of the Z line, we can see the subsidence development trend of the goaf outside the observation area; that is, the maximum subsidence should be between 1.3 and 1.5 m.

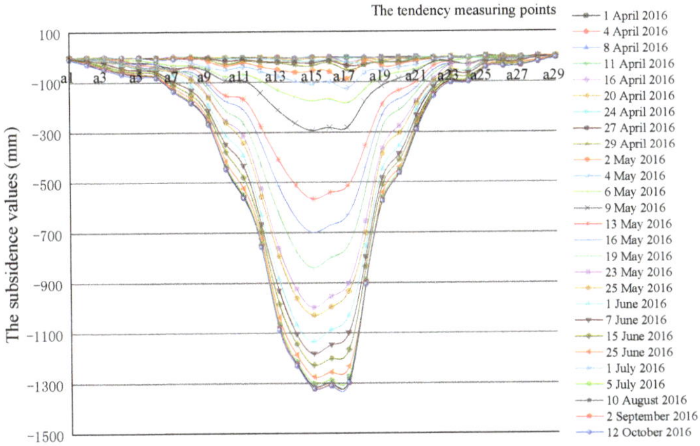

Figure 4. Dynamic subsidence curve of dip observation line.

The point Z13 on the strike observation line is observed, and three stages can be observed from the start of the movement of the surface to the stop of the movement: the initiation stage, the active stage and the decay stage. The maximum sinking rate at this point is reached only after the working surface has pushed through the point for a certain distance.

Table 2 shows that the sinking rate at this point is less than 10 mm/d during the start and decay phases and between 10 mm/d and 90 mm/d during the active phase, with a maximum sinking rate of 67.5 mm/d. It can also be concluded from the data in the table that the amount of sinking at this point during the active phase exceeds 80% of the total amount of sinking, thus indicating that the sinking of the ground surface primarily occurs during the sinking. This shows that the surface subsidence mainly occurs in the active period of subsidence. The change in the subsidence speed and the cumulative subsidence value of measuring point Z13 with time is shown in Figure 5.

Table 2. Comparison of subsidence rates and subsidence values for point Z13.

Date	Sinking Rate v (mm/d)	Subsidence Values ω (m)
1 April 2016–4 April 2016	−0.33	−1.00
4 April 2016–8 April 2016	−0.50	−2.00
8 April 2016–11 April 2016	−0.33	−4.00
11 April 2016–16 April 2016	−0.60	−5.00
16 April 2016–20 April 2016	−1.75	−8.00
20 April 2016–24 April 2016	−2.50	−15.00
24 April 2016–27 April 2016	−2.67	−25.00
27 April 2016–29 April 2016	−13.50	−33.00
29 April 2016–2 May 2016	−18.00	−60.00
2 May 2016–4 May 2016	−42.50	−114.00
4 May 2016–6 May 2016	−45.00	−199.00
6 May 2016–9 May 2016	−55.33	−289.00
9 May 2016–13 May 2016	−67.50	−455.00
13 May 2016–16 May 2016	−36.67	−725.00
16 May 2016–19 May 2016	−36.67	−835.00
19 May 2016–23 May 2016	−29.00	−945.00
23 May 2016–25 May 2016	−12.00	−1061.00
25 May 2016–1 June 2016	−10.43	−1085.00
1 June 2016–7 June 2016	−6.21	−1158.00
7 June 2016–15 June 2016	−4.66	−1195.25
15 June 2016–25 June 2016	−3.73	−1232.50
25 June 2016–1 July 2016	−2.48	−1269.75
1 July 2016–5 July 2016	−3.35	−1284.65
5 July 2016–10 August 2016	−0.05	−1298.06
10 August 2016–2 September 2016	−0.18	−1299.85
2 September 2016–12 October 2016	−0.07	−1304.02

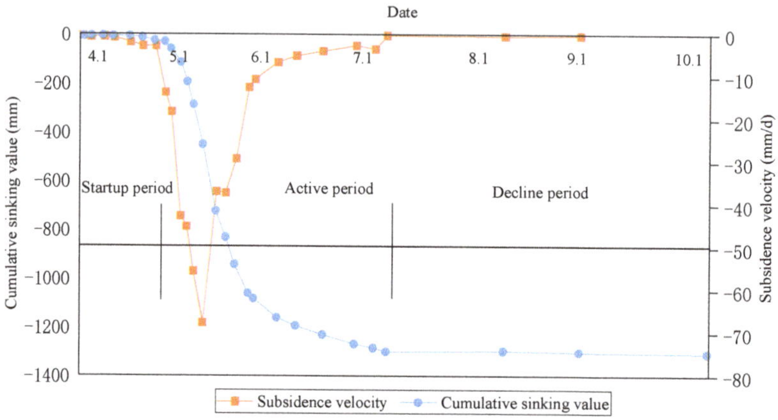

Figure 5. Subsidence velocity and cumulative sinking value of measuring point Z13.

From the overall analysis, the subsidence curves of the two observation lines reflect the initial stage, active stage, and recession stage of the subsidence of each measuring point on the surface, which is in line with the general law of surface subsidence.

3. Mining Subsidence Prediction Theory

For one or more working faces that are intended to be mined, the work of pre-calculating the surface movement and deformation affected by the mining, according to its geological coal mining conditions and the selected predicted functions and parameters, is called "surface movement mining subsidence prediction". The mining subsidence prediction is one of the core contents of mining subsidence and protection, and it is of great significance to the theoretical research and production practice of mining subsidence [4].

According to the different ways of establishing prediction methods, they can be divided into three types: empirical method, theoretical model method and influence function method. The most widely used methods in China are the probability integration method, typical curve method, profile function method, similar material simulation and numerical calculation method [33,34].

3.1. Probability Integral Method

The probability integral method regards the rock mass as a random medium. According to the random medium theory, the subsidence basin of the surface unit caused by the unit mining is normally distributed and is consistent with the distribution of the probability density [9]. Therefore, the subsidence profile equation caused by the whole mining can be expressed as the integral formula of the probability density function. I surface movement and deformation caused by the whole mining are calculated by the superposition principle [35,36].

The random particle medium model is shown in Figure 6a [37]. It takes the assumption that these particles are small spheres of the same size and uniform mass and are packed in squares of the same size and uniform arrangement. When the ball in square a1, in the first layer, is removed, one of the balls in squares b1 and b2 in the second layer will roll into this square due to gravity, and the probability that they roll in both are 1/2; if the ball in square b1 rolls into a1, then b1 will be occupied by the ball from square c1 or c2 in the third layer. Similarly, if the ball in square b2 rolls into a1, then b2 will be occupied by the ball rolling from the c2 or c3 square in the third layer. According to the multiplication and addition theorem in probability, the probability that the ball in the a1 grid is removed and the c1, c2 and c3 grids are emptied are 1/4, 1/2 and 1/4, respectively; the probability of emptying caused by the removal of a1 is written in the corresponding grid, and its probability distribution histogram can be formed, as shown in Figure 6b. If the size of the grid is very small, the histogram will approximate a smooth normal distribution probability density curve [38].

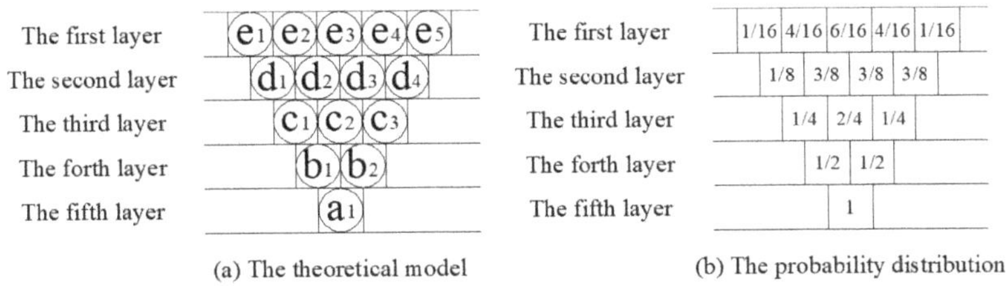

Figure 6. Random granular medium model.

Since the probability integration method is based on the random medium theory, the rock formation movement is regarded as a random process that obeys statistical laws. From a statistical point of view, the entire mining area is decomposed into the mining of infinite tiny unit mining. The impact of the surface is equal to the sum of the impact of each unit of mining on the rock formation and the surface. The subsidence profile equation caused by mining can be expressed as the integral formula of the probability density function. On the basis of the subsidence profile equation, other movement and deformation equations of the measuring points are derived. Therefore, it can be said that the subsidence profile equation is the basis of the probability integral method [35].

3.2. Knothe Time Function Model

From the statistical point of view, in order to infer the formation process of the unit basin, it can be considered that the unit mining is completed in an instant, but when the unit body is excavated, a final unit subsidence basin is not immediately formed on the surface; a certain period of time and process has to be experienced. In order to describe the formation process of the subsidence basin of the unit, it can be represented by a subsidence time function.

In 1952, in order to describe the complex time-space process of surface subsidence caused by underground mining with mathematical methods, Knothe made the following assumptions: the instantaneous subsidence speed $v(t)$ of the surface point is proportional to the difference between the instantaneous subsidence value $w(t)$ and the final subsidence value w_{max} of the point, the difference between the subsidence and the final subsidence value is proportional, and the following differential equation is established to describe the subsidence time course of the surface point [16,39]:

$$\frac{dw(t)}{dt} = c[w_{max} - w(t)] \tag{2}$$

where c is the time coefficient; w_{max} is the maximum subsidence value when the surface subsidence is stable, mm; $w(t)$ is the instantaneous subsidence value at time t, mm.

Substituting the initial condition $t = 0$ and $w(0) = 0$, the solution of Equation (2) can be obtained:

$$w(t) = w_{max}\left(1 - e^{-ct}\right) \tag{3}$$

Equation (3) is the final expression of the Knothe time function. The Knothe time function is the earliest "theoretical model" to study the dynamic subsidence law of the ground surface and is considered to be able to approximately describe the whole process of the subsidence of the ground point [16] (Figure 7).

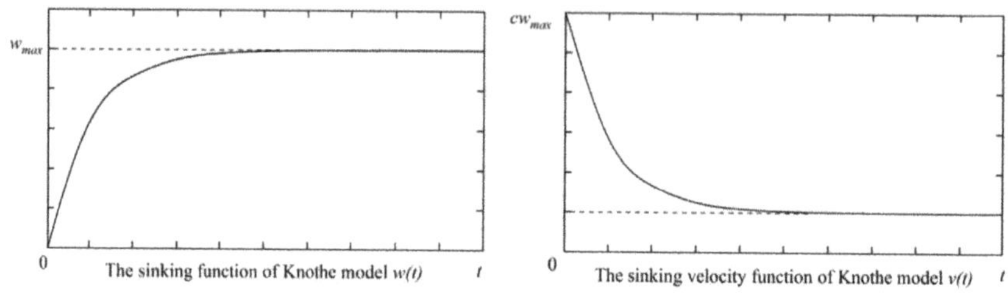

Figure 7. Knothe time function model.

4. Strip Unit Mining Model

4.1. Establishment of the Strip Unit Mining Model

Coal seam mining movement is always a dynamic process in space, so it is necessary to establish a three-dimensional dynamic prediction model, in order to be able to intuitively and accurately analyze the surface movement caused by coal seam mining. Based on this, this paper has made a bold attempt on the establishment of the three-dimensional dynamic prediction model of coal seam mining. The whole modeling idea mainly has the following points:

(1) At present, among the theories on the surface movement caused by mining, the stochastic medium theory is the most mature and widely used. It has been recognized and applied by many scholars for its rigorous reasoning and high accuracy conclusions. It is completely feasible to study the surface movement caused by mining with stochastic medium theory.

(2) There are two speeds in the advancing process of the working face: one is the speed of mining a knife along the working face; the other is the speed of the whole working face. These two speeds are different and related, but in terms of mining, the first speed is generally relatively fast, and it can be considered that during this process, the overlying rock layer and the surface have not been affected. Therefore, the first speed can be ignored and only the effect of the second velocity is considered.

(3) The mining of the working face is not carried out unit by unit, but along the working face. It can be imagined that if the strip unit mining model along the entire working face is established, it will be closer to the actual situation. The strip unit mining of the working face can be regarded as the result of the superposition of many unit body mining [40].

(4) After the mining of each working face strip unit, the surface does not immediately form the final strip subsidence basin, but must go through a certain amount of time and process. The subsidence time function is introduced into the strip unit mining model to establish a strip unit mining model with time factor [41].

(5) The mining time of each strip unit and the elapsed time after mining are inconsistent. These two times are both different and related. Considering the influence of the "double time" factors, it can better reflect the subsidence of the surface. It can realize the dynamic three-dimensional prediction of the surface movement caused by coal seam mining.

(6) On the time issue of the dynamic movement of the surface, the unit is generally "day (d)". Although the subsidence process is a continuous time function, if it is accurate to "hours" or "seconds", it will inevitably bring many mathematical problems to the model. Therefore, in the prediction of surface movement, the time step can be defined as day; that is, the strip unit mined per day is a unit subsidence factor

The text continues here.

4.2. Strip Unit Model without Time Factor

First, considering strip unit mining per unit thickness; assuming that the incline length of the working face is D, the average buried depth of the mining coal seam is H, and selecting projection point of intersection point of strike axis line and coal wall plane in surface water plane as coordinate origin point, it can be obtained:

$$w_{e'}(x,y) = \int_{-\frac{D}{2}}^{\frac{D}{2}} \frac{1}{r^2} exp\left[-\frac{\pi\left(x^2+(y-\xi)^2\right)}{r^2}\right] d\xi \tag{4}$$

Using MATLAB software (Matlab Version 2015a under the license of education version of Dalian University of Technology), the result of the above formula can be obtained as:

$$w_{e'}(x,y) = \frac{exp\left(-\frac{\pi x^2}{r^2}\right)}{2r} \cdot \left[erf\left(\frac{\sqrt{\pi}(2y+D)}{2r}\right) - erf\left(\frac{\sqrt{\pi}(2y-D)}{2r}\right)\right] \tag{5}$$

where: $erf(x) = \frac{2}{\sqrt{\pi}} \int_0^x e^{-t^2} dt$ is a built-in function in MATLAB.

D is the inclination length of the working face, m; r is the major influence radius, m.

Equation (4) is the expected subsidence basin expression for the mining of the strip unit per unit thickness of the working face, without considering the time factor. Since the thickness of the mining coal seam is generally small, usually within 10 m, this thickness is very small compared with the buried depth of the coal seam. Therefore, when establishing the model of the working face inclined to the full-height strip unit mining, the full-height strip unit mining of the working face is regarded as the superposition of the unit thickness strip unit mining; that is, the thickness of a mined coal seam is multiplied on the basis of the original expression, as shown in the following formula:

$$w_{e''}(x,y) = m \cdot \frac{exp\left(-\frac{\pi x^2}{r^2}\right)}{2r} \left[erf\left(\frac{\sqrt{\pi}(2y+D)}{2r}\right) - erf\left(\frac{\sqrt{\pi}(2y-D)}{2r}\right)\right] \tag{6}$$

where m is the thickness of the mined coal seam, m.

Considering the fragmentation of the rock and soil in the subsidence basin of the strip unit, the space of the goaf will not be completely filled; that is, the maximum subsidence of the surface will not exceed the mining height of the coal seam m. If the subsidence coefficient of the subsidence basin of the full-height strip unit be η_e, then Equation (6) is transformed into:

$$w_{e''}(x,y) = m \cdot \eta_e \cdot \frac{exp\left(-\frac{\pi x^2}{r^2}\right)}{2r} \left[erf\left(\frac{\sqrt{\pi}(2y+D)}{2r}\right) - erf\left(\frac{\sqrt{\pi}(2y-D)}{2r}\right)\right] \tag{7}$$

where η_e is the subsidence coefficient.

Equation (7) is the final model for full-height strip unit mining, without considering the time factor. However, it is worth noting that in the influence range of the goaf and the filling rock blocks in the goaf, the force of each spatial position is also different, which makes the bulking coefficient of each strip unit different after mining; that is, the subsidence coefficient η_e of each strip unit is different. However, the geological structure is relatively simple, and each rock layer is relatively horizontal mining conditions, and the subsidence coefficient η_e can be approximately considered equal.

4.3. *Strip Unit Model Considering Time Factor*

4.3.1. Consider the Time Factor of Full-Thickness Strip Unit Mining Experience, *T*

After each strip unit of the working face is excavated, the surface does not immediately form the final subsidence basin of the unit. Generally, it will go through several steps of "start subsidence → maximum subsidence speed → subsidence speed gradually decreases → subsidence stabilizes". This process can be described by the dynamic subsidence time function of the surface.

The classic Knothe subsidence time function model is used here. The characteristic of this model is that the subsidence speed reaches the maximum at the beginning of the subsidence, which is similar to the failure law of the overlying rock and soil layer of the shallow coal seam; that is, during the first subsidence, the surface had a sudden change in subsidence, and the stratum was completely fractured and collapsed within a short period of time. During the subsequent surface movement, the subsidence speed gradually decreased and, finally, stabilized. Therefore, choosing the Knothe time function can better explain the surface movement law of shallow coal seams.

Equation (3) suggests that the time function of the time factor T experienced by strip unit mining is:

$$\varphi(T) = 1 - exp(-c \cdot T) \tag{8}$$

where T is the mining time of the strip unit, d.

Combining Equations (6) and (7), we can obtain the subsidence prediction expression considering the time factor T of the strip unit mining experience:

$$w_{e''}(x,y,T) = m \cdot \eta_e \cdot [1 - exp(-cT)] \cdot \frac{exp\left(-\frac{\pi x^2}{r^2}\right)}{2r} \cdot \left[erf\left(\frac{\sqrt{\pi}(2y+D)}{2r}\right) - erf\left(\frac{\sqrt{\pi}(2y-D)}{2r}\right) \right] \tag{9}$$

4.3.2. Considering the Time Factor of Full-Thickness Strip Unit Advancement, t

Analyzing the actual mining process of the working face, it can be seen that the time experienced by each strip unit after mining is inconsistent with the advancing time of the working face. It is reasonable and necessary to take the mining speed into consideration in the strip unit mining model.

The x axis is parallel to the advancing direction of the working face, and the w axis is perpendicular to the working face. The advancing speed of the working face along the positive direction of the x axis is v (m/d), and the advancing distance over time t is S. Subsequently, the dynamic subsidence basin within the mining distance is obtained from the integral of Equation (9):

$$w_{e''}(x,y,T,t) = \frac{m \cdot \eta_e \cdot [1 - exp(-cT)]}{2r} \left[erf\left(\frac{\sqrt{\pi}(2y+D)}{2r}\right) - erf\left(\frac{\sqrt{\pi}(2y-D)}{2r}\right) \right] \cdot \int_0^s exp\left(-\frac{\pi(x-s)^2}{r^2}\right) ds \tag{10}$$

Since $s = vt$, $ds = vdt$, Equation (9) can be transformed into:

$$w_{e''}(x,y,T,t) = \frac{m \cdot \eta_e \cdot [1 - exp(-cT)]}{2r} \left[erf\left(\frac{\sqrt{\pi}(2y+D)}{2r}\right) - erf\left(\frac{\sqrt{\pi}(2y-D)}{2r}\right) \right] \cdot \int_0^t exp\left(-\frac{\pi(x-vt)^2}{r^2}\right) vdt \tag{11}$$

where v is the speed of advancing along the positive direction of the x-axis, m/d.

The time experienced by each strip unit after mining is different. The strip unit mined first has a relatively long time, and the strip unit mined later has a relatively short time. If one strip unit is used as a factor, it will inevitably bring a lot of inconvenience to the calculation. Here, the time "day (*d*)" is used as the unit: the strip mined every day is used as a calculation factor, and the time experienced by each factor is considered. The surface subsidence is regarded as the result of the superposition of various strip factors after different mining times. Then the dynamic subsidence model of strip unit factor mining with $k = (1, 2, 3, \cdots\cdots)$ on the k day is determined by:

$$w_{e''}(x,y,T,k) = \frac{m \cdot \eta_e \cdot [1 - exp(-cT)]}{2r} \left[erf\left(\frac{\sqrt{\pi}(2y+D)}{2r}\right) - erf\left(\frac{\sqrt{\pi}(2y-D)}{2r}\right) \right] \cdot \int_{k-1}^k exp\left(-\frac{\pi(x-vt)^2}{r^2}\right) vdt \tag{12}$$

4.3.3. The final Three-Dimensional Dynamic Subsidence Model of the Surface

For underground mining, the surface begins to subside only when the mining reaches a certain distance. The mining distance when the surface begins to subside is called the starting length, which is represented by d. Here, the corresponding time may be called the starting time, which is represented by T_d. Then, from the above assumptions, we have $T_d = \frac{d}{v}$. Therefore, the following improvements can be made to the Knothe subsidence time function:

$$\varphi(T) = \begin{cases} 0 & T \leq \frac{d}{v} \\ 1 - exp\left[-c\left(T - \frac{d}{v}\right)\right] & T > \frac{d}{v} \end{cases} \tag{13}$$

where d is the starting length, m.

In the strip mining within the starting length d, the surface does not subside; that is, the strip mining units within the mining distance range d are synchronized in time, so the strip unit mining of this distance segment can be regarded as a whole, that is:

$$\begin{cases} 0 & T \leq \frac{d}{v} \\ \frac{m \cdot \eta_e \cdot [1 - exp(-cT)]}{2r} \left[\begin{array}{c} erf\left(\frac{\sqrt{\pi}(2y+D)}{2r}\right) - \\ erf\left(\frac{\sqrt{\pi}(2y-D)}{2r}\right) \end{array} \right] \cdot \int_0^d exp\left(-\frac{\pi(x-s)^2}{r^2}\right) ds & T > \frac{d}{v} \end{cases} \tag{14}$$

It can be calculated by MATLAB:

$$\int_0^d exp\left(-\frac{\pi(x-s)^2}{r^2}\right) ds = \frac{H}{2} \left[\begin{array}{c} erf\left(\frac{\sqrt{\pi}x}{r}\right) - \\ erf\left(\frac{\sqrt{\pi}(x-d)}{r}\right) \end{array} \right] \tag{15}$$

Substitute Equation (14) into Equation (15), we obtain:

$$\begin{cases} 0 & T \leq \frac{d}{v} \\ \frac{m \cdot \eta_e \cdot [1 - exp(-cT)]}{4} \left[\begin{array}{c} erf\left(\frac{\sqrt{\pi}(2y+D)}{2r}\right) - \\ erf\left(\frac{\sqrt{\pi}(2y-D)}{2r}\right) \end{array} \right] \cdot \left[\begin{array}{c} erf\left(\frac{\sqrt{\pi}x}{r}\right) - \\ erf\left(\frac{\sqrt{\pi}(x-d)}{r}\right) \end{array} \right] & T > \frac{d}{v} \end{cases} \tag{16}$$

When the advancing distance of the working face exceeds the starting length, the ground surface begins to subside. The strip unit mining formula for each day thereafter is as follows:

$$\begin{aligned} w_{e''}(x, y, T, k) &= \frac{m \cdot \eta_e \cdot [1 - exp(-cT)]}{2r} \left[\begin{array}{c} erf\left(\frac{\sqrt{\pi}(2y+D)}{2r}\right) - \\ erf\left(\frac{\sqrt{\pi}(2y-D)}{2H}\right) \end{array} \right] \cdot \int_{k-1}^{k} exp\left(-\frac{\pi(x-d-vt)^2}{r^2}\right) vdt \\ &= \frac{m \cdot \eta_e \cdot [1 - exp(-cT)]}{4} \left[\begin{array}{c} erf\left(\frac{\sqrt{\pi}(2y+D)}{2r}\right) - \\ erf\left(\frac{\sqrt{\pi}(2y-D)}{2r}\right) \end{array} \right] \cdot \left[\begin{array}{c} erf\left(\frac{\sqrt{\pi}(vk-x+d)}{r}\right) - \\ erf\left(\frac{\sqrt{\pi}[v(k-1)-x+d]}{r}\right) \end{array} \right] \end{aligned} \tag{17}$$

Calculating based on the day when the surface moved, the surface subsidence on the k day after the start length d should be the superposition of the previous surface subsidence and the subsidence of the strip unit on each subsequent day, which can be determined by Equation (18):

$$w_{e''}(x,y,k) = \frac{m\cdot\eta_e\cdot[1-exp(-c\cdot k)]}{4}\begin{bmatrix} erf\left(\frac{\sqrt{\pi}(2y+D)}{2r}\right)- \\ erf\left(\frac{\sqrt{\pi}(2y-D)}{2r}\right)\end{bmatrix}\cdot\begin{bmatrix} erf\left(\frac{\sqrt{\pi}x}{r}\right)- \\ erf\left(\frac{\sqrt{\pi}(x-d)}{r}\right)\end{bmatrix} + \begin{bmatrix} erf\left(\frac{\sqrt{\pi}(2y+D)}{2r}\right)- \\ erf\left(\frac{\sqrt{\pi}(2y-D)}{2r}\right)\end{bmatrix}\cdot\begin{bmatrix} erf\left(\frac{\sqrt{\pi}(\mathrm{vi}-x+d)}{r}\right)- \\ erf\left(\frac{\sqrt{\pi}[v(i-1)-x+d]}{r}\right)\end{bmatrix}\sum_{i=0}^{k}\frac{m\cdot\eta_e\cdot[1-exp(-c\cdot(k-i))]}{4} \quad (18)$$

Equation (18) is the final expression of the three-dimensional surface dynamic subsidence prediction, and the expression of the dynamic predicted subsidence velocity can be obtained by partial differential of time variable of Equation (18).

4.4. Determination of Prediction Parameters of Surface Movement

When using the probability integral method to predict the movement and deformation of the mining surface, five predicted parameters should first be determined: subsidence coefficient η, tangent value of main influence angle $tan\beta$, horizontal movement coefficient b, inflection point offset d^* and propagation angle of extraction θ. Through the analysis of the measured data, these parameters can be obtained by calculation [42].

The most important step for obtaining parameters is to determine the maximum subsidence value. Under the condition of critical mining (ensuring that full subsidence is achieved in both strike and dip directions, it can be called critical mining), the maximum subsidence value of each survey line is obtained through the observation data of mobile observation station.

Under the condition of full mining of nearly horizontal coal seam, the maximum subsidence value, maximum horizontal displacement and maximum inclination value in strike and dip are approximately equal. In this paper, the same value is taken in strike and dip for calculation.

(1) Subsidence coefficient η

The subsidence coefficient is determined by the ratio of the maximum subsidence value on the surface to the average mining thickness under critical mining conditions:

$$\eta = \frac{w_{max}}{m \cos \alpha} \quad (19)$$

where α is the dip angle of the coal seam. The calculation results are shown in the Table 3.

Table 3. Calculation results of subsidence coefficient.

Results	The Maximum Subsidence Value (mm)	Mining Height (m)	Dip Angle of Coal Seam α (°)	Subsidence Coefficient η
A Line	1327	3.38	0.5	0.393

(2) The tangent value of main influence angle $tan\ \beta$

The angle between the boundary point connecting the main influence range and the mining boundary line and the horizontal line is called the tangent value of main influence angle [43], and the calculation formula is shown in Equation (20):

From $r = H/tan\beta$, $r = \omega_{max}/i_{max}$, we have:

$$tan\beta = \frac{H}{r} \quad (20)$$

where r is the main influence radius, m; H is the average buried depth of coal seam, mm; ω_{max} is the maximum subsidence value of the surface, mm; i_{max} is the maximum tilt value of the observation point, mm/m. The calculated results are shown in Table 4.

Table 4. Calculation results of tangent of main influence area angle ($tan\beta$).

Result/Unit	Maximum Subsidence (mm)	Maximum Inclination (mm/m)	Radius of Influence (m)	Average Mining Depth (m)	The Tangent of the Main Influence Angle
A Line	1327	15.851	83.276	316	3.795

(3) Propagation angle of extraction θ.

The propagation angle of extraction represents the offset degree of the surface subsidence curve to the downhill direction, under the condition of inclined coal seam mining. The value is the angle between the inflection point of the subsidence curve and the calculation boundary of the same side goaf (considering the offset distance of the inflection point) and the horizontal line on the downhill side. Its value is primarily related to the coal seam dip angle and lithology [43]. In general, the influence propagation angle of extraction is equal to the maximum subsidence angle. It is calculated by:

$$\theta = 90^\circ - c\alpha \tag{21}$$

where c is the coefficient related to lithology, α is the dip angle of the coal seam, $^\circ$.

The dip angle of the working face studied in this study is 0.5°, therefore, the propagation angle of extraction can be approximately 90°.

(4) Deviation of inflection point d^*

Under critical/supercritical mining conditions, the following points on the measured subsidence curve are used to determine the position of the inflection point: point with a subsidence value of $1/2w_{max}$; and point when i_{max} and u_{max} occurs. The distance from the position of the inflection point to the nearest coal wall is known as the deviation of inflection point d^* [44,45]. According to the data of observation A line, the position of the inflection point of 0.5 ω_{max} can be calculated, and the deviation of inflection point d^* = 70 m = 0.222 H can be obtained.

(5) Horizontal movement coefficient b

The horizontal movement coefficient is a coefficient that reflects the relationship between the maximum horizontal movement value and the maximum subsidence value [45]. Under critical/supercritical mining conditions, it is calculated by:

$$b = \frac{u_{max}}{w_{max}} \tag{22}$$

where u_{max} is the maximum value of the horizontal movement of the surface.

According to the observation data of the two survey lines, the horizontal movement factor can be calculated (Table 5).

Table 5. Calculation results of the horizontal movement coefficient.

Results	Maximum Subsidence (mm)	Maximum Horizontal Movement (mm)	Horizontal Movement Coefficient
A Line	1327	712	0.537

Several groups of surface movement parameters in different observation periods are selected below (Table 6). It is clear that when the goaf range is small, the fluctuation of each parameter is large, indicating that the change is unstable. With the expansion of the mining range, the parameters gradually become stable, indicating that the surface movement at this time has also stabilized. From this result, it can be inferred that with the continuous increase of the goaf, the deformation of the ground surface will not change much, and the range of surface subsidence behind the goaf will not change much. The last few measurements show that the subsidence has stabilized.

Table 6. Parameters of probability integral method in different observation periods of line A.

Results	Subsidence Coefficient, η	The Tangent of the Main Influence Angle, $tan\beta$	Deviation of Inflection Point (m)	Horizontal Movement Coefficient
2016/5/6	0.053	3.88	75	1.12
2016/5/19	0.249	3.3	74	0.41
2016/8/10	0.39	3.14	73	0.55
2016/10/12	0.393	3.13	72	0.54

5. Performance of the Probability Integral Theory

A large number of measured data have confirmed that the surface movement in mountainous areas has certain rules to follow. However, the surface deformation caused by mining subsidence has a positive relationship with the topography and landforms, and the regularity is very poor under general conditions. Figures 8 and 9 are the predicted parameters of the surface probability integral, based on the measured surface movement and deformation, comparing the measured surface subsidence curve caused by mining with the theoretical predicted curve, in order to analyze the selected theoretical prediction model and parameters and the measured fitting.

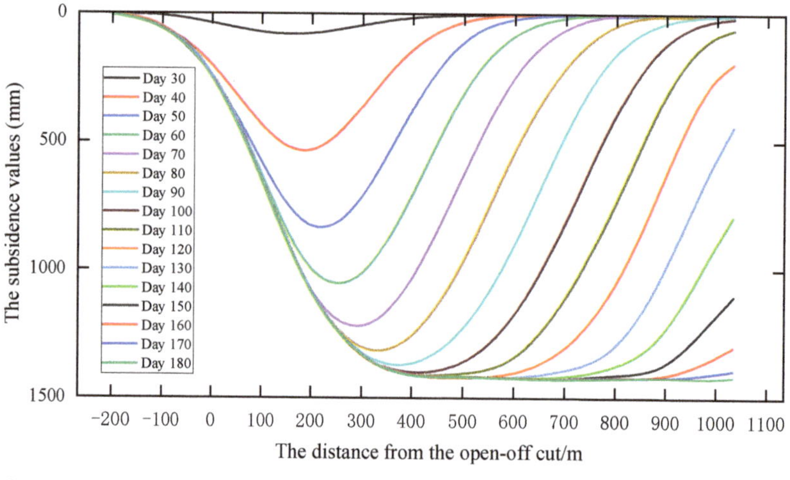

Figure 8. Calculation results of dynamic surface subsidence curve of strike observation line Z.

Since the entry measurement work started after 540 m of mining at the working face, in order to better reflect the law of surface movement in the early stage of mining, the Knothe subsidence time function model was used to numerically calculate the surface movement in the early stage of mining.

Figure 8 shows that the surface activity near the open-off cut is active at the initial stage of mining, and the active subsidence period of each measuring point changes with the advancement of the working face. The theoretical prediction of maximum surface subsidence reaches approximately 1.4 m, which is basically consistent with the measured value. The subsidence to the strike observation line gradually expands forward with the advancement of the working face. On the strike main section, the surface points subsidence sequentially, from near to far. When the working face advances to 400 m, the goaf will have a certain range. When the working face advances to 400 m, the subsidence of the surface point within a certain range of the goaf is stable. As the working face continues to advance, the surface points in front of the working face cycle the previous subsidence process in turn, which is consistent with the actual measurement.

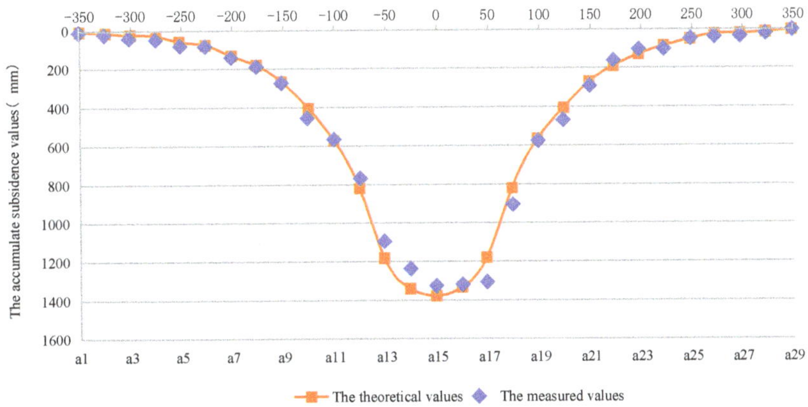

Figure 9. Comparison between measured surface subsidence curve and theoretical prediction curve of dip observation line A.

Figure 9 shows that after the critical mining is achieved, the measured values are greatly affected by the terrain, and the values of local points jump. But on the whole, the subsidence trend of the theoretical value and the measured value is basically consistent. The maximum subsidence value of the surface is basically consistent with the theoretical prediction value, reaching approximately 1.4 m. Therefore, the measured surface subsidence value of the dip observation line in the influence range of the working face has a good fit with the theoretically predicted curve.

6. Conclusions

This study conducted surface movement observations at the first mining face in the Weiqiang coalfield, obtained the dynamic subsidence curves, and proposed a three-dimensional dynamic prediction model to obtain the predicted deformation of surface movement, and compares the measured values of subsidence with the predicted values. The conclusions of this paper are as follows.

(1) According to the mining environment of the first working face of the Weiqiang coalfield, the design of surface movement observation was carried out. A total of 2 systematic observations and 27 daily observations were carried out in the observation area, and 1508 sets of reliable observation data were obtained. Through data analysis, it can be seen that the surface movement of each measuring point in the observation area has experienced the start-up stage, the active stage, and the decline stage. The duration of the three stages varies at each measuring point, and the last few observations show that the surface movement has basically stabilized.

(2) A new three-dimensional dynamic prediction model is proposed: the strip unit model. Considering the influence of dual time factors, the probability integration method is used to predict the deformation of the main section of the surface mobile basin during semi-infinite mining. The estimated parameters suitable for the probability integration method are obtained.

(3) The measured surface subsidence curve is fitted by the theoretical prediction formula. The fitting results show that the parameters of the probability integration method determined according to the measured data meet the requirements of the geological mining conditions in the mining area, and the predicted results are in positive agreement with the measured results.

Author Contributions: P.L.: Data curation, methodology, formal analysis, original draft. Y.L.: Supervision, Data curation, methodology, review and editing. Q.L.: Resources, methodology, review and editing. Z.C.: Methodology, review and editing. L.W.: Data curation, methodology, review and editing. All authors have read and agreed to the published version of the manuscript.

Funding: This research was funded by the Open Fund of State Key Laboratory of Water Resource Protection and Utilization in Coal Mining (Grant No. GJNY-18-73.5).

Institutional Review Board Statement: Not applicable.

Informed Consent Statement: Not applicable.

Data Availability Statement: Not applicable.

Acknowledgments: This work is supported by the Open Fund of State Key Laboratory of Water Resource Protection and Utilization in Coal Mining (Grant No. GJNY-18-73.5).

Conflicts of Interest: We declare that we have no conflict of interest associated with this work.

References

1. Liu, Z.; Liu, B.; Wang, R.; Zhang, P. Main research methods and application of mining subsidence in China. *West. Explor. Eng.* **2011**, *23*, 176–178.
2. Wang, J.; Liu, X.; Liu, X. Dynamic prediction model of mining subsidence. *J. Coal Sci. Eng.* **2015**, *40*, 516–521.
3. Wang, J. Study on some theoretical and technical problems of mining subsidence. *Mine Surv.* **2003**, *3*, 1–5.
4. Cui, X.; Deng, K. Review on theory and method of mining subsidence prediction. *Coal Sci.* **2017**, *45*, 160–169.
5. Cheng, J.; Zhao, G.; Sa, Z.; Zheng, W.; Wang, Y. Prediction model and engineering application of overlying strata movement and deformation caused by mining. *J. Min. Strat. Control. Eng.* **2020**, *2*, 20–29.
6. Tong, L.; Liu, S.; Qiu, Y.; Fang, L. Research status and progress of goaf under expressway at home and abroad. *J. Rock Mech. Eng.* **2004**, *7*, 1198–1202.
7. Berry, D.S.; Sales, T.W. An elastic treatment of ground movement due to mining—II. Transversely isotropic ground. *J. Mech. Phys. Solids* **1961**, *9*, 52–62. [CrossRef]
8. Wang, J. Study on Plastic Damage Structure Theory of Mining Subsidence and Grouting Filling Technology of Coal Gangue Gap. Ph.D. Thesis, General Institute of Coal Research, Taiyuan, China, 2001.
9. Liu, B.; Dai, H. Origin and research progress of probability integral method. *Coal Min.* **2016**, *21*, 1–3.
10. Wei, Z.; Xiong, W. Numerical calculation and analysis of mining subsidence prediction by probability integral method. *Surv. Mapp. Eng.* **2019**, *28*, 35–40.
11. Zhou, J.; Dong, J.; Zhu, Y.; Zou, Y. Numerical calculation and analysis of multi-layer mining subsidence prediction. *Sichuan J. Geol.* **2020**, *40*, 255–257.
12. Li, Y.; Cui, F.; Yang, W.; Yuchi, X. Dynamic migration law and its control of roof in fully mechanized top coal caving mining in extremely steep and thick coal seams. *J. Min. Strat. Control. Eng.* **2020**, *4*, 043538.
13. Cai, Y.; Li, X.; Deng, W.; Xiao, W.; Deng, W. Simulation of surface movement and deformation rules and detriment key parameters in high-strength mining. *J. Min. Strat. Control. Eng.* **2020**, *24*, 043511.
14. Peng, S.S. *Surface Subsidence Engineering*; Society for Mining, Metallurgy, and Exploration, Inc.: Littleton, CO, USA, 1992.
15. Cui, X.; Wang, J.; Liu, Y. Prediction of progressive surface subsidence above longwall coal mining using a time function. *Int. J. Rock Mech. Min. Sci.* **2001**, *38*, 1057–1063. [CrossRef]
16. Chang, Z.; Wang, J. Study on time function of surface subsidence—The improved Knothe time function. *Chin. J. Rock Mech. Eng.* **2003**, *22*, 1496–1499.
17. Kowalski, A. Surface subsidence and rate of its increments based on measurements and theory. *Arch. Min. Sci.* **2001**, *46*, 391–406.
18. Kwinta, A.; Hejmanowski, R.; Sorka, A. A time function analysis used for the Prediction of rock mass subsidence. In Proceedings of the International Symposium on Mining Science and Technology, Xuzhou, China, 16 October 1996; pp. 419–424.
19. Zhengshuai, W.; Kazhong, D. Richards model of surface dynamic subsidence prediction in mining area. *Rock Soil Mech.* **2011**, *32*, 1664–1668.
20. Gonzalez-Nicieza, C.; Alvarez-Fernandez, M.I.; Menendez-Diaz, A.; Alvarez-Vigil, A.E. The influence of time on subsidence in the Central Austrian Coalfield. *Bull. Eng. Geol. Environ.* **2007**, *66*, 319–329. [CrossRef]
21. Guangyi, Z.; Hongxia, S.; Liguo, W. Study of dynamic prediction function of surface movement and deformation. *Chin. J. Rock Mech. Eng.* **2011**, *30*, 1889–1895.
22. Cheng, J.; Liu, F.; Li, S. Model for the prediction of subsurface strata movement due to underground mining. *J. Geophys. Eng.* **2017**, *14*, 1608–1623. [CrossRef]
23. Zhang, B.; Cui, X. Optimization of segmented Knothe time function model for dynamic prediction of mining subsidence. *Rock Soil Mech.* **2017**, *38*, 541–548.
24. Mirsalari, S.E.; Fatehi Marji, M.; Gholamnejad, J.; Najafi, M. A boundary element/finite difference analysis of subsidence phenomenon due to underground structures. *J. Min. Environ.* **2017**, *8*, 237–253.

25. Abdollahi, M.S.; Najafi, M.; Yarahmadi Bafghi, A.R.; Fatehi Marji, M. A 3D numerical model to determine suitable reinforcement strategies for passing TBM through a fault zone, a case study: Safaroud water transmission tunnel, Iran. *Tunn. Undergr. Space Technol.* **2019**, *88*, 186–199. [CrossRef]
26. Sepehri, M.; Apel, D.B.; Hall, R.A. Prediction of mining-induced surface subsidence and ground movements at a Canadian diamond mine using an elastoplastic finite element model. *Int. J. Rock Mech. Min. Sci.* **2017**, *100*, 73–82. [CrossRef]
27. Alejano, L.R.; Ramírez-Oyanguren, P.; Taboada, J. FDM predictive methodology for subsidence due to flat and inclined coal seam mining. *Int. J. Rock Mech. Min. Sci.* **1999**, *36*, 475–491. [CrossRef]
28. Cheng, G.; Chen, C.; Li, L.; Zhu, W.; Yang, T.; Dai, F.; Ren, B. Numerical modelling of strata movement at footwall induced by underground mining. *Int. J. Rock Mech. Min. Sci.* **2018**, *108*, 142–156. [CrossRef]
29. Li, Y.; Du, X.; Ji, Y. Prediction of the transitional normal stress of rock joints under shear. *Int. J. Rock Mech. Min. Sci.* **2022**, *159*, 105203. [CrossRef]
30. Yu, X.; Wang, Z.; Yang, Y.; Mao, X. Movement rule of overburden in fully mechanized caving mining with thick depth and high mining height based on 3DEC. *J. Min. Strat. Control. Eng.* **2021**, 3, 013533.
31. Li, P.; Tan, Z.; Deng, K. Analysis of related factors of calculation parameters of surface movement probability integral method. *Coal Min.* **2011**, *16*, 14–18.
32. Ye, W.; Xu, L. Analysis of surface movement and deformation law due to repeated mining. *Met. Mine* **2020**, *8*, 188–194.
33. Deng, K.; Ma, W.; He, G. Study on stratification effect in mining subsidence. *J. Coal Sci. Eng.* **1995**, *4*, 380–384.
34. Zhu, Q.; Li, H.; Yang, X.; Shen, Y. Structural evolution characteristics of mining overburden and its influence on surface subsidence. *J. Coal Sci. Eng.* **2019**, *44*, 9–17.
35. Chen, Y.; Zhang, Z.; Zhang, Y. Prediction of ground subsidence based on probability integral method. *J. Liaoning Univ. Eng. Technol.* **2007**, *S2*, 143–145.
36. Díez, R.R.; Álvarez, J.T. Hypothesis of the multiple subsidence trough related to very steep and vertical coal seams and its prediction through profile functions. *Geotech. Geol. Eng.* **2000**, *18*, 289–311. [CrossRef]
37. Jiang, Y.; Misa, R.; Li, P.; Yuan, X.; Sroka, A.; Jiang, Y. A review of the development of mining subsidence theory. *Met. Mines* **2019**, *10*, 1–7.
38. Sun, C.; Bo, J.; Sun, Y. Research history and current situation of goaf subsidence. *J. Coll. Disaster Prev. Technol.* **2008**, *10*, 128–131.
39. Cui, X.; Miao, X.; Zhao, Y.; Jin, R. On the time function of surface movement process. *J. Coal Sci. Eng.* **1999**, *5*, 3–5.
40. Xu, J.; You, Q.; Zhu, W.; Li, X.; Lai, W. Theoretical study on controlling mining subsidence by strip filling. *J. Coal Sci. Eng.* **2007**, *2*, 119–122.
41. Cui, X.; Yang, S. Discussion on rheological model of mining subsidence. *China Min.* **1996**, *2*, 52–55.
42. Hao, Y.; Wu, L.; Chen, S. Statistical law and influencing factors of rock movement parameters. *Coal Mine Saf.* **2000**, *5*, 30–31.
43. Tan, Z.; Deng, K. Comprehensive analysis and application research on surface deformation prediction parameters of fully mechanized top coal caving face. *J. Rock Mech. Eng.* **2007**, *5*, 1041–1047.
44. Hao, Y.; Wu, L. Influence factors and formation mechanism of inflection point offset. *Zhongzhou Coal* **2000**, *1*, 6–7.
45. Xie, W.; Da, J.; Zhang, L.; Cheng, S.; Huang, S. Influence of geological and mining factors on deflection of inflection point. *Mine Press. Roof Manag.* **1998**, *1*, 3–5.

sustainability

MDPI

Article

On-Site Measurement on Surface Disturbance Law of Repeated Mining with High Relief Terrain

Weitao Yan [1,2], Junjie Chen [1,*], Wenfu Yang [2,3], Xiaosong Liu [2,3], Wenwen Wang [2,3] and Wenkai Zhang [2,3]

1 State Collaborative Innovation Center of Coal Work Safety and Clean-Efficiency Utilization, Henan Polytechnic University, Jiaozuo 454003, China; yanweitao@hpu.edu.cn

2 Key Laboratory of Monitoring and Protection of Natural Resources in Mining Cities, Ministry of Natural Resources, Jinzhong 030600, China; wcyymf@163.com (W.Y.); xunjieweifeng@163.com (X.L.); gouzai607@126.com (W.W.); zwk1619@163.com (W.Z.)

3 Coal Geological Geophysical Exploration Surveying and Mapping Institute of Shanxi Province, Jinzhong 030600, China

* Correspondence: chenjj@hpu.edu.cn; Tel.: +86-139-3910-0612

Abstract: Underground coal mining in high relief terrain often causes problems such as ground collapse and mountain sliding. This produces disasters such as mountain collapse and landslides, which seriously endangers the ecological environment and human settlements in the coal mining area and causes a series of public security events. To improve and solve the related problems, it is necessary to master the impact law of underground mining on ground surface. Therefore, taking the working face 1310 of Jincheng coal industry as an example, this paper adopts a new and improved RTK measurement technology for field measurement and analyzes the surface disturbance law by using field data. The results show that under the action of repeated mining in high relief terrain areas, there is a sudden subsidence in the flat bottom area of the surface moving basin, the strike influence scope is small, the dip influence scope in the downhill direction is larger, the dip horizontal movement curve loses its antisymmetry, and the dip subsidence curve loses its left-right symmetry. Through the analysis, it is found that the high reverse slope weakens the original disturbance and reduces the influence range of surface mining. On the contrary, repeated mining enhances the original disturbance, expands the influence range of surface mining, and flattens the movement deformation curve. Research results can provide a technical reference for the formulation and implementation of environmental remediation and treatment technologies in mining areas under similar conditions.

Keywords: high relief terrain; repeated mining; actual measured law; surface disturbance in mining area

1. Introduction

China is the world's largest coal producer, accounting for 37% of the world's total coal output. The exploitation of underground coal resources often leads to large-scale overburden movement and surface disturbance. Surface disturbance has destroyed the original terrain, landform, and natural landscape of the mine, causing a lot of damage to land resources, water resources, and vegetation resources, and seriously damaged the living environment and ecological environment of the mining area [1–4]. This situation is particularly serious in mining areas with high relief terrain areas. The topographic relief degree of the mining area is often expressed by slope. The slope can be calculated by Formula (1).

$$\alpha = \arctan(\Delta H / D) \quad (1)$$

where: ΔH is the height difference and D is the horizontal distance.

According to the slope, the terrain can be divided into flat land, hilly land, mountainous land, and alpine land. The slope of flat land is less than 2°, that of hilly land is 2–6°, that of mountainous land is 6–25°, and that of alpine land is more than 25°. Mountains and alpine land are collectively referred to as high relief terrain, that is, the slope of the high relief terrain area is more than 6°.

In high relief terrain mining areas, underground mining often induces large-scale mountain sliding, large collapse cracks, and tensile cracks, which seriously damage mountains, mountain roads, buildings, structures, and water bodies. As shown in Figure 1.

(a) (b)

Figure 1. Surface disturbance and damage under the influence of mining. (**a**) Mountain crack; (**b**) mountain road slip crack.

At present, China is pushing the concept of "green and eco-friendly coal mining". To promote "green and eco-friendly coal mining", coal mining needs to be carried out within the ecological carrying capacity of the mining area. The key and prerequisite for the realization of this concept is to master the disturbance law of coal mining on the surface of the mining area.

At present, many people have conducted research on the surface subsidence monitoring in the mining area using new measurement methods, including UAV [5], 3D laser scanning [6], InSAR [7–11], etc., but these methods have the disadvantage that the accuracy cannot meet the engineering requirements. Therefore, traditional field measurement methods, such as leveling, traverse, and GPS are still the main methods for subsidence monitoring in the mining area.

The disturbance of coal mining to the surface is a very complex time-space movement problem under the combined action of multiple geological and topographic mining factors [12–14]. The same mining will show different mining effects under different geological, topographic, and mining conditions. Scholars and experts have conducted a lot of research on the mining-induced surface disturbance law under different geology, different terrain, and different mining conditions [15–18]. Among them, the more typical is the large-scale mining of coal resources under the condition of repeated mining in high relief terrain areas.

From the geological structure of mining areas, high relief terrain mines in China are widely distributed in Shanxi, Henan, Guizhou, and other provinces, accounting for more than 1/3 of the total number of mines in China. Scholars and experts have conducted in-depth research and analysis on the disturbance law of mining on the surface in high relief terrain areas.

Wang [19] and others combined AutoCAD, Surfer, and FLAC3D software to build a numerical simulation model for mining different inclined coal seams in exposed bedrock high relief terrain areas, and analyzed the influence of coal seam inclination on surface movement and deformation; Wang [20], Han [21], Lian [22], Zhao [23] and others deduced and established an improved algorithm for predicting surface subsidence in high relief terrain areas; Liu [24] constructed the surface subsidence basin model of four types of surface mountains based on the small deflection bending problem in the thin plate bending deformation theory; Dai [25], Bai [26], and Lan [27] studied the law of surface movement and deformation in high relief terrain areas in combination with model test or numerical simulation; Hu [28] and Yi [29] analyzed the damage law and mechanism of environment, ecology, ground fissures and slopes under mining conditions in high relief terrain areas.

The mining conditions in high relief terrain areas vary greatly, and the disturbance laws are also different. Limited by the mining plan and relevant policies of the mining area, the working faces in high relief terrain areas are often mined in sequence. The overburden above the working face is affected by the high relief terrain and repeated at the same time. The disturbance law of the surface above the working face is different from the traditional situation. Which not only leads to abnormal laws in the scope, degree, and location of surface disturbance, but also increases the difficulty of subsequent disturbance prediction.

Therefore, taking working face 1310 as an example, combined with the field measurement and data analysis results, this paper reveals the distribution law of surface disturbance under the comprehensive action of repeated mining in high relief terrain area.

2. Engineering Background

2.1. Basic Information of Working Face 1310

Working face 1310 is subordinate to Zhaozhuang coal mine and is located at the junction of Gaoping city and Changzi county, Shanxi province, China. The longitude is 112°48′00″~112°58′00″ E and the latitude is 35°54′00″~36°03′00″.

Coal seam 3# is the main mining coal seam of working face 1310. The ground elevation is 1012.40–1193.20 m, the maximum ground fall is nearly 200 m, the elevation of the coal seam floor is 334.00–444.00 m, the average mining depth is 760 m, the average coal seam dip angle is 3°, and the average thickness of coal seam is 4.88 m. The coal seam structure is simple, the minability coefficient is 1.0, the variation coefficient is 5.13%, and the coal seam is stable. The strike length of the working face is 2741 m and the dip length is 219.27 m. The roof control adopts all caving method. Mining method is the one-time full height mechanized mining method. The characteristics of the roof and floor strata of the working face 1310 are shown in Table 1.

Table 1. Characteristics of roof and floor strata of coal seam 1310.

Name of Roof and Floor	Rock Name	Thickness/m	Lithologic Characteristics
Main roof	Fine grained sandstone	6.10	Gray, mainly composed of rock debris, intercalated with black organic stripes
Direct roof	Sandy mudstone	10.29	Black sandy mudstone, thin layer, with coal line at the top, rich in plant fossils
Direct floor	Sandy mudstone	7.34	Grayish black, intercalated with lenticular siltstone, shear joint, unfilled, occasionally vegetated
Main floof	Siltstone	10.60	Dark gray, thin layer, horizontal bedding, intercalated with a small amount of argillaceous strips and a small amount of animal fossils

Working face 1310 has the following characteristics:

① High relief terrain

The surface above the working face 1310 fluctuates greatly, as shown in Figure 2. The surface slope is steep with large drop, the maximum vertical drop can reach 200 m, and the mountain slope is about 12°. It belongs to the typical high relief terrain.

Figure 2. Topography of study area.

② Repeated mining

The mining sequences of the working faces in the mining area are 1308–1309–1310. The north side of working face 1310 is the planned fully mechanized mining face 1311, and the south side are fully mechanized mining faces 1308 and 1309 whose mining has been completed. In the early mining stage of working face 1310, it is affected by the residual deformation of 1308 and 1309. At the same time, working faces 1308 and 1309 are affected by the repeated mining of working face 1310, and the repeated mining disturbance is obvious.

2.2. Layout of Observation Stations

According to the terrain above the working face and the code for coal mine survey, the layout of surface movement observation stations is carried out. The observation station is arranged on the side of the stopping line. The point spacing is 30 m, and half of the strike and one dip line are arranged. The information of the observation stations is shown in Figure 3 and Table 2. A total of 49 monitoring points are arranged along the strike observation line, numbered from A1 to A49, with a total length of 1440 m. A total of 66 monitoring points are arranged on the dip observation line, numbered from B1 to B66 from the downhill direction, with a total length of 1950 m

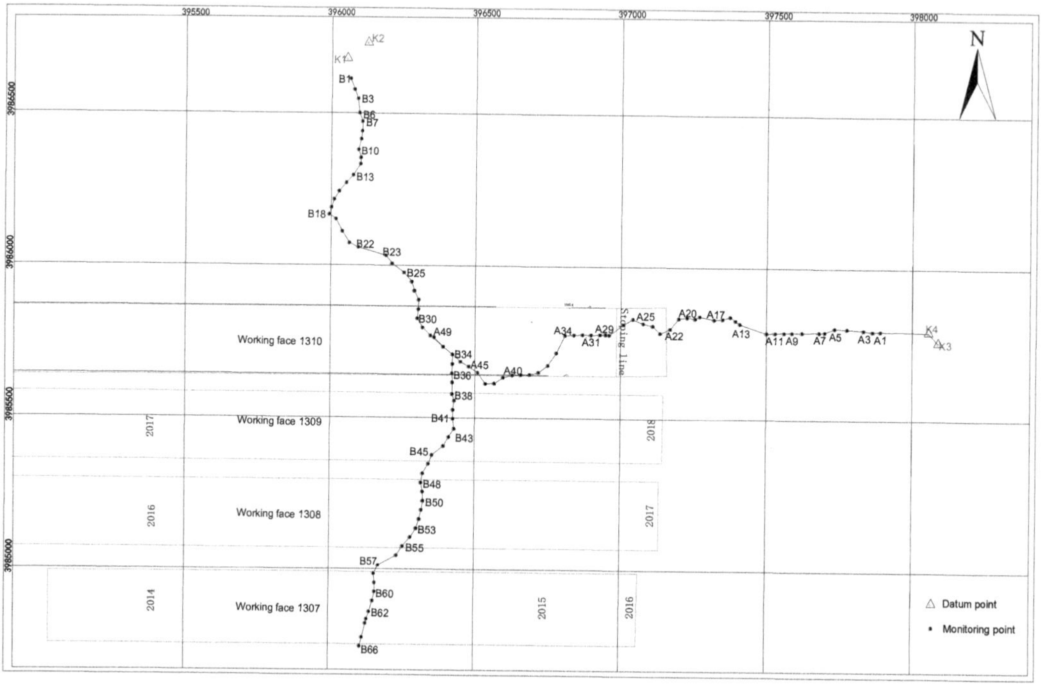

Figure 3. Actual layout point map of observation station.

Table 2. Summary of observation station layout.

	Number of Monitoring Points	Length/m
Strike line	49	1440
Dip line	66	1950
In total	115	3390

2.3. *Topographic Analysis of Main Section*

The topographic profile of strike and dip according to the point layout of the working face 1310 is shown in Figure 4.

Along the strike direction, surface monitoring points A1–A49 are arranged from the side of the coal pillar to the goaf. Point A22 is roughly above the stopping line. The terrain along the strike line from point A1 to point A34 generally shows an upward slope trend, with a large surface slope of about 14°. The terrain from point A34 to A49 generally shows a downhill trend. A34–A37 and A46–A49 are two steep downhill areas, with a mountain slope of about 12°.

Along the dip direction, surface monitoring points B1-B66 are set from downhill to uphill. B30 is located directly above the downhill mining boundary and B37 is located directly above the uphill mining boundary. The dip line from point B1 to point B47 generally shows an upward trend, with a slope of about 10°. The terrain from point B47 to point B66 generally shows a downward trend with a small slope of about 7°.

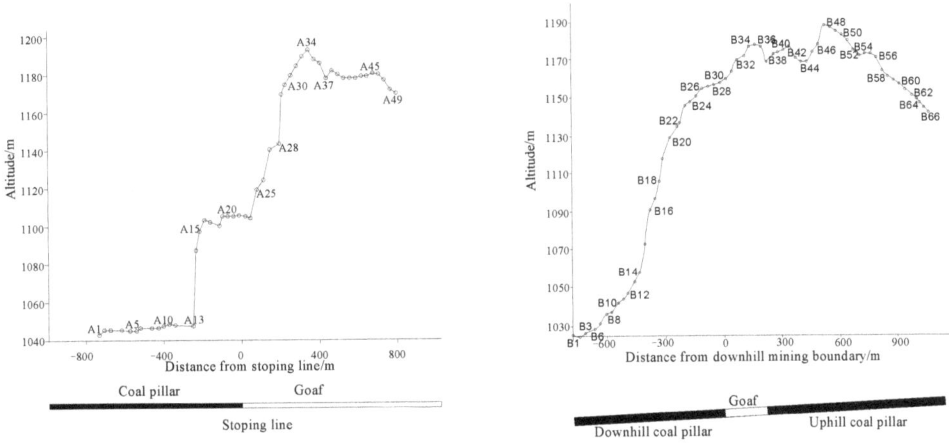

(a) (b)

Figure 4. Topographic map. (**a**) Strike line direction; (**b**) dip line direction.

3. Measurement Method and Data Acquisition

3.1. Monitoring Content

The disturbance degree and scope of the surface caused by underground mining are often described by five indexes: subsidence, inclination, curvature, horizontal movement, and horizontal deformation. The subsidence and horizontal movement are the basic quantities, and the inclination, curvature and horizontal deformation are the derived quantities. The inclination and curvature can be deduced from the subsidence by Equation (2):

$$\begin{aligned} i_{1-2} &= \frac{w_2-w_1}{l_{1-2}} \\ k_{1-2-3} &= \frac{i_{2-3}-i_{1-2}}{0.5(l_{1-2}+l_{2-3})} \end{aligned} \tag{2}$$

where: w_1 and w_2 are the subsidence values of observation points 1 and 2, respectively. i_{1-2} and i_{2-3} are the average slopes of line segments 1–2 and 2–3, respectively. l_{1-2} and l_{2-3} are the horizontal lengths of segments 1–2 and 2–3, respectively. k_{1-2-3} represents the curvature between observation points 1–2–3.

The horizontal deformation can be deduced from the horizontal movement by Equation (3):

$$\varepsilon_{1-2} = \frac{u_2 - u_1}{l_{2-3}} \tag{3}$$

where: ε_{1-2} is the horizontal deformation between observation points 1 and 2. u_1 and u_2 are the horizontal movement values of observation points 1 and 2, respectively.

The subsidence and horizontal movement are mainly calculated by the following formula:

$$w_i = h_{i0} - h_{in} \tag{4}$$

$$u_i = l_{i0} - l_{in} \tag{5}$$

where, h_{i0} and h_{in} are the elevation values of observation point i during the first and n times observations, respectively. l_{i0} and l_{in} are the horizontal distance from point i to the control point during the first and n times observations, which can be obtained by the following formula:

$$\begin{aligned} l_{i0} &= \sqrt{(x_{i0} - x)^2 + (y_{i0} - y)^2} \\ l_{in} &= \sqrt{(x_{in} - x)^2 + (y_{in} - y)^2} \end{aligned} \tag{6}$$

where: (x, y) are the plane coordinates of the control point, x_{i0} and x_{in} are the x coordinates of point i during the first observation and n times observations, respectively. y_{i0} and y_{in} are the y coordinates of point i during the first observation and n times observations, respectively.

In conclusion, if the elevation h and plane coordinate values (x, y) of the observation points are measured and obtained, the subsidence and horizontal movement values can be calculated; the inclination, curvature and horizontal deformation values can be deduced to analyze the subsidence law. Therefore, the content of field monitoring mainly includes (x, y) and H.

3.2. Measuring Method

According to the accuracy requirements of mine survey regulations, RTK technology is often used to collect the plane coordinate information of observation points (x, y) and the fourth-order leveling is used to collect the H elevation data information of observation points. According to the relevant requirements, if in the area with large topographic relief, the triangular elevation method can be used to collect elevation data. Both leveling and triangular elevation require good intervisibility between observation points in the survey area. However, in the survey area, the surface topography fluctuates too much, the vegetation is dense, and the intervisibility conditions between observation points are poor, so the traditional triangular elevation and leveling cannot be carried out. Therefore, in the process of observation implementation, an improved RTK elevation acquisition method supported by cores network is proposed. The instrument adopts Southern Galaxy 6 receiver for observation.

To meet the accuracy of elevation observation. In the process of observation, the method of multiple observations is adopted, and the observation process is as follows:

① Use the triangular forced centering device to erect RTK and collect the data in the fixed solution mode. Set the sampling interval to 1 s and the sampling to 60 s to obtain the first set data of the observation point;

② Invert the RTK receiver to make it lose contact with the satellite, then put it back in the right direction and reconnect the satellite;

③ Collect the second set of data in the fixed solution mode;

④ Turn the RTK receiver upside down to make it lose contact with the satellite again, and then put it right back to reconnect the satellite;

⑤ Collect the third set data in fixed solution mode.

Then compare and analyze the three-set data. If the difference between the maximum and minimum elevation data exceeds 2 cm, remeasure it. If it is less than 2 cm, take its average value as the final value.

To verify the accuracy of this method, a verification experiment was carried out on a mountain road. The slope of the mountain road is about 10°. Select a distance of 50 m on the mountain road, take point G1 as the datum point and use the RTK survey data of point G1 as the datum elevation data to carry out leveling and RTK elevation survey. The spacing of experimental sampling points is 50 m and the number is A1–A10 (Figure 5).

This method and the fourth-order leveling method are used to measure the elevation of the experimental sampling points. The measurement results are as follows.

According to the relevant specifications, the tolerance requirements for fourth-order leveling are as follows:

$$\varepsilon = 12\sqrt{n} \tag{7}$$

The number of measured stations from each monitoring point to the datum point measured in the field is listed in Table 3. The tolerance requirement for the elevation survey of each monitoring point obtained are also listed in Table 3. By comparing the relationship between the absolute value of the elevation difference measured by the two methods and the tolerance, the following conclusions can be drawn.

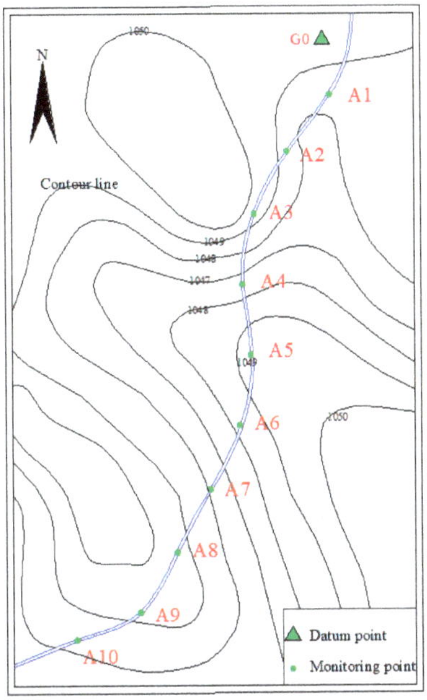

Figure 5. Layout of the test area and points.

Table 3. Experimental result.

Point Number	Elevation of Leveling/m	RTK Survey Elevation/m	Absolute Value of Elevation Difference/mm	From the Base Point, the Cumulative Number of Measured Stations	Tolerance Requirements for Elevation Survey/mm
A1	1048.386	1048.362	24	2	17
A2	1048.774	1048.748	26	4	24
A3	1049.431	1049.456	25	6	29
A4	1047.401	1047.405	−4	9	36
A5	1049.108	1049.139	31	11	40
A6	1048.097	1048.112	15	12	42
A7	1047.995	1047.968	27	13	43
A8	1049.208	1049.213	5	15	46
A9	1049.099	1049.076	23	16	48
A10	1048.214	1048.242	28	17	49

(1) The monitoring errors of monitoring points A1 and A2 (within 100 m) slightly exceed the limit;
(2) The monitoring errors of other monitoring points are less than the tolerance and meet the accuracy requirements.

According to the requirement of mine survey code, datum points are generally buried 150–200 m outside the mining affected area. Therefore, the distance between monitoring points and datum point is greater than 100 m, that is, the improved RTK can be used for elevation measurement, and the data meet the accuracy requirements.

This new and improved RTK measurement method not only solves the problem of non-intervisibility of monitoring points in high undulating terrain area but also greatly improves the measurement speed. At the same time, the original three-person mode of leveling is changed into a one-person mode of RTK, which greatly reduces the labor intensity.

3.3. Data Acquisition

The time and contents of surface disturbance monitoring are shown in Table 4.

Table 4. Monitoring information.

Observation Order	Observation Content	Observation Time	Advancing Distance of Working Face
1	Coordinate survey +	5 Septembet 2019	10~15 days before the mining
2	elevation survey	11 Septembet 2019	impact of working face
3	Elevation survey	22 Septembet 2019	Working face advancing 1345 m
4	Elevation survey	12 October 2019	Working face advancing 1414 m
5	Elevation survey	3 November 2019	Working face advancing 1603 m
6	Elevation survey	24 November 2019	Working face advancing 1750 m
7	Coordinate survey + elevation survey	22 December 2019	Working face advancing 1844 m
8	Elevation survey	4 January 2020	Working face advancing 1898 m
8	Elevation survey	27 February 2020	Working face advancing 2163 m
10	Elevation survey	18 March 2020	Working face advancing 2274 m
11	Elevation survey	22 April 2020	Working face advancing 2442 m
12	Elevation survey	17 May 2020	Working face advancing 2559 m
13	Coordinate survey + elevation survey	21 June 2020	End of working face advance
14	Elevation survey	19 July 2020	End of working face advance
15	Coordinate survey + elevation survey	23 August 2020	End of working face advance
16	Coordinate survey + elevation survey	9 October 2020	End of working face advance

4. Results

4.1. Surface Disturbance Law along Strike Direction

The data obtained in each stage of strike are processed to obtain the subsidence value and horizontal movement value. The drawing is as follows.

With the advancement of the working face, it can be seen from Figure 6a that the surface subsidence gradually increases. By 9 October 2020, a flat bottom appears within the observation range, indicating that the strike direction has reached full mining. The edge point of the flat bottom is near A35, and the maximum subsidence value is 1747 mm. However, due to the influence of high relief terrain on the inner side of the flat bottom, the flat bottom part of the surface basin has slipped again, showing serious inequality, and the maximum surface subsidence has increased to 2436 mm.

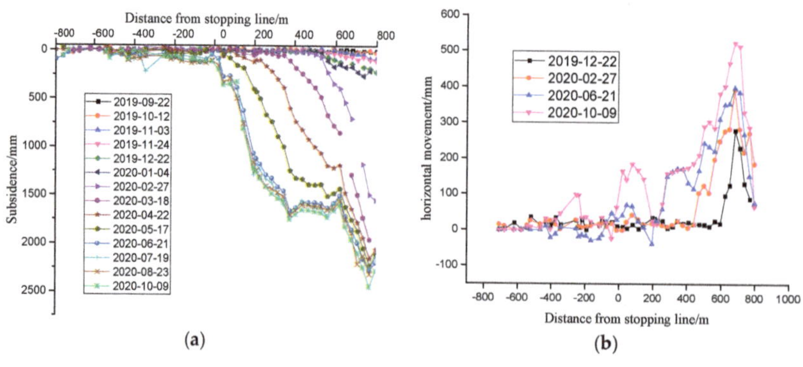

Figure 6. Dynamic moving curve of strike direction. (**a**) Subsidence; (**b**) horizontal movement.

It can be seen from the Figure 6b that with the advance of the working face, the maximum value of horizontal movement in the strike direction increases and the influence range expands. However, due to the influence of high relief terrain, there are two positive extreme values of horizontal movement, among which the extreme value above the goaf is large, which is mainly the influence of high relief terrain in mountainous area.

4.2. Surface Subsidence Law along Dip Direction

The obtained subsidence values and horizontal movement values of each period in the dip direction are plotted as follows.

Figure 7a shows the subsidence distribution at different times in the dip direction. It can be seen from the Figure 7a that the surface subsidence value and subsidence range gradually increase with the advance of the working face, the position of the maximum subsidence point remains unchanged, near point B33. By 23 August 2020, the maximum surface subsidence will reach 2307 mm. However, due to the repeated disturbance near the old mining faces on the south sides, the dip subsidence shows the following special laws. Compared with the downhill subsidence curve, the uphill subsidence curve changes more gently and has a wider range.

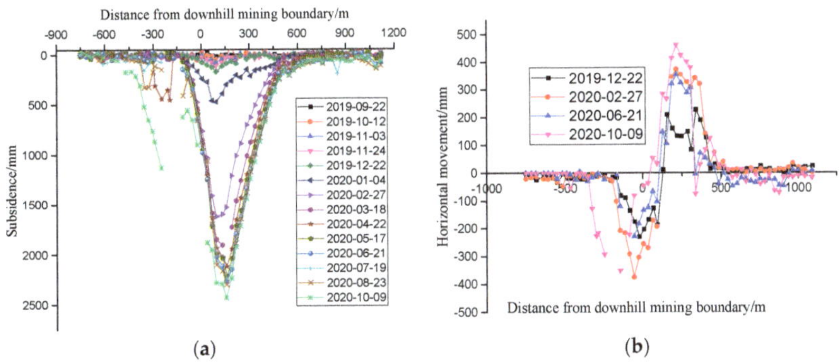

Figure 7. Dynamic moving curve of dip direction. (**a**) Subsidence; (**b**) horizontal movement.

Figure 7b shows the horizontal movement of the trend line at different times. Traditionally, the horizontal movement curve should be an antisymmetric curve. It can be seen from the figure that the horizontal movement gradually increases with the advance of the working face, but it does not form an inverse symmetry.

4.3. Parameter Analysis of Disturbance Range

The surface disturbance range caused by underground mining is often expressed by boundary angle and movement angle. The boundary angle delineates the outermost disturbance range of the surface. The movement angle delineates the scope of surface damage. The smaller the angle value, the larger the delineated range. According to the characteristics of surface movement in high relief terrain areas, the minimum boundary angle between horizontal movement and subsidence of 10 mm is taken as the final boundary angle. The final movement angle is obtained according to the outermost critical deformation value points of tilt (3 mm/m), curvature (0.2 mm/m^2), and horizontal deformation (2 mm/m). The obtained boundary angle and movement angle are shown in Table 5.

Table 5. Boundary angle and movement angle.

	Strike Direction/°	Dip Direction/°		Average
	Reverse-Positive Slope	Downhill (Reverse Slope)	Uphill (Reverse Slope)	
Boundary angle	63.1	50.8	57.7	57.2
Movement angle	72.8	63.5	70.6	69.0

Traditionally, the boundary angle is about 55–60°and the moving angle is about 65–70° [30]. It can be seen from Table 5 that the boundary angle and movement angle of the strike line are large, and the influence range of the surface is small. However, the boundary angle and movement angle in the downhill direction are small, and the surface influence range is large.

5. Discussion

The following mainly analyzes the special law of mining-induced surface disturbance from two aspects: surface topography and repeated mining.

5.1. High Relief Terrain

According to the spatial correspondence between working face distribution and surface topography, high relief terrain can be divided into positive slope and reverse slope. When the ground surface slope inclines to the goaf, the slope is called the positive slope, as shown in Figure 8a. On the contrary, when the ground surface slope tends to the side of the coal pillar, the slope is called reverse slope, as shown in Figure 8b.

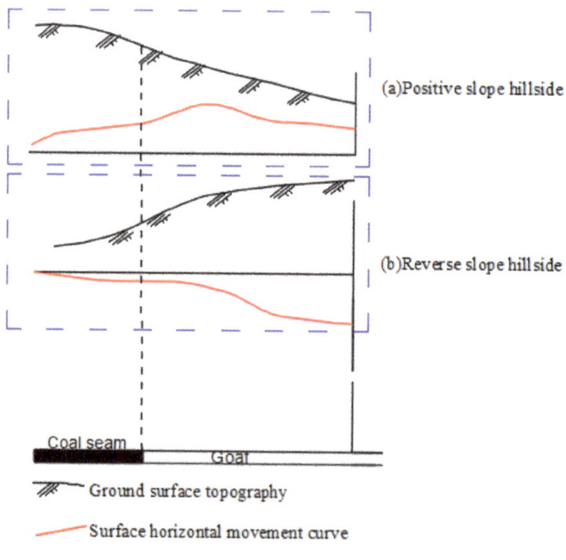

Figure 8. Positive slope hillside and reverse slope hillside. (**a**) Positive slope hillside; (**b**) reverse slope hillside.

According to the general theory of mining subsidence, under the condition of flat terrain, all surface points in the mining affected area move to the goaf, which is called mining-induced surface movement. In addition, under the influence of external action, the slope of high relief terrain will slip itself. Therefore, when the surface terrain is highly relief terrain, the surface movement is the superposition of mining-induced surface movement and surface slope sliding.

(1) When the terrain directly above the goaf is a positive slope, the sliding direction of the slope itself is consistent with the movement direction of the mining surface. The slope slip will aggravate the surface movement, aggravate the damage degree of surface mining and increase the influence range of mining.
(2) When the terrain directly above the goaf is an inverse slope, the sliding direction of the slope itself is opposite to the mining surface movement direction. The slope slip will weaken the surface movement and reduce the surface damage degree and the influence range.

The A1–A34 terrain of the working face 1310 is a reverse slope, which reduces the surface damage degree and the influence range of mining, which explains the reason the boundary angle and movement angle of the strike direction are large and the influence range is small. In addition, A35–A49 terrain is a positive slope. When the sliding of the broken body and the movement of the mining surface exceed the anti-sliding ability of the broken body, a large-scale sudden landslide will occur in the broken body, resulting in a sudden increase of the subsidence and horizontal movement in this area. This is the reason for the sudden increase of the subsidence and horizontal movement value at the last few points of the strike observation line.

5.2. Repeated Mining

Working face 1310 is the continuous working face of working face 1309 and 1308. Working face 1310 is mined after 1309 and 1308 have been mined, but the rock stratum movement has not been completed.

In general, the duration of overburden and surface movement after the mining of working face is 1–2 years. The dip observation line of working face 1310 is arranged above the non-mining area in the downhill direction and the mined working faces 1309 and 1308 in the uphill direction. When working face 1310 is mined, the working faces 1309 and 1308 are affected, and the separation layer and gap in the overlying strata are recompressed and transmitted to the surface, resulting in secondary disturbance of the surface. It leads to a wider range of surface disturbance and a larger degree of disturbance in this area. Finally, the boundary angle and movement angle in the downhill direction are small, and the subsidence curve of the inclined observation line is smoother and wider in the downhill direction than that in the uphill direction, which is the reason for the loss of anti-symmetrical characteristics of the dip horizontal movement curve.

6. Conclusions

(1) In the mining area with high relief terrain, the intervisibility between surface monitoring points is poor, and it is inconvenient to obtain data by traditional measurement methods. This paper presents an improved RTK elevation measurement method, which not only meets the observation accuracy but also improves the observation efficiency.
(2) Based on the measured data, the special surface disturbance law of the working face 1310 is analyzed. The strike boundary angle and movement angle are large, the boundary angle and movement angle in the downhill direction are small, the dip horizontal movement curve loses its antisymmetry, and the dip subsidence curve loses its left-right symmetry.
(3) The special surface disturbance law is the result of the combined action of high relief terrain and repeated mining. The high reverse slope reduces the influence range of surface mining and increases the corresponding boundary and movement angle. Landslides occur on high positive slopes, resulting in a sudden increase in subsidence. Repeated mining leads to slower convergence of movement deformation curve, expansion of influence range, and increase of disturbance value.

Author Contributions: Conceptualization, W.Y. (Weitao Yan) and W.Z.; methodology, J.C.; software, W.Y. (Wenfu Yang); validation, W.Y. (Weitao Yan); formal analysis, W.Y. (Wenfu Yang); investigation, X.L.; resources, W.W.; data curation, W.W.; writing—original draft preparation, W.Y. (Weitao Yan) and W.Y. (Wenfu Yang); writing—review and editing, J.C.; visualization, J.C.; supervision, W.Y. (Weitao Yan); project administration, J.C.; funding acquisition, J.C. All authors have read and agreed to the published version of the manuscript.

Funding: This research was funded by the Henan Scientific and Technological Projection grant number (212102310012, 212102310399), the National Natural Science Foundation of China grant number [U21A20108, 51974105, U1810203], the Open Fund of State Key Laboratory of Water Resource Protection and Utilization in Coal Mining grant number (WPUKFJJ2019-20, WPUKFJJ2019-17), the Fundamental Research Funds for the Universities of Henan Province grant number (NSFRF200314), the Youth Backbone Teacher Support Program of Henan Polytechnic University (2019XQG-07), and Doctoral Fund Program of Henan Polytechnic University (B2017-07, B2019-03).

Institutional Review Board Statement: Not applicable.

Informed Consent Statement: Not applicable.

Data Availability Statement: Not applicable.

Conflicts of Interest: The authors declare no conflict of interest.

References

1. Bell, F.G.; Stacey, T.R.; Genske, D.D. Mining subsidence and its effect on the environment: Some differing example. *Environ. Geol.* **2000**, *40*, 135–152. [CrossRef]
2. Saha, S.; Pattanayak, S.; Sills, E. Under-mining health: Environmental justice and mining in India. *Health Place* **2011**, *17*, 140–148. [CrossRef] [PubMed]
3. Sasaoka, T.; Takamoto, H.; Shimada, H.; Oya, J.; Hamanaka, A.; Matsui, K. Surface subsidence due to underground mining operation under weak geological condition in Indonesia. *J. Rock Mech. Geotech. Eng.* **2015**, *7*, 337–344. [CrossRef]
4. Sinha, S.; Bhattacharya, R.N.; Banerjee, R. Surface iron ore mining in eastern India and local level sustainability. *Resour. Policy* **2007**, *32*, 57–68. [CrossRef]
5. Edyta, P.; Wojciech, G.; Ćwiąkała, P.; Wojciech, M. Application of UAV-based orthomosaics for determination of horizontal displacement caused by underground mining. *ISPRS J. Photogramm. Remote Sens.* **2021**, *174*, 282–303.
6. Anders, K.; Marx, S.; Boike, J.; Herfort, B.; Wilcox, E.; Langer, M.; Marsh, P.; Höfle, B. Multitemporal terrestrial laser scanning point clouds for thaw subsidence observation at Arctic permafrost monitoring sites. *Earth Surf. Process. Landf.* **2020**, *45*, 1589–1600. [CrossRef]
7. Guillaume, M.; Doubre, C.; Masson, F. Time evolution of mining-related residual subsidence monitored over a 24-year period using InSAR in southern Alsace, France. *Int. J. Appl. Earth Obs. Geoinf.* **2021**, *102*, 102392.
8. Fan, H.; Wang, L.; Wen, B.; Du, S. A New Model for three-dimensional Deformation Extraction with Single-track InSAR Based on Mining Subsidence Characteristics. *Int. J. Appl. Earth Obs. Geoinf.* **2020**, *94*, 102223. [CrossRef]
9. Kim, J.; Lin, S.; Singh, R.; Lan, C.; Yun, H. Underground burning of Jharia coal mine (India) and associated surface deformation using InSAR data. *Int. J. Appl. Earth Obs. Geoinf.* **2021**, *103*, 102524. [CrossRef]
10. Carnec, C.; Delacourt, C. Three years of mining subsidence monitored by SAR interferometry, near Gardanne, France. *J. Appl. Geophys.* **2000**, *43*, 43–54. [CrossRef]
11. Essica, M.W. Application of DInSAR for short period monitoring of initial subsidence due to longwall mining in the mountain west United States. *Int. J. Min. Sci. Technol.* **2020**, *30*, 33–37.
12. Tang, F.-Q. Research on mechanism of mountain landslide due to underground mining. *J. Coal Sci. Eng.* **2009**, *15*, 351–354. [CrossRef]
13. Gao, C.; Xu, N.; Sun, W. Dynamic surface subsidence prediction model based on Bertalanffy time function. *J. China Coal Soc.* **2020**, *45*, 2740–2748.
14. Dai, H. Mining subsidence variables and their time-space relationship description. *J. China Coal Soc.* **2018**, *43*, 450–459.
15. Cui, X.; Gao, Y.; Yuan, D. Sudden surface collapse disasters casused by shallow partial mining in Datong coalfield, China. *Nat. Hazards* **2014**, *74*, 911–929. [CrossRef]
16. Li, Q.; Wang, J.; Yang, Q. Application and research on prediction method for mining subsidence in shallow buried deep coal seam. *Coal Sci. Technol.* **2019**, *47*, 175–181. [CrossRef]
17. Zhang, H. Study on equivalent calculation method of surface subsidence in inclined coal seam mining. *Coal Sci. Technol.* **2020**, *48*, 119–123.
18. Jia, X.; Song, G.; Chen, K. Study on influence of mining face advancing velocity on progressive surface subsidence and deformation. *Coal Sci. Technol.* **2019**, *47*, 208–214.

19. Wang, Q.; Li, T.; Guo, G. Study on seam inclination variation affected to ground surface movement law in base rock exposed mountain area. *Coal Sci. Technol.* **2016**, *44*, 155–160. [CrossRef]
20. Wang, L.; Guo, G.; Wang, M. New method of updating design for model and its parameter to prediction surface movement in mountainous mining. *J. China Coal Soc.* **2014**, *39*, 1070–1076.
21. Han, K.; Kang, J.; Wang, Z. Uniform prediction parameters for ground movement model in mountain area caused by coal mining. *J. Min. Saf. Eng.* **2013**, *30*, 107–111.
22. Lian, X.; Hu, H.; Guo, B. Regularities of surface dynamic movement and deformation induced by mining in mountainous area. *Met. Min.* **2016**, *9*, 151–156.
23. Zhao, B.; Liang, N.; Wu, C. Prediction system for mining subsidence of steeply inclined coal seam based on matlab in mountainous area. *Met. Min.* **2020**, *03*, 159–167.
24. Liu, Y.; Dai, H. A surface subsidence model of coal seam mining in mountain area based on plate bending deformation. *Mec. Eng.* **2019**, *41*, 559–564.
25. Dai, H.; Zhai, J.; Hu, Y. Testing study of surface displacement of mountainous region with similar material. *Chin. J. Rock Mech. Eng.* **2000**, *04*, 501–504.
26. Bai, G.; Zhang, D.; Zhang, J. Law of surface movement and deformation of mining subsidence in mountain area. *J. Liaoning Tech. Univ.* **2017**, *36*, 684–688.
27. Lan, H.; Zhang, H.; Yao, J. Numerical smiulation of mining subsidence predict of mountain surface. *J. China Coal Soc.* **2007**, *9*, 912–916.
28. Hu, H.; Lian, X.; Cai, Y. Study on ecological environment damage and restoration for coal mining-subsided area in loess hilly area of Shanxi Province. *Coal Sci. Technol.* **2020**, *48*, 70–79.
29. Yi, Z.; Liu, M.; Liu, X.; Wang, Y.; Wu, L.; Wang, Z.; Zhu, L. Long-term Landsat monitoring of mining subsidence based on spatiotemporal variations in soil moisture: A case study of Shanxi Province, China. *Int. J. Appl. Earth Obs. Geoinf.* **2021**, *102*, 102447. [CrossRef]
30. Yan, W.; Chen, J.; Tan, Y.; Zhang, W.; Cai, L. Theoretical Analysis of Mining Induced Overburden Subsidence Boundary with the Horizontal Coal Seam Mining. *Adv. Civ. Eng.* **2021**, 3, 6657738. [CrossRef]

sustainability

MDPI

Article

Quantitative Study on the Law of Surface Subsidence Zoning in Steeply Inclined Extra-Thick Coal Seam Mining

Yueguan Yan, Yanjun Zhang *, Yuanhao Zhu, Jinchi Cai and Junyao Wang

College of Geoscience and Surveying Engineering, China University of Mining and Technology-Beijing, Beijing 100083, China; yanyueguan@cumtb.edu.cn (Y.Y.); zqt1800204097g@student.cumtb.edu.cn (Y.Z.); zqt2100204107@student.cumtb.edu.cn (J.C.); 2010200306@student.cumtb.edu.cn (J.W.)
* Correspondence: bqt2100204049@student.cumtb.edu.cn

Abstract: The damage of overlying strata and ground surface caused by the one-time mining space is relatively severe in steeply inclined extra-thick coal seams. The unique law of surface subsidence at these conditions is still missing. Taking Huating Dongxia Coal Mine as the research background, this paper reveals the law-governing effects on rock strata and surface movement and deformation caused by steeply inclined extra-thick coal seam mining with different coal seam dip angles and coal thicknesses by using the methods of surface measurement, theoretical analysis, and numerical simulation. Based on the characteristics of the surface inclination deformation, the surface is divided into four areas along the tendency section line—namely, an outcrop discontinuous deformation area, an overall subsidence area, a gradual subsidence area, and a slight subsidence area. The results show that the influence of the coal seam dip angle on surface subsidence zoning in steeply inclined and thick coal seams is mainly reflected in the affected area range and the form of damage. Coal thickness has a weak effect on the form of rock strata damage and surface movement. Utilizing the influence of the coal seam dip angle and coal seam thickness on the change in the surface subsidence zoning, the calculation formulas for each area range and zoning angle in relation to the coal seam dip angle, coal thickness, mining depth, and vertical stage height are established. The research results can provide a reference to evaluate the influence of mining, especially in steeply inclined extra-thick coal seams.

Keywords: steeply inclined extra-thick coal seam; numerical simulation; discontinuous deformation; surface subsidence zoning; influence range; zoning angle

1. Introduction

Generally speaking, a coal seam with a dip angle greater than 45° is a steeply inclined coal seam. The coal quality of steeply inclined coal seams is excellent, the degree of joint development and metamorphism is high, and the mining value is also high. Steep coal seams are widely distributed in China, covering many mining areas such as Huating, Beijing, Huainan, and Urumqi [1,2]. Compared with horizontal and gently inclined coal seams, steeply inclined coal seams have complex occurrence conditions and numerous influencing factors, and they undergo extensive changes. The outcrop area can become seriously damaged and it is easy for step cracks to form, as shown in Figure 1. There are thus many factors that lead to complex surface subsidence laws and rock strata control problems [3–5].

In recent years, a large number of scholars have produced a series of studies on rock damage, surface movement, and deformation laws in steep seam mining. The following two aspects can be noted: (1) the damage laws, mechanical mechanisms, rock strata control, and prediction methods for steep seam mining are studied based on field experience, mechanical theory, and numerical calculation methods; (2) the surface movement and deformation laws and surface subsidence zoning of steeply inclined coal seam mining are studied based on the analysis of measured data, similar material model tests, and numerical simulation.

Figure 1. Surface steps and cracks in a steeply inclined coal seam mine. (**a**) Surface steps; (**b**) Surface cracks.

From the 1970s to 1985, the former Soviet Union carried out systematic research on steeply inclined coal seams and formulated the "Rules for the protection of natural objects and buildings in the mining area", which summed up the typical predicted parameters for steeply inclined coal seam mining in the Dunbas mining area [6]. Afsari [7] used FLAC software to simulate the surface movement caused by the mining of steeply inclined coal seams and proposed a method to simulate the mining of inclined and steeply inclined coal seams, including a combination method utilizing both a macroscopic isotropic elastic model and a global node model. Alejano et al. [8] defined a rock behavior model and characteristics, used FLAC software to simulate the rock behavior characteristics, and proposed a prediction method for a surface subsidence trough basin in horizontal and inclined coal seam mining. Asadi [9] proposed a new section function method to predict the surface subsidence of inclined coal seam mining, and the results showed that the predicted value was in good agreement with the measured value. Ma et al. [10] used FLAC software to study the variation law for the stress and displacement of overlying strata caused by mining of steeply inclined coal seams under the influence of faults in Zhaozhuang. The results showed that due to the influence of a weak fault surface, the stress was concentrated around the roof and fault after mining. With an increase in the mining depth, the stress displacement around the fault changed obviously. Wang et al. [11] used numerical simulation to study the law determining overlying strata damage and coal pillar stability in a steeply inclined coal seam group, providing the technical basis for coal mining under buildings. Yang Fan [12] proposed a "factory"-type moving arch model of rock strata movement in steeply inclined coal seams utilizing the characteristics of rock strata movement and the force transmission mechanism and analyzed the stability of "factory"-type moving arch structures in steeply inclined coal seams, which provided a new theoretical basis for the prediction and control of mining subsidence in steeply inclined coal seams. Yi et al. [13] identified subsidence laws such as those that produce large surface movement ranges, severe deformations and easy formation of step cracks in the horizontal slicing mining of steeply inclined and extra-thick coal seams. They also proposed a calculation method for the surface movement angle parameter in horizontal slicing mining. Dai et al. [14,15] revealed the zoning characteristics of surface subsidence in space in steeply inclined coal seams based on the measured data accumulated over many years in Majiagou Mine, Tangshan. Utilizing the surface subsidence and its average change rate can divide the subsidence area into a slight subsidence area, a subsidence growth area, and an overall subsidence area. Yan et al. [16–18] used a similar material simulation to reveal the laws governing surface movement and deformation, surrounding rock collapse, and damage mechanisms. The surface subsidence basin can be divided into an outcrop subsidence area, an overall subsidence area, a gradual subsidence area, and a slight subsidence area.

The above research results combined with field practice, the characteristics of rock strata movement, and deformation under steeply inclined coal seam mining are studied. However, the stress condition and mechanical transmission mechanism of the rock strata caused by the mining of the steeply inclined coal seam are complex, and the research on the mining subsidence law of the steeply inclined coal seam in the relevant literature is not sufficient. At present, the research on the surface subsidence zoning is only qualitative research after mining the thin and medium-thickness coal seam, and the research on the surface subsidence zoning of the steeply inclined extra-thick coal seam is still in a relatively blank state. Because of this, in order to maintain the efficient and sustainable development of mining mines in steeply inclined extra-thick coal seams, it is necessary to study the subsidence law and surface subsidence zoning of steeply inclined extra-thick coal seams. This paper takes the steeply inclined extra-thick coal seam mining in Huating Dongxia Coal Mine as the research object, and the deformation and damage of rock strata, as well as the characteristics and laws of surface subsidence zoning in steeply inclined extra-thick coal seam mining, are analyzed and quantitatively studied. By changing the geometric conditions of coal seams, the formation, development, and deformation mode of surface subsidence zoning pattern is studied. The formation conditions and variation law of outcrop discontinuous deformation area are emphatically analyzed. The purpose is to provide guidance for the protection of buildings (structures) and land in the mining area and provide relevant theoretical guidance for preventing and controlling different subsidence areas in the mining area.

2. Materials and Methods

2.1. Overview of Study Area

The Huating mining area is located in Pingliang District of Gansu Province, which is jointly controlled by Huating County and Chongxin County. Figure 2 shows the geographical location and distribution of the Huating mining area, including Huayan, Shanzhai, Dongxia, and other mining areas. Dongxia Coal Mine is located in the southeast of Huating coalfield. The terrain in the mining area is complex, and the gully is vertical and horizontal. The surface is almost covered by loess, and the loess hilly landform is the main landscape. The highest altitude is +1701 m, and the lowest altitude is +1525 m. The general terrain is high in the southeast and low in the northwest.

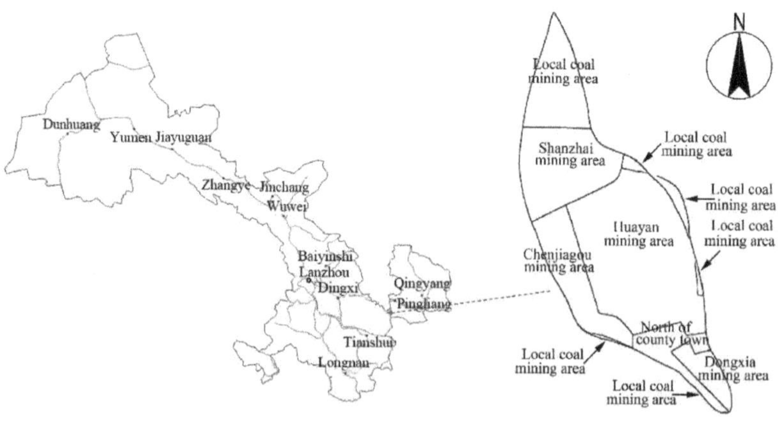

Figure 2. Geographical location and distribution of mining area.

The strata in the Dongxia mine are the Upper Triassic Yanchang Group, the Lower Jurassic Huating Group, the Middle Jurassic Zhiluo Group, the Lower Cretaceous Zhidan Group, the Upper Tertiary Gansu Group, and Quaternary. The direction of strata is generally 20° north to west (inclined to the southwest), the dip angle is 28°–44°, the shallow dip angle is larger than the deep dip angle, and the northern dip angle is larger than the southern dip angle. The coal seam structure is simple in the north and complex in the south. The coal seam roof is sandy mudstone and mudstone, followed by carbonaceous mudstone; the north of the coal seam floor is mudstone and argillaceous sandstone, the middle is shale, and the south is carbonaceous shale.

The 220 working face is located in the west wing expansion area of the mine field. The average coal seam dip angle is 50°, the average coal thickness is 30 m, and the mining rate is 85%. The inclined length of the working face is 60 m, and the stage vertical height is 50 m. The mining depth of the lower trough is about 520 m, and the horizontal section is adopted. The working face adopts the longwall backward mining method. All caving methods manage the roof. The overlying strata can collapse automatically by gravity, and there are no additional measures to control surface subsidence. After the mining of the upper section is completed, and the next section of the working face is followed. The daily mining progress is 2.4 m, and the annual mining progress is 792 m. Figure 3 shows the layout of the working face.

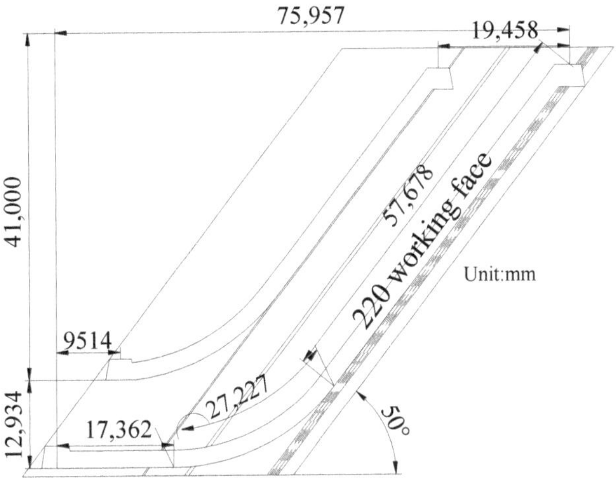

Figure 3. The layout of working face.

According to the terrain distribution of the 220 working face, an inclination observation line was set up. The inclination observation line is perpendicular to the strike of the working face. A measuring line is 820 m in length, including 28 monitoring points (3 control points and 25 monitoring points). The monitoring points numbered as D1, D2, D3, Q1, Q2...Q22, respectively, and the distance between each measuring points is 30 m. Three control points, K1, K2, and K3, respectively, with spaces of 50 m, are located outside the influence range of the working face (on the side of the floor). The layout of observation line is shown in Table 1, whereas the layout of measuring points and working face is shown in Figure 4.

Table 1. The layout of observation line.

Line Name	Time of Observation	Length (m)	Points Number	Observation Number	Remark
Q line	2012.7~2014.5	820	28	7	Vertical to the working face

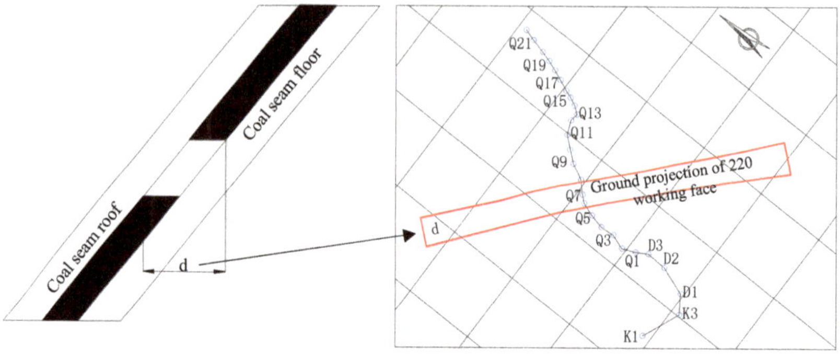

Figure 4. The field layout of measuring points and working face.

2.2. Numerical Simulation

2.2.1. Introduction of 3DEC Numerical Simulation Software

In the field of geotechnical engineering, there are mainly numerical analysis methods, such as the finite element method (FEM), the finite difference method (FDM), the discrete element method (DEM), and the boundary element method (BEM). According to different principles, different numerical simulation software is generated, such as FLAC, PFC, and 3DEC [19].

FLAC is a finite difference software, mainly used through analyzing stress and plastic zone changes to express rock strata movement and deformation, and cannot directly see the surface discontinuous deformation. PFC is a particle flow software based on the discrete element. The numerical model is composed of many particles, which can intuitively see the development of cracks. However, it is difficult to model because of the large amounts of calculation involved. The analysis software 3DEC is based on the discrete element method. Because of the small amount of calculation required, simple modelling, and the ability to simulate discontinuous deformation, it is widely used in the engineering field. Based on the advantages and disadvantages of the above software, this paper selects the 3DEC software to carry out the related research on the law of surface subsidence zoning [20,21].

Furthermore, 3DEC software is a numerical analysis program used for discontinuous simulation developed by ITASCA based on two-dimensional discrete element software UDEC. This software has great application in rock engineering design and analysis, and is suitable for large deformation calculation, rotation calculation, and other problems. The application scope of 3DEC software covers the simulation analysis and calculation of rock slope instability, underground engineering excavation, rock joints, faults, and bedding structure effects. It can effectively solve the mechanical problems of potential instability caused by discontinuous characteristics [22].

2.2.2. Scheme Design and Parameter Determination

(1) Scheme design

In order to study the surface subsidence zoning law, the surface movement, and the deformation law of steeply inclined thick coal seam mining, the 3DEC numerical simulation program was used to simulate coal seams with different geometric conditions under the same geological conditions. Taking the geological and mining conditions of Dongxia Mine as the prototype, a numerical model with a 50° dip angle and 30 m coal thickness was established. At the same time, the numerical models of different coal seam dip angles and coal thicknesses were established by changing the coal seam dip angle and coal thickness, namely the models of coal seam dip angles (45°, 50°, 55°, 60°, 70°, and 80°) (coal thickness of 30 m) and the models of coal thickness (20 m, 30 m, 40 m, and 50 m) (coal seam dip angle of 50°). The specific simulation scheme is shown in Table 2.

Table 2. Numerical simulation scheme.

Number	Coal Seam Thickness (m)	Dip Angle (°)	Stage Mining Depth (m)	Stage Number	Mining Depth (m)
1	30	50	50	10	20~520
2	30	45	50	10	20~520
3	30	55	50	10	20~520
4	30	60	50	10	20~520
5	30	70	50	10	20~520
6	30	80	50	10	20~520
7	20	50	50	10	20~520
8	40	50	50	10	20~520
9	50	50	50	10	20~520

The numerical model is based on the steeply inclined extra-thick coal seam mining in the inclination direction of the Dongxia Coal Mine. The coal seam dip angle is 50°, the coal thickness is 30 m, the upper boundary of the goaf is 20 m, and the stage mining depth is 50 m. The model size is 1200 m × 600 m × 1 m, and the thickness of the loose layer is 10 m. The Mohr–Coulomb model is selected as the constitutive model. According to different lithology, the mesh size of the model is divided into 10 m, 15 m, and 20 m, across a total of 3805 zones. The lateral displacement of the left and right boundaries of the model is constrained, and the bottom is clamped. The upper boundary is free and is loaded on the model by gravity (the gravity acceleration is 9.8 m/s^2).

There are seven layers of model rock strata. Except for the surface layer, the dip angle of the model rock strata is the same as that of the coal seam, and the names and thickness of each rock strata from left to right are as follows: (1) fine sandstone, with a thickness of 120 m; (2) siltstone, with a thickness of 100 m; (3) sandy mudstone, with a thickness of 50 m; (4) sandy mudstone and sandstone, with a thickness of 30 m; (5) coal seam, with a thickness of 30 m; and (6) sandy shale, with a thickness of 100 m. The surface layer is covered by loess, and the thickness is 10 m. The stratigraphic distribution is shown in Figure 5.

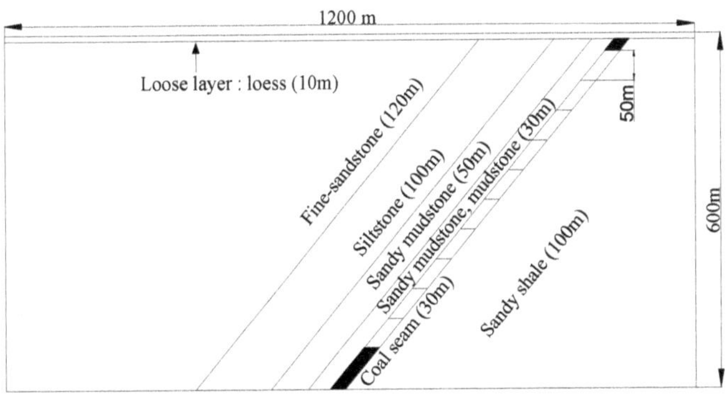

Figure 5. Numerical simulation design.

(2) Parameter determination

The mechanical parameters are selected according to the geological conditions of the coal seam and the condition of the overlying strata, mainly including density, bulk modulus, shear modulus, friction angle, cohesion, and tensile strength. The relationship between bulk modulus K, shear modulus G, elastic modulus E, and Poisson's ratio v is as follows.

$$K = \frac{E}{3(1-2v)} \tag{1}$$

$$G = \frac{E}{2(1+v)} \tag{2}$$

The block division of the discrete element should be based on the actual development and distribution of rock strata combined with the physical and mechanical properties of rock strata. The specific physical and mechanical parameters are shown in Tables 3 and 4. The selection of joint parameters is obtained by the empirical formula in [23,24].

Table 3. Physical and mechanical parameters of the 3DEC model.

Lithology	Thickness (m)	Volume Weight (kg/m^3)	Bulk Modulus (GPa)	Shear Modulus (GPa)	Cohesion (MPa)	Friction Angle (°)	Tension (MPa)
Sandy shale	100	2400	3.2	1.48	0.60	30	0.12
Coal	30	1350	1.62	0.62	0.28	28	0.10
Sandy mudstone and sandstone	30	2400	3.04	1.58	0.60	30	0.13
Sandy mudstone	50	2450	4.30	2.60	0.56	35	0.20
Siltstone	100	2300	2.80	1.28	0.48	34	0.24
Fine sandstone	120	2200	2.48	1.02	0.24	23	0.21
Loess	10	1600	0.16	0.06	0.20	17	0.10

Table 4. Physical and mechanical parameters of rock joints.

Lithology	Normal Stiffness (GPa)	Tangential Stiffness (GPa)	Friction Angle (°)	Cohesion (MPa)	Tension (MPa)
Sandy shale	3.2	1.48	22	0.21	0.12
Coal	1.62	0.62	20	0.13	0.10
Sandy mudstone and sandstone	3.01	1.58	22	0.21	0.13
Sandy mudstone	4.40	2.60	27	0.56	0.20
Siltstone	3.20	1.28	22	0.25	0.24
Fine sandstone	2.40	1.02	21	0.20	0.21
Loess	0.16	0.06	12	0.10	0.10

3. Quantitative Method of Surface Subsidence Zoning

According to the subsidence increase in the adjacent surface points in steeply inclined extra-thick coal seam mining, the surface movement basin can be divided into four areas from the coal seam floor to the roof: the outcrop discontinuous deformation area, the overall subsidence area, the gradual subsidence area, and the slight subsidence area. In order to further reveal the zoning law of surface subsidence in steeply inclined extra-thick coal seam mining, based on the results of field observation, it is determined that the surface subsidence of 10 mm is the outermost boundary in the slight subsidence area. The surface inclination deformation of 6 mm/m is the reference standard for other areas. At the same time, the subsidence curve is comprehensively determined as the division principle of each zoning boundary, as shown in Figure 6. In summary, we define the zoning angle and zoning range of each zone in steeply inclined extra-thick coal seam mining as follows.

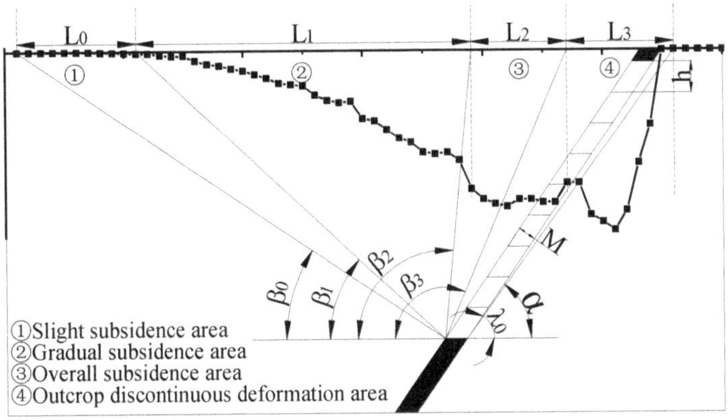

Figure 6. Angle relation of each area.

(1) Floor boundary angle (λ_0)

The angle between the boundary point of the surface movement basin (subsidence of 10 mm) on the inclination main section and the boundary line on the floor side of the goaf and the horizontal line on the floor side.

(2) Roof boundary angle (β_0)

The roof boundary angle is also the zoning angle in the slight subsidence area, which tends to move the basin boundary point (subsidence of 10 mm) on the inclination main section to the angle between the boundary line on the roof side of the goaf and the horizontal line on the roof side.

(3) Zoning angle of the gradual subsidence area (β_1)

The angle between the boundary point on one side of the roof in the gradual subsidence area of the surface movement basin on the main section, the boundary line on the roof side of the goaf, and the horizontal line on the roof side.

(4) Zoning angle of the overall subsidence area (β_2)

The angle between the boundary point of the overall subsidence area of the surface movement basin on the inclination main section, the boundary line on the roof side of the goaf, and the horizontal line on the roof side.

(5) Zoning angle of the outcrop discontinuous deformation area (β_3)

The angle between the boundary point on the side of roof and the boundary line on the roof side of the goaf, and the horizontal line on the side of the roof in the outcrop discontinuous deformation area of the surface movement basin on the inclination main section.

(6) L_3: range of the outcrop discontinuous deformation area; L_2: range of the overall subsidence area; L_1: range of the gradual subsidence area; L_0: range of the slight subsidence area.

In [17,18,25], the authors believe that the main factors affecting the zoning law of surface subsidence are coal seam dip angle, coal thickness, mining depth, and stage vertical height. The change in surface subsidence zoning is mainly reflected in the surface subsidence, scope, and each zoning angle. Therefore, on this basis, this paper establishes the calculation model of zoning scope and zoning angle and various influencing factors, expressed by Formulas (3) and (4).

$$L = f(\alpha, H, M) \tag{3}$$

$$\beta = f(\alpha, h, M) \tag{4}$$

where L is the zoning range, β is the zoning angle, α is the coal seam dip angle, h is the stage vertical height, M is the coal thickness, and H is the mining depth.

The influence range of surface subsidence in the outcrop discontinuous deformation area is:

$$L_3 = H(\cot \lambda_0 - \cot \beta_3) \tag{5}$$

The influence range of surface subsidence in the overall subsidence area is:

$$L_2 = H(\cot \beta_2 - \cot \beta_3) \tag{6}$$

The influence range of surface subsidence in the gradual subsidence area is:

$$L_1 = H(\cot \beta_1 - \cot \beta_2) \tag{7}$$

The influence range of surface subsidence in the slight subsidence area is:

$$L_0 = H(\cot \beta_0 - \cot \beta_1) \tag{8}$$

4. Results and Analysis

4.1. Division of Surface Subsidence Zoning

The subsidence curve is plotted by extracting the surface subsidence values calculated by the numerical model, as shown in Figure 7. In the early stage of working face mining, the subsidence curve is relatively gentle. With the deepening of mining, the influence range of surface movement and deformation is mainly extended to the roof side. The center of the surface subsidence curve shifts to the uphill side, showing obvious asymmetry. The surface influences the degree and range of the roof side, which are greater than those of the floor side. After the end of mining, the maximum surface subsidence is 3655 mm. According to the measured data of the surface, after the mining of the 220 working face, the maximum subsidence value of the surface is 3590 mm. By comparison, the numerical simulation results are evidently not much different from the actual results, indicating that the 3DEC numerical simulation software is reliable and correct for studying the zoning law of the surface subsidence.

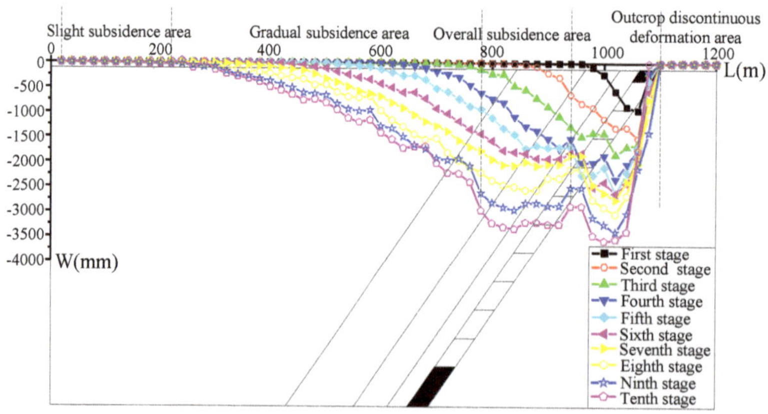

Figure 7. Surface subsidence curve.

According to the representation method of the surface subsidence zoning in Section 3, the subsidence curve is divided, as shown in Figure 8. It can be seen from the figure that the difference of surface subsidence is large. However, the difference of subsidence in the same subsidence area is small, especially in the overall subsidence area and the slight subsidence area, and the amount of deformation is also small. The coal seam outcrop and its vicinity are prone to serious discontinuous deformation, mainly reflected as cracks or collapse pits. In the overall subsidence area, the subsidence is large, but the surface inclination amount is small, and the surface subsidence is overall. The gradual subsidence area connects the slight subsidence area with the overall subsidence area or the outcrop discontinuous deformation area, and the subsidence varies greatly. The subsidence in the slight subsidence area is small, and the change in the surface damage range is stable. The subsidence increment in the subsidence area shows that the subsidence variation in the slight subsidence area is the smallest, followed by the overall subsidence area and the gradual subsidence area, and the subsidence variation in the outcrop discontinuous deformation area is the largest.

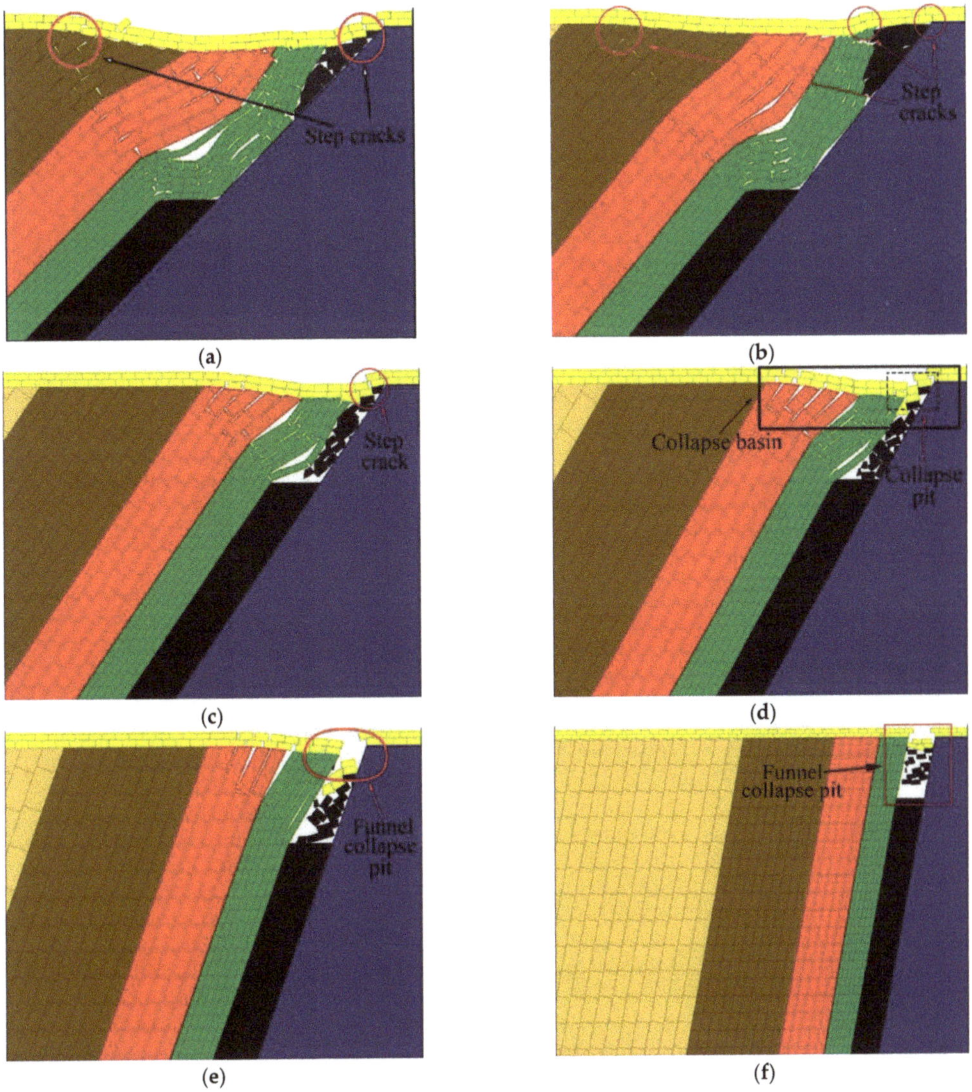

Figure 8. The outcrop discontinuous deformation of shallow mining. (**a**) Coal seam dip angle of 45°, (**b**) coal seam dip angle of 50°, (**c**) coal seam dip angle of 55°, (**d**) coal seam dip angle of 60°, (**e**) coal seam dip angle of 70°, and (**f**) coal seam dip angle of 80°.

According to Formulas (3)–(8), the zoning ranges of surface subsidence can be obtained as follows: L_0 = 120 m; L_1 = 480 m; L_2 = 200 m; and L_3 = 180 m. The total influence range of surface subsidence is 980 m.

4.2. Analysis of the Law of Surface Subsidence Zoning with the Change in the Coal Seam Dip Angle

With an increase in the coal seam dip angle, the degree of surface discontinuous deformation near the coal seam outcrop increases continuously, as shown in Figure 8. When the dip angle of the coal seam is less than 55°, in shallow mining, the coal above the goaf slides to the goaf under the action of its gravity parallel to the floor. At the same time, the roof rock above the goaf bends and collapses to the goaf, and a part of the rock mass squeezes and slides the coal. Under the combined action of the two, the surface near the coal seam outcrop is mainly characterized by the collapsed basin. The surface of the floor side and the roof side appear step cracks. The damage degree of the floor side is greater than that of the roof side. With an increase in the mining depth, the slope of the floor side increases.

When the dip angle of the coal seam is greater than 55°, in the shallow mining process, under the combined effect of the dip angle of the coal seam and the gravity of overlying strata, the coal above the goaf falls to the goaf and fills the goaf, which inhibits the caving of the boundary roof under the goaf. The boundary space above the goaf expands continuously, and the loose layer falls to the boundary space above the goaf, resulting in the formation of collapse pits on the surface. The size and offset distance of the collapse pit are related to the dip angle of the coal seam. The larger the dip angle of the coal seam, the larger the size of the collapse pit, and the smaller the deviation from the top of the coal seam outcrop. Because the coal above the goaf fills the goaf, which inhibits the collapse of the roof at the lower boundary of the goaf, a cantilever beam structure is formed at the upper boundary of the goaf. The larger the dip angle of the coal seam, the smaller the bending degree of the cantilever beam, the smaller the horizontal movement of the surface, and the deeper the funnel collapse pit formed on the surface. At the same time, the larger the coal seam dip angle, the longer the collapse pit exists. This is because the cantilever beam formed by the roof after coal seam mining is perpendicular to the side of the floor, and the cantilever beam needs a certain length to collapse to the goaf. When the coal seam with the dip angle of 80° is mined to the fourth stage (the mining depth is 200 m), the cantilever rock mass is filled in the goaf, and the funnel collapse pit gradually disappears.

Figure 9 shows the variation curve of the range and zoning angle of the outcrop discontinuous deformation area with different coal seam dip angles. It can be seen from the figure that the range and zoning angle change more obviously during the shallow mining (50 m~200 m). When the coal seam dip angle is less than 55°, the range of the outcrop discontinuous deformation area decreases with an increase in the coal seam dip angle; when the coal seam dip angle is greater than 55°, the range of the outcrop discontinuous deformation area increases with an increase in the coal seam dip angle. During deep mining, the range of the outcrop discontinuous deformation area of each outcrop is no longer significantly changed, because the formation of the outcrop discontinuous deformation area is mainly caused by the collapse of coal above the goaf and the collapse of roof strata during shallow mining. When the coal seam dip angle is less than 55°, after the coal is mined out, the roof strata of the goaf bend to the goaf, which inhibits the surface movement above the goaf, and the surface subsidence tends to the downhill direction. With an increase in the coal seam dip angle, the coal caving in the upper part of the goaf increases, making it easier to form collapse pits above the goaf. The damage range of the outcrop discontinuous deformation area is more concentrated in the upper part of the goaf. When the coal seam dip angle is greater than 55°, after the coal is mined out, the upper coal of the goaf is filled in the goaf first, and the surface collapse pit appears. Due to the effect of the coal seam dip angle, the rock mass on the roof side will bend to the goaf in the form of a cantilever beam, thus filling the goaf. During the shallow mining, the collapse pits appear, while in deep mining, large cracks and steps appear on the surface, and the overall subsidence area no longer appears on the surface, all of which are manifested as the outcrop discontinuous deformation area, increasing the range of the discontinuous deformation area. During deep mining, the area scope remains stable, and the change in the zoning angle is stable.

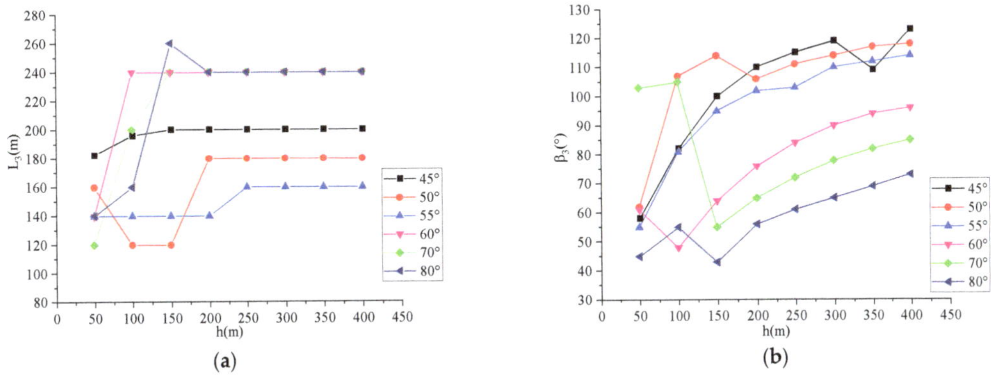

Figure 9. The variation curve of the range and zoning angle of the outcrop discontinuous deformation area with different coal seam dip angles. (**a**) The variation curve of the range, (**b**) the variation curve of the zoning angle.

Figure 10 shows the variation curve of the range and zoning angle of the overall subsidence area with different coal seam dip angles. The overall subsidence area shows large subsidence, but the change in inclination is small. In the shallow mining of steeply inclined extra-thick coal seam, the surface mainly demonstrates discontinuous damage, and there is no overall subsidence area on the surface. When deep mining is carried out, due to a gradual increase in the range of internal rock movement, a part of the surface originally belonging to the gradual subsidence area forms an overall subsidence area. It can be seen from the figure that the overall subsidence area mainly occurs below the dip angle of the coal seam 55°, and it is easy to occur in deep mining. When the dip angle of the coal seam is greater than 55°, the "integrity" of the overall subsidence area no longer appears, but is presented in discontinuous deformation (large cracks and large steps). With the deepening of mining, the discontinuous deformation intensifies.

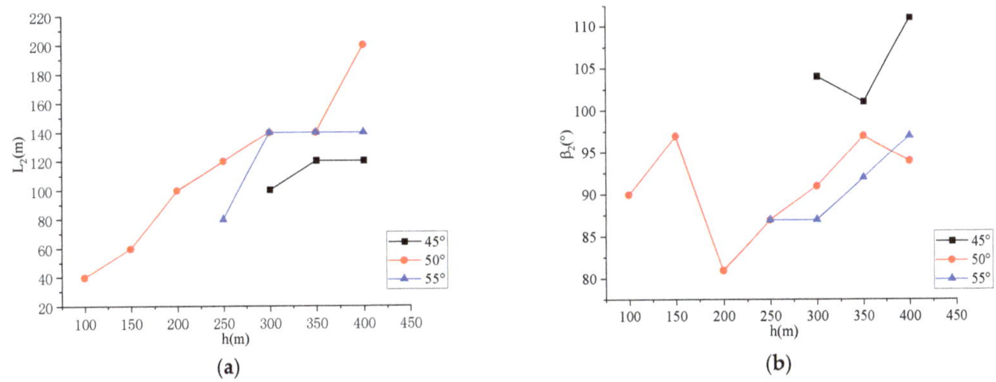

Figure 10. The variation curve of the range and zoning angle of the overall subsidence area with different coal seam dip angles. (**a**) The variation curve of the range, (**b**) the variation curve of the zoning angle.

Figure 11 shows the variation curve of the range and zoning angle of the gradual subsidence area with different coal seam dip angles. It can be seen from the figure that with an increase in the mining depth, the range of gradual subsidence area increases continuously, and the growth rate remains stable. At the same time, with an increase in the coal seam dip angle, the range of gradual subsidence area is smaller. This is because with an increase in the coal seam dip angle, the influence range of the surface is more concentrated, and the surface deformation shifts more significantly to the side of the floor, and the horizontal movement is larger. When the dip angle of the coal seam is greater than 55°, the surface subsidence shows "mutation", that is, from the discontinuous deformation area with large subsidence to the slight subsidence area with small subsidence, the range of intermediate gradual subsidence area is small, and the excessive area is not obvious. When the coal seam dip angle is less than 55°, the surface subsidence curve deviates significantly from the downhill side, and the surface subsidence changes more gently, and the range of gradual subsidence area is larger.

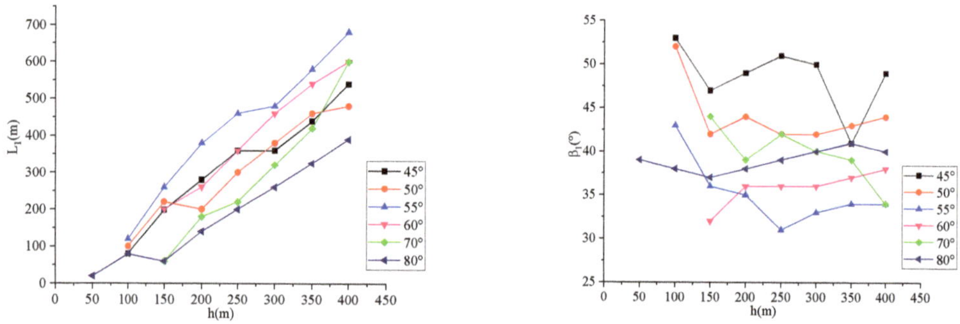

(a) (b)

Figure 11. The variation curve of the range and zoning angle of the gradual subsidence area with different coal seam dip angles. (**a**) The variation curve of the range, (**b**) the variation curve of the zoning angle.

Figure 12 shows the variation curve of the range and zoning angle in the slight subsidence area with different coal seam dip angles. The slight subsidence area is the area with small surface subsidence and shows a certain integrity. The surface horizontal movement and inclination deformation are small. It can be seen from the figure that in shallow mining, a larger coal seam dip angle correlates with a smaller range in the slight subsidence area. This is because a smaller coal seam dip angle causes a larger influence range. The larger the coal seam dip angle, the more concentrated the damage is. With an increase in the mining depth, the range in the slight subsidence area decreases. At this time, the larger the dip angle of the coal seam, the larger the range in the slight subsidence area. This is because with a larger dip angle of the coal seam, the damage gradually concentrates in the outcrop area, and the surface subsidence shows a sudden change. From the outcrop discontinuous deformation area with very serious damage to the overall subsidence area, the excessive effect of the intermediate gradual subsidence area is no longer obvious, so the gradual subsidence area is small, and the slight subsidence area is large. When the dip angle of the coal seam is smaller, the surface area which belongs to the slight subsidence area is easier to change into the gradual subsidence area, and the change in the surface subsidence area is gentler, which makes the gradual subsidence area increase and the slight subsidence area decrease. Ultimately, the larger the dip angle of the coal seam, the larger the range in the slight subsidence area.

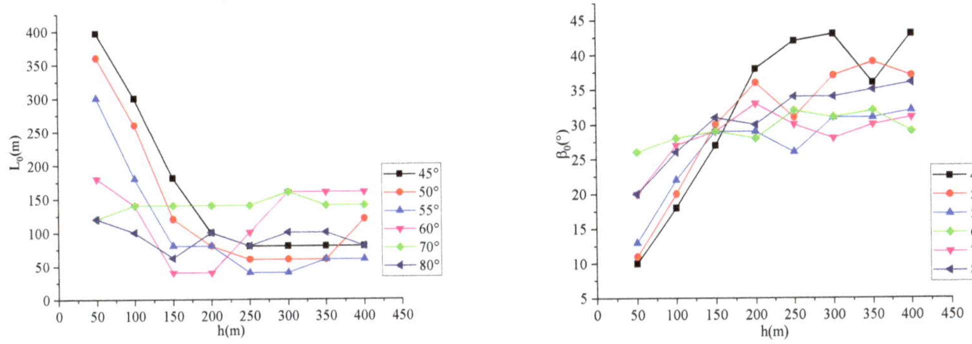

(**a**) (**b**)

Figure 12. The variation curve of the range and zoning angle in the slight subsidence area with different coal seam dip angles. (**a**) The variation curve of the range, (**b**) the variation curve of the zoning angle.

Figure 13 presents the variation curve of the surface subsidence range with different coal seam dip angles. It can be seen from the figure that with the coal seam dip angle of 55° as the dividing line, the range of the surface subsidence basin shows different trends. The larger the coal seam dip during the shallow mining, the smaller the range of the surface subsidence basin. With the deepening of mining, when the coal seam dip angle is less than 55°, a larger coal seam dip angle correlates with a larger range in the subsidence basin. When the coal seam dip angle is greater than 55°, a larger coal seam dip angle correlates with a smaller influence range in the subsidence basin.

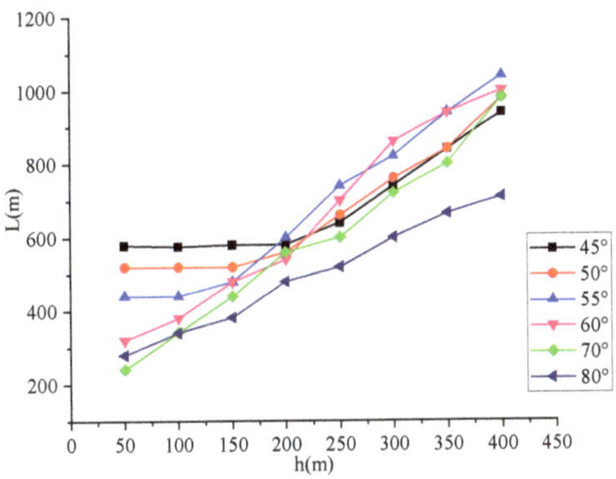

Figure 13. The variation curve of the surface subsidence range with different coal seam dip angles.

The above analysis shows that with the change in the coal seam dip angle, the surface subsidence zoning changes show different forms, and the influence of the coal seam dip angle on the mining zoning law is very obvious. When the outcrop discontinuous deformation area is shallow mining, a larger dip angle of the coal seam correlates with a smaller range of the outcrop discontinuous deformation area. However, in deep mining, a larger dip angle of the coal seam leads to a larger range, and the surface damage forms are mainly collapsed pits and step cracks. The overall subsidence area is characterized by large surface subsidence and small inclination deformation. With a change in the coal seam dip angle, the form is different: a larger coal seam dip angle increases the range, and the overall subsidence area is presented in the form of collapse basins. When the dip angle is greater than 55°, the overall subsidence area no longer appears. Most of it is integrated with the outcrop discontinuous deformation area. The surface demonstrates serious discontinuous deformation. The gradual subsidence area shows that a larger coal seam dip angle reduces the range. When the slight subsidence area is shallow mining, a larger coal seam dip angle reduces the range. However, in deep mining, a larger coal seam dip angle increases the range. At the same time, for the mining of steeply inclined extra-thick coal seam with the same dip angle, with an increase in the mining depth, the range of the outcrop discontinuous deformation area and the overall subsidence area becomes larger, while the range in the slight subsidence area becomes smaller.

4.3. Analysis of the Law of Surface Subsidence Zoning with the Change in the Coal Thickness

The influence of coal thickness on surface subsidence is mainly reflected in the degree of surface damage and subsidence. For coal seams with the same dip angle and different coal thicknesses, the damage mechanisms of the rock strata and the surface are basically the same, so the overall surface damage form and damage range have no obvious changes. However, a larger coal thickness leads to a greater outcrop discontinuous deformation in the surface, and increases the overall subsidence. Figure 14 shows the discontinuous deformation near the surface outcrop of shallow mining in a steeply inclined coal seam when the coal thickness is 20 m, 40 m, and 50 m (coal seam dip angle of 50°). It can be seen from the figure that the change om coal thickness does not directly change the discontinuous deformation form near the surface outcrop, which is mainly presented as a collapse basin. The greater the coal thickness, the greater the degree of discontinuous deformation, making the damage near the surface outcrop more serious. With deep mining, the damage near the coal seam outcrop is more serious, and the slope of the floor is steeper.

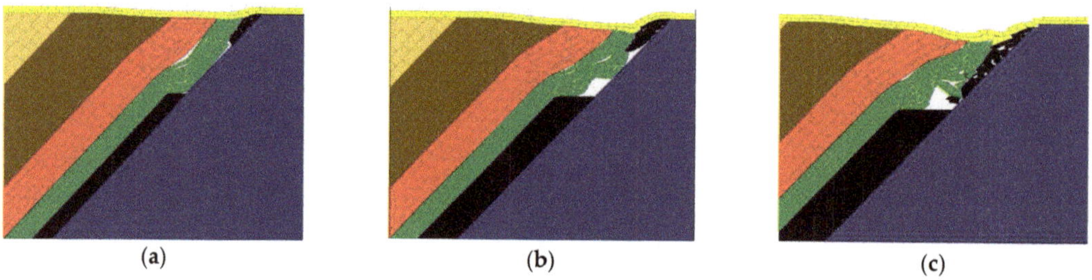

Figure 14. The discontinuous deformation near the surface outcrop of shallow mining in steeply inclined coal seam. (**a**) Coal thickness of 20 m, (**b**) coal thickness of 40 m, (**c**) coal thickness of 50 m.

Figure 15 shows the variation curve of the range and zoning angle of the outcrop discontinuous deformation area with different coal thicknesses. It can be seen from the figure that the coal thickness continuously increases the range of the outcrop discontinuous deformation area, and the surface is more likely to form discontinuous damage. With the same dip angle of the coal seam, when the steeply inclined coal seam with a different thickness is mined in the shallow mining, the discontinuous damage range of the outcrop area expands rapidly, while the range is basically unchanged when deep mining is carried out.

Figure 16 shows the variation curve of the range and zoning angle of the overall subsidence area with different coal thicknesses. With an increase in the mining depth, the range of the overall subsidence area continues to increase, and the expansion trend of the overall subsidence area with different coal thicknesses is basically consistent. For steeply inclined coal seams with the same dip angle and different coal thicknesses, a smaller coal thickness makes it easier for an overall subsidence area to form. A larger coal thickness makes the discontinuous deformation of the surface more serious, increases the influence range of the outcrop discontinuous deformation, and reduces the range of the overall subsidence. On the whole, there is little difference in the overall subsidence range corresponding to different coal thicknesses.

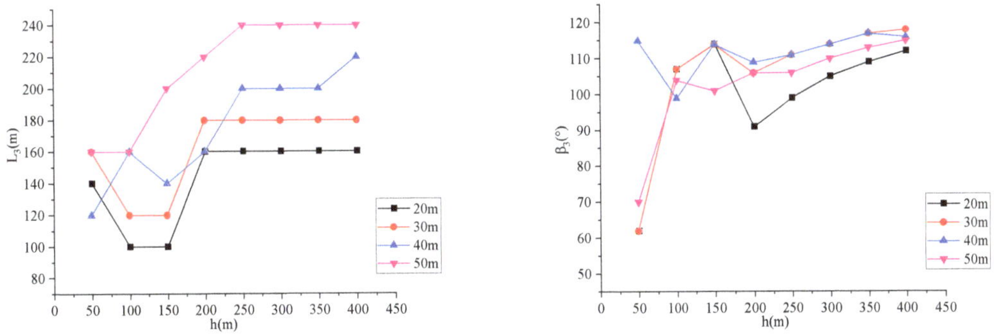

(**a**) (**b**)

Figure 15. The variation curve of the range and zoning angle of the outcrop discontinuous deformation area with different coal thickness. (**a**) The variation curve of the range, (**b**) the variation curve of the zoning angle.

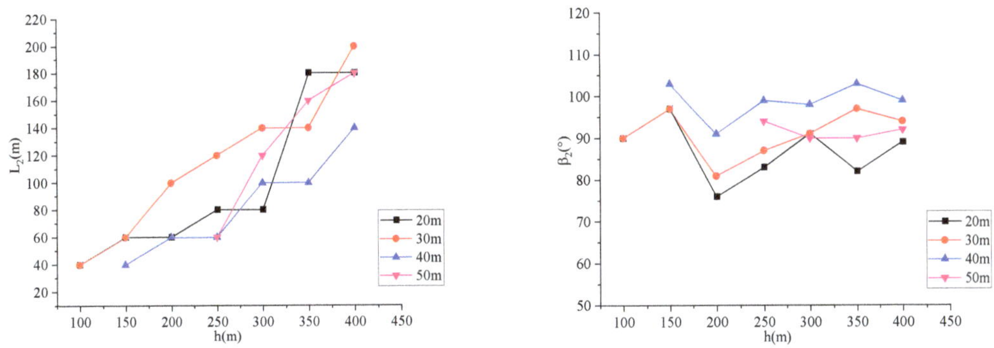

(**a**) (**b**)

Figure 16. The variation curve of the range and zoning angle of the overall subsidence area with different coal thickness. (**a**) The variation curve of the range, (**b**) the variation curve of the zoning angle.

Figure 17 shows the variation curve of the range and zoning angle of the gradual subsidence area with different coal thicknesses. It can be seen from the figure that the change in the gradual subsidence area is basically the same in steeply inclined coal seam mining with the same coal seam dip angle and different coal thickness. From the overall trend, a larger coal thickness increases the gradual subsidence area, but the coal thickness has little effect on the change in the gradual subsidence area.

Figure 18 shows the variation curve of the range and zoning angle in the slight subsidence area with different coal thicknesses. It can be seen from the figure that an increase in the mining depth changes the range of slight subsidence area corresponding to different coal thicknesses from large to small. This is because in the process of deep mining, the range of the original slight subsidence area changes into the gradual subsidence area due to an increase in the subsidence, so the range in the slight subsidence area shrinks and the gradual subsidence area expands. At the same time, the variation trend in the slight subsidence area corresponding to different coal thicknesses is consistent. There is no significant difference in the range in the slight subsidence area between different coal thicknesses. On the whole, a larger coal thickness increases the range in the slight subsidence area.

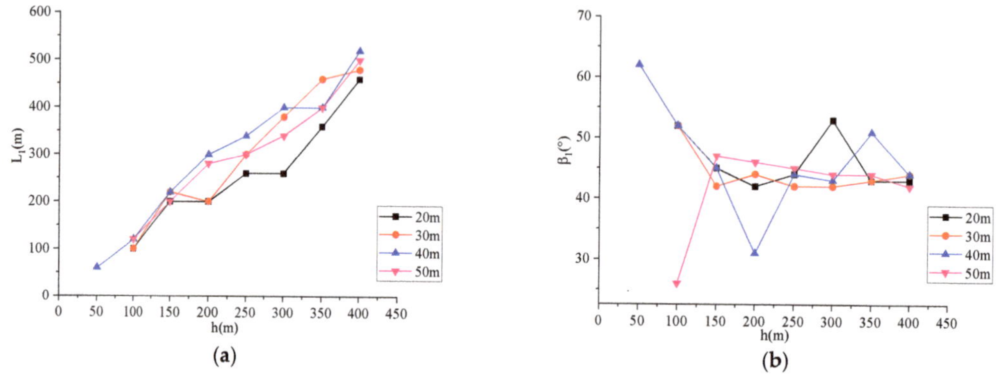

Figure 17. The variation curve of the range and zoning angle of the gradual subsidence area with different coal thickness. (**a**) The variation curve of the range, (**b**) the variation curve of the zoning angle.

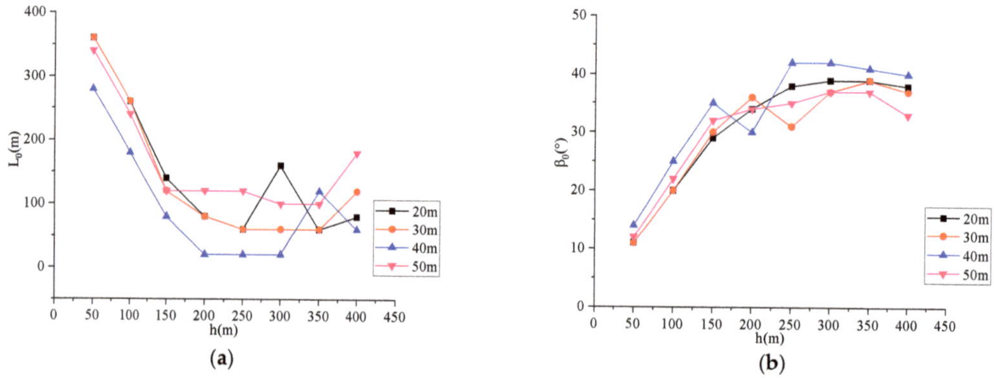

Figure 18. The variation curve of the range and zoning angle in the slight subsidence area with a different coal thickness. (**a**) The variation curve of the range, (**b**) the variation curve of the zoning angle.

Figure 19 shows the variation curve of the surface subsidence range with different coal seam thicknesses. It can be seen from the figure that when in the shallow mining, the range of the surface subsidence basin is not significantly different. With the deepening of mining, the range of the surface subsidence basin grows steadily, and the range continues to expand. At the same time, a larger coal thickness increases the range of the surface subsidence basin, but the influence of different coal thicknesses on the range of the surface subsidence basin is not obvious.

The above analysis shows that the influence of coal thickness on the surface subsidence zoning is mainly reflected in the subsidence of each area and has no significant influence on the range of each area. The influence range of each area corresponding to different coal thicknesses is basically the same, but the damage degree is quite different. On the whole, the difference in the influence of coal thickness on the range of the surface subsidence basins is mainly reflected in the range of the outcrop discontinuous deformation area. A larger coal thickness increases the outcrop discontinuous deformation area, making the damage more serious, and making the surface more likely to form cracks and steps. The range of other areas keeps the same trend as the mining depth increases.

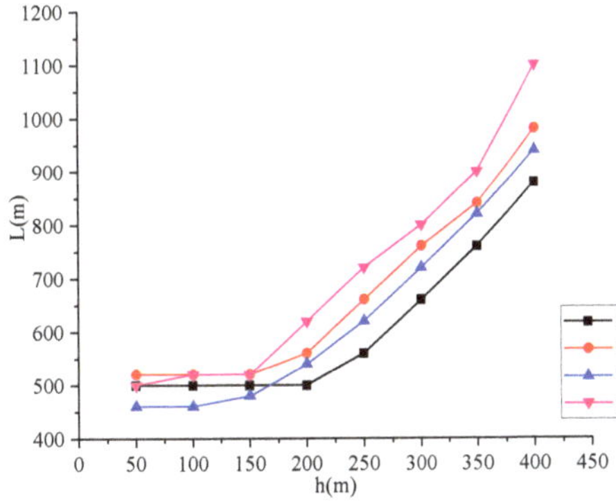

Figure 19. The variation curve of the surface subsidence range with different coal seam thickness.

5. Discussion

5.1. Variation Law of the Outcrop Discontinuous Deformation Area

Figure 20 shows the relationship between the coal seam dip angle and the zoning angle of the discontinuous deformation area, where SSE represents the sum of squares of the error term. A smaller SSE value correlates with a smaller error between the fitting curve and the sample value, thus improving the fitting degree. The figure shows a negative linear relationship between the cotangent value of the coal seam dip angle and the cotangent value of the zoning angle of the discontinuous deformation area. With an increase in the cotangent value of the coal seam dip angle, the cotangent value of the zoning angle of the outcrop discontinuous deformation area becomes smaller and smaller.

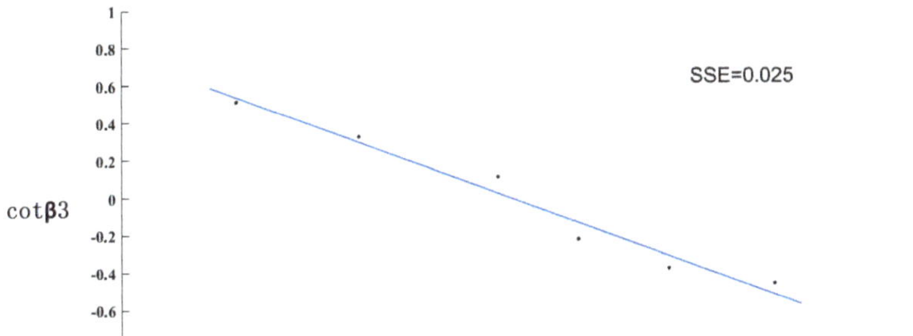

Figure 20. Relationship between the coal seam dip angle and the zoning angle of the discontinuous deformation area.

Figure 21 shows the relationship between the depth–thickness ratio (ratio of stage vertical height to coal seam thickness) and the zoning angle of the discontinuous deformation area. It can be seen from the figure that the cotangent value of the zoning angle of the outcrop discontinuous deformation area is negatively correlated with the depth–thickness ratio.

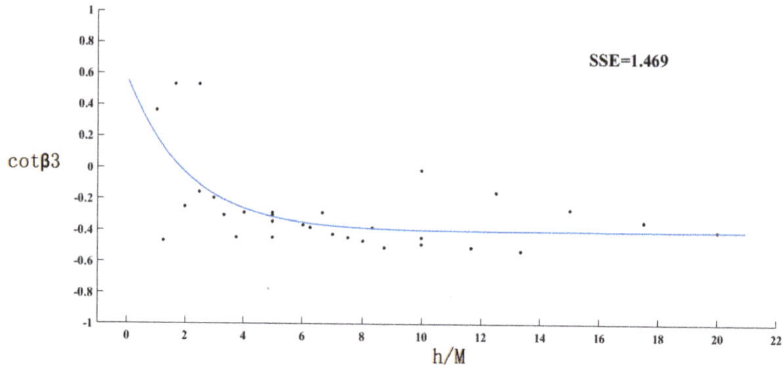

Figure 21. Relationship between the depth–thickness ratio and the zoning angle of the discontinuous deformation area.

Comprehensive analysis of the relationship between the zoning angle of the outcrop discontinuous deformation area and the coal seam dip angle and the depth–thickness ratio is shown in Formula (9):

$$\cot\beta_3 = ae^{-b\frac{h}{M}} + c\cot\alpha + d \tag{9}$$

By using the indirect adjustment method to obtain the parameters of Formula (9), the fitting formula can be obtained as follows:

$$\beta_3 = \mathrm{arccot}(0.492e^{-0.482\frac{h}{M}} + 0.643\cot\alpha + 0.179)\ \left(\begin{array}{c} 45^\circ < \alpha < 80^\circ \\ h > 0m \end{array}\right) \tag{10}$$

5.2. Variation Law of the Overall Subsidence Area

The coal seam dip angle greatly influences the overall subsidence area. When the coal seam dip angle is greater than 55°, the overall subsidence area of the surface no longer appears and becomes the outcrop discontinuous deformation area.

Figure 22 shows the relationship between the depth–thickness ratio and the zoning angle of the overall subsidence area. It can be seen from the figure that the cotangent value of the zoning angle of the overall subsidence area is linearly and positively correlated with the depth–thickness ratio. The overall trend of the curve shows that a greater depth–thickness ratio leads to a greater cotangent value of the zoning angle of the overall subsidence area. When the SSE is small, the fitting degree is good.

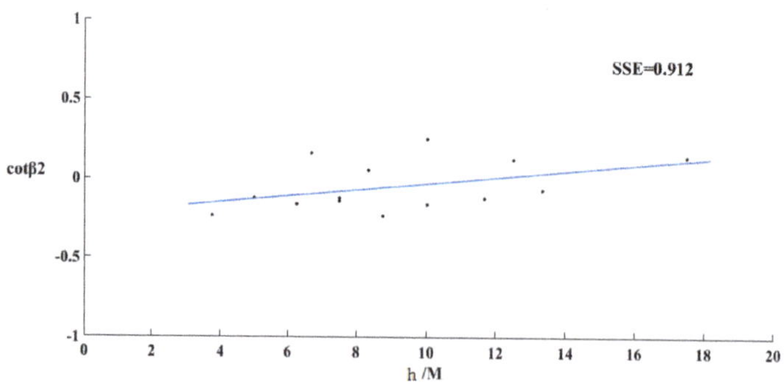

Figure 22. Relationship between the depth–thickness ratio and the zoning angle of the overall subsidence area.

Through comprehensive analysis of the relationship between the depth–thickness ratio and the zoning angle of the overall subsidence area, the relationship is shown in Formula (11):

$$\cot\beta_2 = a\frac{h}{M} + b \tag{11}$$

By using the indirect adjustment method to obtain the parameters of Formula (11), the fitting formula can be obtained as follows:

$$\beta_2 = \operatorname{arccot}(0.012\frac{h}{M} - 0.158)\begin{pmatrix} 45^\circ < \alpha < 55^\circ \\ h > 250m \end{pmatrix} \tag{12}$$

5.3. Variation Law of the Gradual Subsidence Area

Figure 23 shows the relationship between the coal seam dip angle and the zoning angle of the gradual subsidence area. It can be seen from the figure that the cotangent value of the coal seam dip angle and the cotangent value of the zoning angle of the gradual subsidence area have a negative linear relationship. The larger the cotangent value of the coal seam dip angle, the smaller the zoning angle of the gradual subsidence area.

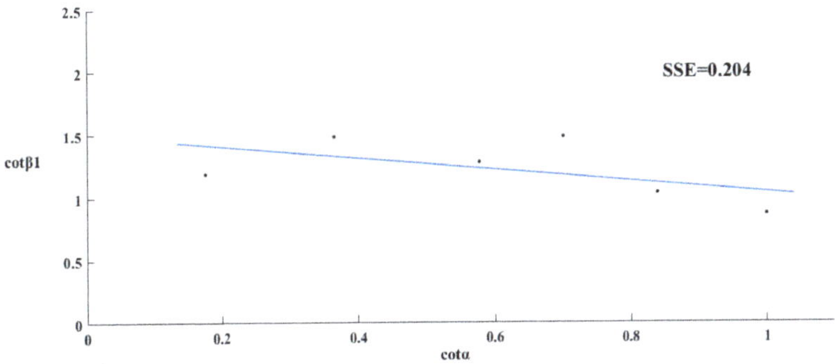

Figure 23. Relationship between the coal seam dip angle and the zoning angle of the gradual subsidence area.

Figure 24 shows the relationship between the depth–thickness ratio and the zoning angle of the gradual subsidence area. It can be seen from the figure that the cotangent value of the zoning angle of the gradual subsidence area is positively correlated with the depth–thickness ratio. The curve fitting degree is good.

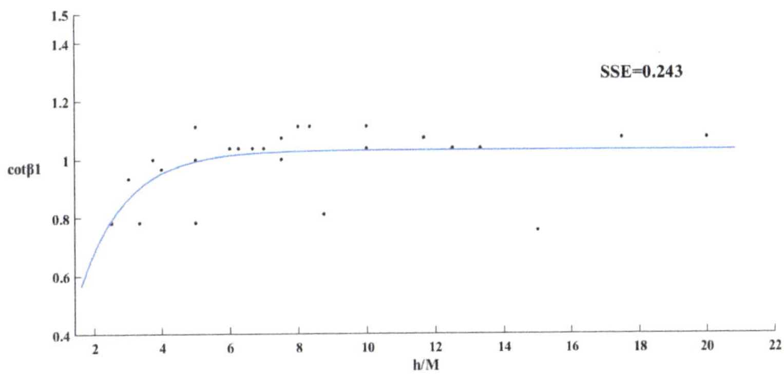

Figure 24. Relationship between the depth–thickness ratio and the zoning angle of the gradual subsidence area.

Comprehensive analysis of the relationship between the zoning angle of the gradual subsidence area and the coal seam dip angle and the depth–thickness ratio is shown in Formula (13):

$$\cot\beta_1 = ae^{-b\frac{h}{M}} + c\cot\alpha + d \quad (13)$$

By using the indirect adjustment method to obtain the parameters of Formula (13), the fitting formula can be obtained as follows:

$$\beta_1 = \text{arccot}(-0.778e^{-0.742\frac{h}{M}} - 0.216\cot\alpha + 0.765)\begin{pmatrix} 45^\circ < \alpha < 80^\circ \\ h > 100m \end{pmatrix} \quad (14)$$

5.4. Variation Law of the Slight Subsidence Area

Figure 25 shows the relationship between the coal seam dip angle and the zoning angle in the slight subsidence area. There is a negative linear relationship between the cotangent value of the coal seam dip angle and the cotangent value of the zoning angle in the slight subsidence area, and the curve fitting is good.

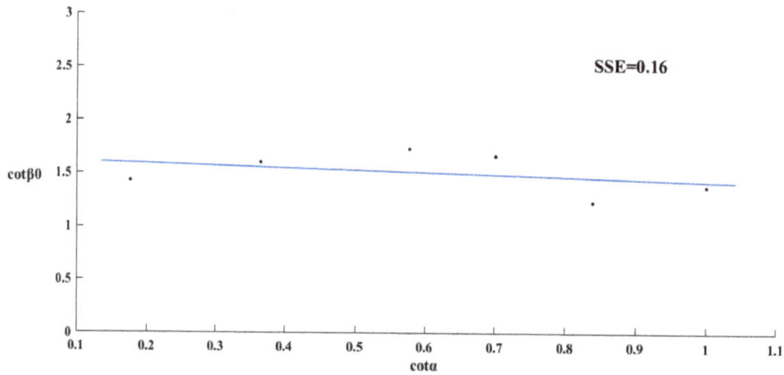

Figure 25. Relationship between the coal seam dip angle and the zoning angle i the slight subsidence area.

Figure 26 shows the relationship between the depth–thickness ratio and the zoning angle in the slight subsidence area. It can be seen from the figure that the depth–thickness ratio and the cotangent value of the zoning angle in the slight subsidence area have a negative exponential relationship, and the curve fitting is in good condition.

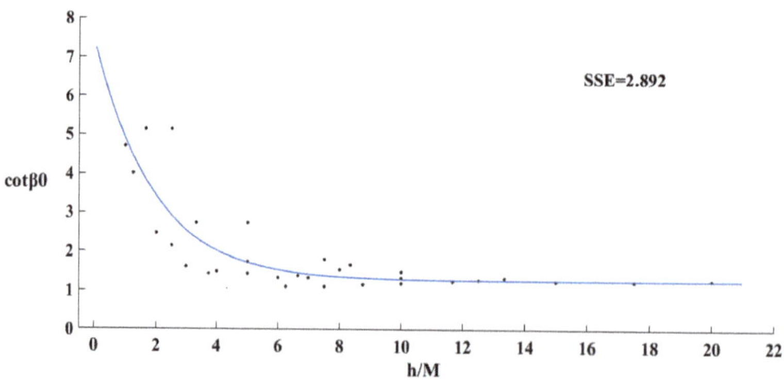

Figure 26. Relationship between the depth–thickness ratio and the zoning angle in the slight subsidence area.

The comprehensive analysis of the relationship between the zoning angle in the slight subsidence area and the coal seam dip angle and the depth–thickness ratio is shown in Formula (15):

$$\cot\beta_0 = ae^{-b\frac{h}{M}} + c\cot\alpha + d \quad (15)$$

By using the indirect adjustment method to obtain the parameters of Formula (15), the fitting formula can be obtained as follows:

$$\beta_0 = \text{arccot}(3.067e^{-0.52\frac{h}{M}} - 0.103\cot\alpha + 1.446)\left(\begin{array}{c}45^\circ < \alpha < 80^\circ \\ h > 0m\end{array}\right) \quad (16)$$

In summary, the zoning angle calculation model of each surface moving area is :

$$\begin{cases} \beta_3 = arc\cot(0.492e^{-0.482\frac{h}{M}} + 0.643\cot\alpha + 0.179) & (45^\circ < \alpha < 80^\circ, h > 0m) \\ \beta_2 = arc\cot(0.012\frac{h}{M} - 0.158) & (45^\circ < \alpha < 55^\circ, h > 250m) \\ \beta_1 = arc\cot(-0.778e^{-0.742\frac{h}{M}} - 0.216\cot\alpha + 0.765) & (45^\circ < \alpha < 80^\circ, h > 100m) \\ \beta_0 = arc\cot(3.067e^{-0.52\frac{h}{M}} - 0.103\cot\alpha + 1.446) & (45^\circ < \alpha < 80^\circ, h > 0m) \end{cases} \quad (17)$$

Then, the range calculation model of each surface moving area is:

$$\begin{cases} L_3 = H(\cot\lambda_0 - \cot\beta_3) \\ L_2 = H(\cot\beta_2 - \cot\beta_3) \\ L_1 = H(\cot\beta_1 - \cot\beta_2) \\ L_0 = H(\cot\beta_0 - \cot\beta_1) \end{cases} \quad (18)$$

Some scholars took the shallow-buried thick coal seam mining area as the research background and divided the different development areas of ground cracks through field measurement and similar material physical tests. The surface was divided into a crack-producing area, crack transfixion development area, and a stable crack area [26,27]. However, due to the severe surface movement and deformation caused by the mining of steeply inclined coal seam, the above research results are not applicable to this geological condition.

Some scholars' research on the law of surface subsidence zoning after mining in a thin and medium-thickness coal seam is only qualitative, lacking quantitative representation [16–18]. Moreover, with an increase in the coal thickness, the law of surface subsidence zoning becomes more complex, and the previous research results are no longer applicable.

For steeply inclined thick coal seam mining, the surface subsidence law of the trend main section is similar to that of the horizontal coal seam, and some scholars have carried out relevant research on this [28,29]. Therefore, this paper aims to study the surface subsidence zoning law of the inclination main section in the mining of steeply inclined extra-thick coal seams.

In this paper, a change in the surface tilt is used as the criterion to determine the area boundary. However, this criterion needs further research in the future to determine a more accurate criterion. At the same time, the zoning law of the whole subsidence basin has not been studied in detail in this paper. The next step is to extend the zoning law of the profile to the whole basin. The research results can be applied to the design and implementation of coal mining schemes under buildings (structures) in steeply inclined thick coal seams, which provides a basis for the safe mining of steeply inclined thick coal seams under buildings (structures).

6. Conclusions

In this paper, a calculation formula for the variation of each area range and zoning angle with a coal seam dip angle, coal thickness, mining depth, and stage vertical height area is established. The characteristics and deformation laws of rock strata and surface movement in the mining of steeply inclined extra-thick coal seam in Dongxia Coal Mine are validated using the methods of field measurement, numerical simulation, and theoretical analysis. Based a systematic analysis, the following conclusions can be drawn:

(1) Surface subsidence basins can be divided into four areas: the outcrop discontinuous deformation area, the overall subsidence area, the gradual subsidence area, and the slight subsidence area;
(2) The influence of the coal seam dip angle on surface subsidence zoning of steeply inclined extra-thick coal seam mining is mainly reflected in the range of the affected area and the form of damage;
(3) The influence of coal thickness on surface subsidence zoning of steeply inclined thick coal seam mining is mainly reflected in the surface subsidence and the degree of damage. On the whole, a larger coal thickness can increase the range of surface subsidence basins and subsidence zoning.

Author Contributions: Conceptualization, Y.Y.; methodology, Y.Y., Y.Z. (Yuanhao Zhu) and Y.Z. (Yanjun Zhang); software, Y.Y. and Y.Z. (Yuanhao Zhu); validation, Y.Z. (Yuanhao Zhu); formal analysis, Y.Z. (Yuanhao Zhu) and Y.Z. (Yanjun Zhang); investigation, Y.Z. (Yuanhao Zhu) and J.C.; resources, J.C.; data curation, J.C.; writing—original draft preparation, Y.Y. and Y.Z. (Yanjun Zhang); Writing—review & editing, Y.Y. and Y.Z. (Yanjun Zhang); visualization, J.W.; supervision, J.W.; project administration, Y.Y.; funding acquisition, Y.Y. All authors have read and agreed to the published version of the manuscript.

Funding: This research was funded by the National Natural Science Foundation of China (grant number: 51404272), and the Fundamental Research Funds for the Central Universities (grant number: 2022YJSDC20). The APC was funded by 2022YJSDC20.

Conflicts of Interest: The authors declare that they have no competing interests.

References

1. Lai, X.P.; Dai, J.J.; Li, C. Analysis on hazard characteristics of overburden structure in steeply inclined coal seam. *J. China Coal Soc.* **2020**, *45*, 122–130.
2. Sun, C.; Chen, D.X.; Cheng, Y.H.; Lu, J.X. Study on collapse rule and control of hard roofs in steeply inclined coal seams. *Chin. J. Rock Mech. Eng.* **2019**, *38*, 1647–1658.
3. Wang, J.C.; Yang, S.L.; Li, L.H. Toppling-slumping failure mode in horizontal sublevel top-coal caving face in steeply-inclined seam. *J. China Univ. Min. Technol.* **2018**, *47*, 1175–1184.
4. Zhang, J.W. The fracture mechanism of main roof stratum and strong mine pressure control method in long wall mining of steeply inclined coal seam in Wangjiashan colliery. *Chin. J. Rock Mech. Eng.* **2018**, *37*, 1776.
5. Huang, X.C. Study on stress and fracture evolution law of steeply inclined coal seam. *Coal Sci. Technol.* **2018**, *46*, 93–96.
6. Zhong, H.L. Experience of Coal Mining under Buildings in Dunbas Mining Area of Soviet Union. *Coal Min. Technol.* **1991**, *2*, 55–58.
7. Afsari, N. Numerical modelling of surface subsidence arising from longwall mining of steeply inclined coal seams. In Proceedings of the Turkish Coal Congress, Bartm-Amasra, Turkey, 10–12 June 1998.
8. Alejano, L.R.; Ramírez-Oyanguren, P.; Taboada, J. FDM predictive methodology of subsidence due to flat and inclined coal seam mining. *Int. J. Rock Mech. Min. Sci.* **1999**, *36*, 475–491. [CrossRef]
9. Asadi, A.; Shahriar, K.; Goshtasbi, K.; Najm, K. Development of a new mathematical model for prediction of surface subsidence due to inclined coal-seam mining. *J. S. Afr. Inst. Min. Metall.* **2005**, *105*, 15–20.
10. Ma, F.; Sun, L.; Li, D. Numerical simulation analysis of covering rock strata as mining steep-inclined coal seam under fault movement. *Trans. Nonferrous Met. Soc. China* **2011**, *21*, 556–561. [CrossRef]
11. Wang, M.L.; Hu, B.N.; Cheng, X.H.; Zhang, G.Y. Numerical simulation of overburden failure and coal pillar stabilityin steep multi-seam mining. *Coal Min. Technol.* **2005**, *10*, 64–66.
12. Yang, F. Study on Overburden Strata's Movement Pattern and Mechanism of Mining Steeply Inclined Seams. Ph.D. Thesis, Liaoning Technical University, Shenyang, China, 2006.
13. Yi, S.H. The Research on Surface Subsidence Due to Mining Horizontal Slice of Extra-Thick and Steep Coal Seam. Ph.D. Thesis, China University of Mining and Technology, Beijing, China, 2009.
14. Dai, H.Y.; Tiao, X.G.; Zhao, S.L. Rules of surface movement induced by mining steeply inclined coal seam in majiagou colliery. *Coal Min. Technol.* **2010**, *15*, 48–50.
15. Dai, H.Y. Study on seam inclination based mining subsidence model and its visualized application with GIS. *Chin. J. Rock Mech. Eng.* **2002**, *21*, 148.

16. Yan, Y.G.; Guo, S.C.; Liu, J.B.; Zhang, G.G.; Wang, H.L. Principle and rock strata control effect based on mining mixed with backfilling and keeping. *Coal Min. Technol.* **2016**, *21*, 64–68.
17. Yan, Y.G.; Dai, H.Y.; Wang, Z.W.; Jing, S.Q.; Chen, H.Z. Ground subsidence zone and surrounding rock failure mechanism due to steep multiple coal seam mining: A case study at Muchenyjian Datai Mine. *J. China Univ. Min. Technol.* **2013**, *42*, 547–553.
18. Yan, Y.G.; Dai, H.Y.; Wang, Z.W.; Zhu, G.H.; Chen, H.Z.; Xu, C. Study on failure characteristics and normal height of caving zone due to steep seam mining. *Coal Sci. Technol.* **2015**, *43*, 23–26.
19. Zhang, C.; Zhao, Y.X.; Han, P.H.; Bai, Q.S. Coal pillar failure analysis and instability evaluation methods: A short review and prospect. *Eng. Fail. Anal.* **2022**, *138*, 06344. [CrossRef]
20. Zhang, Y.; Yan, Y.; Dai, H.; Zhu, Y.; Wu, T. Stability and force chain characteristics of "inclined step cutting body" in stope. *Appl. Sci.* **2021**, *11*, 10276. [CrossRef]
21. Emad, M.Z.; Mitri, H.; Kelly, C. Dynamic model validation using blast vibration monitoring in mine backfill. *Int. J. Rock Mech. Min. Sci.* **2018**, *107*, 48–54. [CrossRef]
22. Wang, Y.; Ye, H.W.; Li, Y.Z. Numerical simulation on jointed and fractured rock blasting. *Blasting* **2012**, *29*, 20–22.
23. Wang, H.; Gao, Y.T.; Jin, A.B.; Zhang, K. Determination of stiffness parameters of jointed rock masses with 3dec simulations. *Chin. J. Rock Mech. Eng.* **2014**, *33*, 2894–2900.
24. Yan, Q.R.; Feng, J.; Yang, T.; Deng, W.D. Numerical experiment method to determining mechanical parameters of jointed rock mass. *J. Tongji Univ.* **2017**, *45*, 187–194.
25. Yang, F.; Yu, H.F.; Guo, J.T. Mechanism of mining-induced surface crack by numerical simulation method. *J. Liaoning Tech. Univ.* **2016**, *35*, 566–570.
26. Xu, F.Y. Surface Cracks Distribution Characteristics and Development Mechanism in High-Intensity Mining of Nearly Shallow Thick Coal Seam. Ph.D. Thesis, Henan Polytechnic University, Jiaozuo, China, 2019.
27. Guo, W.B.; Xu, F.Y. Numerical simulation of overburden and surface movements for Wongawilli strip pillar mining. *Int. J. Min. Sci. Technol.* **2016**, *26*, 71–76. [CrossRef]
28. Xu, Z.H.; Zhu, R.S.; He, W.R.; Li, X.B. Subsidence model of large working face in shallow buried coal seam with thick loose layer. *J. Min. Saf. Eng.* **2020**, *37*, 264–271.
29. Wang, T.B. Research on the influence range of coal mining in mountainous areas of Guizhou. *Coal Eng.* **2018**, *50*, 141–144.

sustainability

Article

Study on the Mechanism of a Hanging Roof at a Difficult Caving End in a Fully-Mechanized Top Coal Caving Face

Hongtao Liu [1,*], Cheng Hao [1], Zijun Han [1], Qinyu Liu [1], Haozhu Wang [1], Jialu Liang [1] and Dandan Zhang [2]

1 School of Energy and Mining Engineering, China University of Mining and Technology, Beijing 100083, China
2 School of Emergency Management and Safety Engineering, China University of Mining and Technology, Beijing 100083, China
* Correspondence: 108925@cumtb.edu.cn

Abstract: During the mining process of fully-mechanized caving faces, the roof of the roadway behind the working face easily forms an arched-shape hanging roof structure with the working face pushed forward, which results in potential hazards such as gas accumulation and large-scale roof collapse. Based on the actual situation of a hanging roof at a difficult caving end in fully-mechanized top coal caving faces, through borehole exploration, surrounding rock displacement observation, bolt stress monitoring, theoretical formula calculation, and numerical simulation methods, the structure characteristics of the hanging roof at the end of the fully-mechanized caving face are studied. The ultimate failure depth and ultimate break distance of the hanging roof structure at the end of the working face are obtained, and its formation mechanism is analyzed. It is concluded that the hanging structure is formed by the following reasons: the lithology of sandy mudstone and fine sandstone above the top coal of the roadway is strong; the hanging roof structure is less affected by working-face mining; there is a result of insufficient rotary pressure of the upper mudstone while working together with the protective coal pillar and end support the caving step distance of the curved hanging roof structure is 10~13.55 m.

Keywords: end hanging roof; caving mining method; roof caving; numerical simulation

1. Introduction

In coal mining, with the working face forward, the periodic fracture of the roof will form an arc-shaped hanging roof at the end of the working face [1,2]. The existence of a certain range of arc triangular plate structure is conducive to the stability of the roadway end and ensures the safety of the mining face. However, when the hanging roof area is too large, there will be potential safety hazards such as gas accumulation and large-scale roof collapse. The arched-shape hanging roof easily causes high underground stress and the accumulation of gas and dust [3,4]. When the hanging roof structure rotates and breaks, a large area of pressure will be generated, leading to rock burst, coal and gas outburst, or large deformations in roadways. On the other hand, in the process of moving, the end support will inevitably rub and collide with the end of the bolt at the top of the roadway, which may produce sparks and lead to the explosion of gas accumulated at the end, causing unpredictable disasters [5–11].

As one of the common problems in mine disasters, the end-hanging problem has been studied by many scholars.

Wang et al. [12] took the hard roof as the research object and, using a mechanical model and theoretical research methods, obtained a reasonable length of the hanging roof of the working face end roadway under the premise of safe production.

He et al. [13] selected the key strata in the roof of a mined-out area as the research object and simplified it into the rock beam model for analysis. Considering the bending deformation in roof strata, the creep model is used to discuss the characteristics of buckling load time and instability conditions.

Wang et al. [14] discussed the rheological properties of materials under different conditions and obtained three cases of delayed instability of hanging rooves in goaf.

Wang et al. [15] analyzed the structure of the curved triangular plate of the roof after mining, established the mechanical model of the rectangular plate structure, determined the failure position of the hanging plate, and explained the generation mechanism of the curved triangular plate.

Qin et al. [16], based on the thin plate theory of elasticity, analyzed the rectangular plate structure with fixed support on both sides. It is considered that the hanging roof structure can be simplified as a thin plate with two sides fixed. The bending moment calculation formula of the hanging plate structure is obtained by using the generalized simply-supported edge concept and the superposition algorithm.

Luan et al. [17] established the mechanical model of gob-side entry retaining, analyzed the relationship between the characteristics of mine pressure behavior and the arc triangle plate at the end of the retaining roadway, and proposed the position prediction and measurement method of the arc triangle plate.

Liu et al. [18,19] analyzed the basic roof structure of the working face end, used a three-parameter Weibull function to analyze the two-way abutment pressure in the triangular area of the end, and put forward the transfer method of two-way abutment pressure.

Xu [20] studied the coupling relationship between hydraulic support and roof strata by means of theoretical analysis and numerical simulation and analyzed the hanging mechanism of the arc triangular plate at the end of the working face in the Jinjitan coal mine.

Pavlova et al. [21] used a numerical simulation research method to the influence of different length of the hanging roof, and the relationship between the displacement and stress distribution of the working face and the length of the hanging roof is obtained, which provides a basis for formulating preventive measures for the collapse of the hanging roof structure.

Xue et al. [22] used the research method of numerical simulation to discuss the influence of the length of the hanging roof and periodic weighting step on the stress in front of the coal wall, the classification prediction method of rock burst hazard is established, and the countermeasures for different hazard levels of rock burst are proposed.

Dychkovskyi et al. [23] used SolidWorks 2019 software to evaluate stress of a mine field in terms of Lvivvuhillia SE mine and analyzed the properties of the support pressure zone formed in front of the stope and along the length of the pillar. It is considered that, if the stope is in the same plane, its front support pressure area and the wall of the development work area will connect to each other. Based on the analysis of the stress and strain state of rock mass in the process of double unit longwall mining, the effective mining parameters and the control method of rock pressure are determined.

By analyzing and evaluating the geological conditions, technological parameters, and support methods of longwall mining and using methods of statistics, investigation, and analysis of field data, Vu [24] pointed out the causes and laws of spalling and roof fall in longwall fully-mechanized mining faces. In addition, some measures to prevent spalling and roof fall are put forward.

Chen et al. [25], based on the research background of thick and hard hanging roof in Datong mining area, established a mechanical model of roof periodic collapse based on load and instability characteristics, and the relationship between the breaking distance, breaking angle, and the working resistance of the support is studied.

In the previous studies, most of them studied the reasonable length of the hanging roof structure and its impact on the stress of the roadway, or their theoretical calculations, but few of them studied the mechanism of the hanging roof structure from multiple angles, which combines field practice, theoretical analysis, and numerical simulation. The purpose of this paper is to study the collapse form and collapse mechanism of the end arched-shape hanging roof structure. The research results can provide reference for the roof control of the working face end in the process of advancing the working face, so as to ensure the safety of the working face end and ensure the safe production of the working face.

2. Background

2.1. Mining Conditions of the Working Face

As shown in Figure 1, Wangjialing coal mine is located in the Shanxi Province, its area is 119.71 km^2, including 5 layers of recoverable and locally recoverable coal seams. The main mining coal seams are No. 2 coal seams and No. 10 coal seams, which can be mined in the whole area. The dip angle of coal seams is small and does not change much, with an average of 3°~5°. The roof of the coal seam is stable, the geological structure is relatively simple, the hydrogeological conditions are simple, the water inflow is small, the gas content is low, and the mining technical conditions are good.

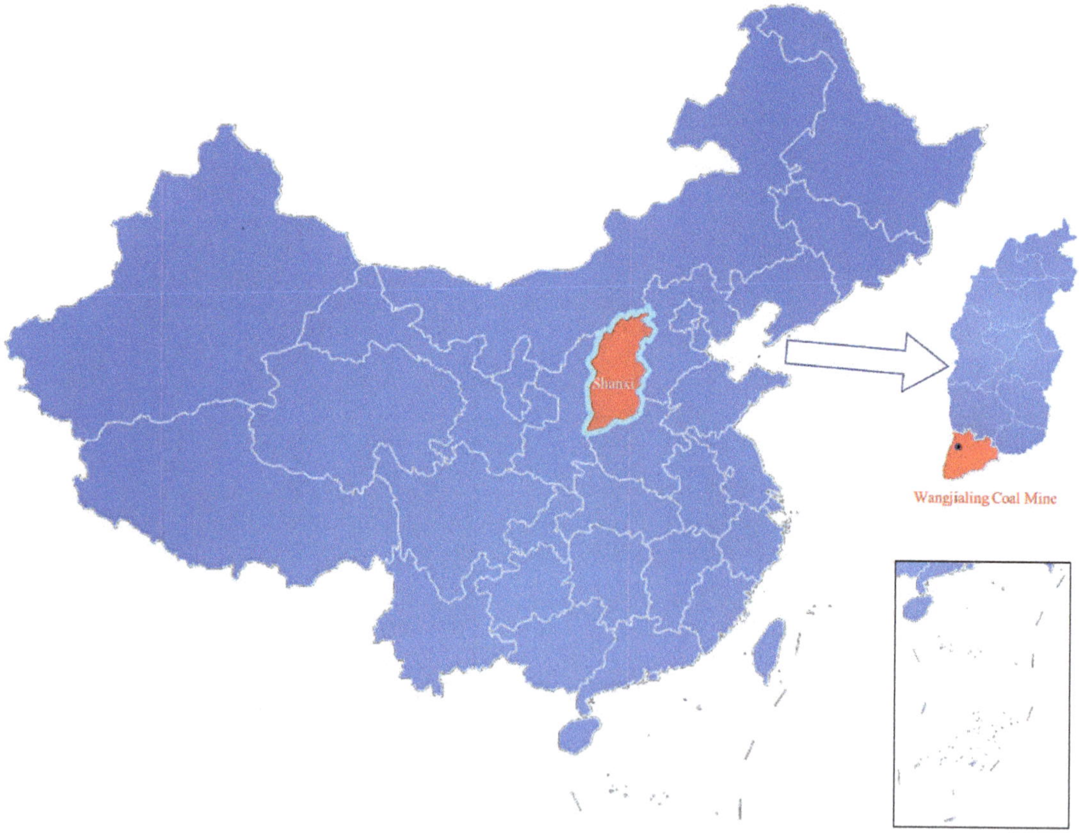

Figure 1. The location of Wangjialing coal mine.

At present, the mine is mining the No. 2 coal seam, with a thickness of 3.09–8.50 m at an average of 6.20 m. The coal seam generally contains 1–2 layers of dirt band. The roof of the coal seam is sand mudstone, above is fine sandstone, the floor is sand mudstone, the comprehensive column is shown in Figure 2a, and the local floor contains fine sandstone and quartz sandstone.

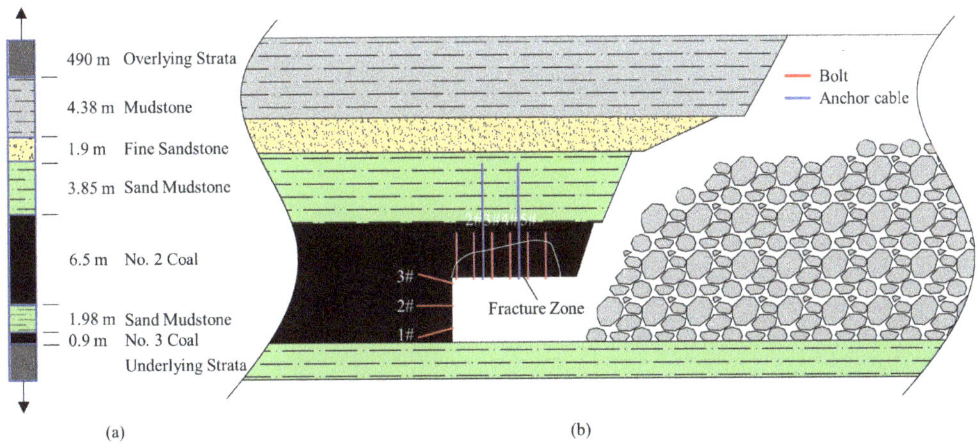

Figure 2. (**a**) Comprehensive column of the roof and floor of the No. 2 coal seam; (**b**) the schematic diagram of the hanging roof at the end of the goaf.

The Platts coefficient of the No. 2 coal seam is about 1.5, the coal seam is soft, the joint fissure is developed, and the top coal caving is good. In the process of working-face mining, a large area of hanging roof appears in the end area of the working face, as shown in Figure 2b, which brings great safety hazard to mine production.

The plan of the working face is shown in Figure 3. The study working face is located in the northern area of its panel. It is arranged in two lanes, which are composed of a haulage roadway, return air roadway, cut hole, and related roundabout and chamber. The average buried depth of the working face is 500 m, the length is 299.6 m, and the advancing length is 3316 m. The working face adopts fully-mechanized top coal caving mining technology. The mining height is 3.1 m, the average height of coal caving is 3.23 m, the mining ratio is 1:1.04, and the cycle progress is 0.85 m. Both sides of the working face are not mined.

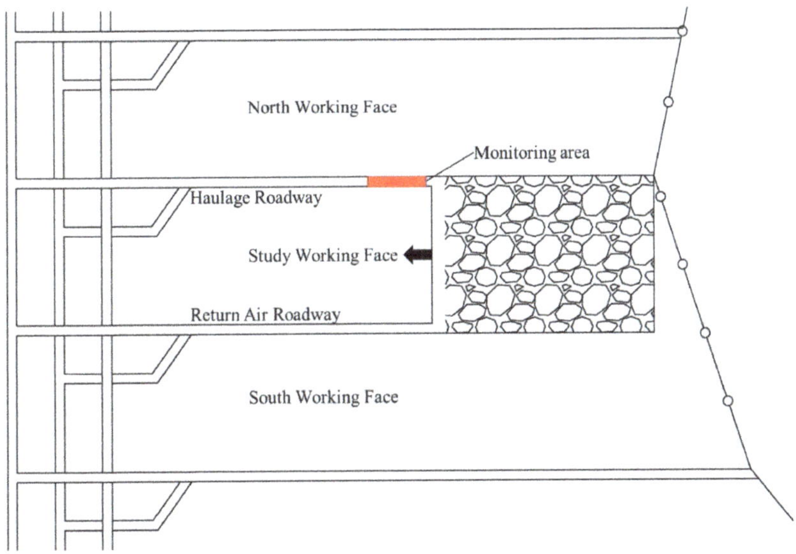

Figure 3. The plan of the working face layout.

2.2. *Roadway Support Parameters*

The working face's haulage roadway has a width of 5.6 m, height of 3.1 m, and uses "anchor-net-cable" support; roadway roof-specific support measures are as follows:

Top bolt: φ20 × 2500 mm left-handed thread steel is used, and the row spacing is 1000 × 900 mm. In addition, φ14 mm steel beam is used to connect the top bolt at the same time in the roadway roof, and the top mesh is φ6 mm steel mesh. The bearing capacity of the bolt is 10 tons.

Top anchor cable: φ17.8 × 6250 mm steel strand, spacing of 2000 × 1800 mm, 2-1-2 arrangement. The bearing capacity of the anchor cable is 18.6 tons.

It can be found from the support conditions of the haulage roadway in the working face that the support strength of the roadway roof is large, and there are 1.5 anchor cables and 6 bolts in the roof of the roadway per meter on average. According to the structural characteristics of the roadway roof strata mentioned above, the supporting strength of the haulage roadway in the working face is too large, and the roadway is only affected by one mining. The influence of mining on the roadway roof is limited, so it is easy to form a hanging structure at the end of the working face during the advance of the working face.

3. Study on the Law of Ground Pressure Behavior in the Working Face

3.1. *Deformation Characteristics of Surrounding Rock in a Mining Roadway*

In order to observe the deformation degree of the surrounding rock of the roadway during the mining process of the working face, the mine pressure monitoring station is arranged at 40 m ahead of the working face of the haulage roadway in the working face, the displacement of the two sides of the roadway and the roof and floor during the advancement of the working face are continuously observed, and the observation results are recorded. The layout of the observation points is shown in Figure 3, and the observation results are shown in Figure 4.

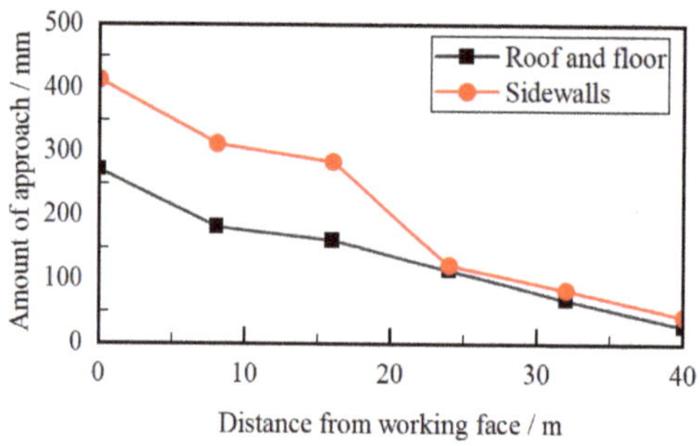

Figure 4. Surrounding rock deformation curve of the haulage roadway.

From Figure 4, it can be concluded that, during the observation period, as the working face advances forward, the displacement of the two sides of the roadway and the roof and floor is increasing, and the rate is growing. When the working face advances to the 5 m measuring point, the displacement of the two sides of the roadway and the roof and floor is the largest, and the maximum displacements are 415 mm and 274 mm, respectively. The roadway is mainly characterized by the bulging of the two sides, and the deformation of the roof and floor is not obvious.

Combined with the actual roof hanging at the end of the working face, under the influence of working-face mining, the maximum subsidence of the roadway roof is 274 mm. At this time, the anchor cable is in a load-bearing state and plays a major role.

3.2. *Roof Failure Characteristics of the Mining Roadway*

In order to find out the damage of the roadway roof in the mining process of the working face and the change of the overlying strata of the roadway roof, five monitoring stations were arranged along the haulage roadway of the working face from 20 m away from the working face, with a spacing of 20 m. The roof peeper was used to study the roof peep, and the peeping results are shown in Figure 5.

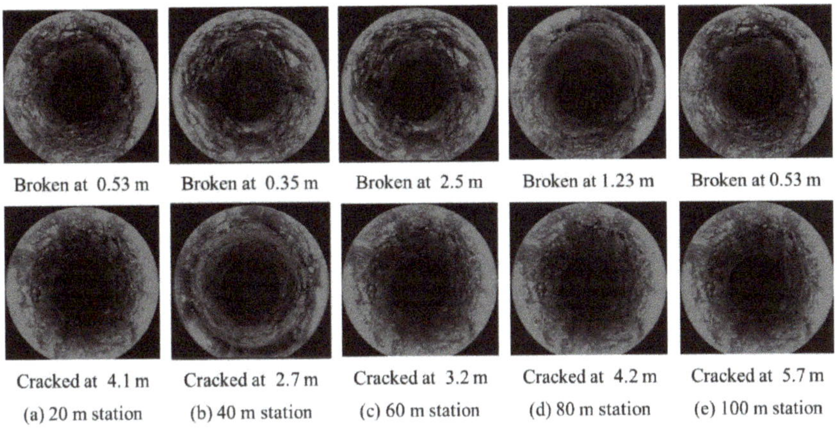

Figure 5. Detecting breaks and cracks at several monitoring stations placed apart from the working face and spaced by 20 m.

From Figure 5, it can be found that, when the distance from the working face is about 20 m, the shallow roof of the roadway is broken at 0.53 m, slight cracks appear at 4.1 m, and the deep rock layer of the roof is relatively complete. When the distance from the working face is 40 m, the roadway roof is broken near 0.35 m and cracks appear at 2.7 m of the roadway roof. From the 40 m station, the deep rock strata of the roadway roof are complete, and there is no obvious fragmentation. With the increase of the distance from the working face, the integrity of the roadway roof is better, and no obvious broken zone and cracks are found in the roof. The shallow-layer damage at 40 m and 100 m away may be caused by the influence of mining on the primary fissure of rock mass. The mine pressure of the roadway is not obvious, and the surrounding rock is in a stable state.

The results of the peep are summarized, and the changes of the rock strata at each peep point are analyzed, as shown in Table 1. Based on this, the rock stratum structure diagram of the haulage roadway roof of the working face is drawn, as shown in Figure 6.

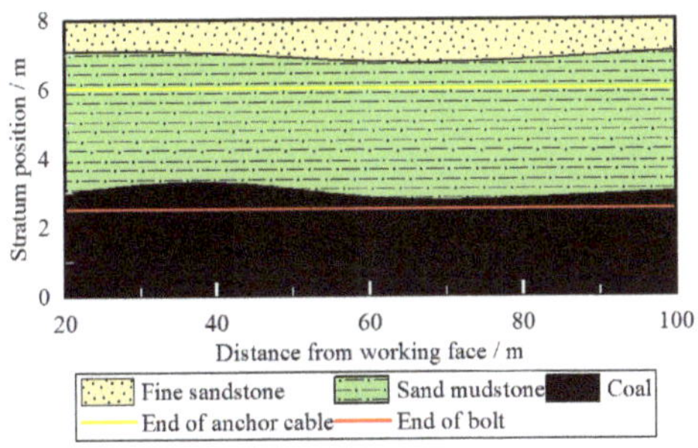

Figure 6. Schematic diagram of roof rock structure of roadway.

Table 1. Summary of roof peeping results.

Rock Stratum	STA [1]-20		STA-40		STA-60		STA-80		STA-100	
	D [2] /m	GT [3] /m	D /m	GT /m	D /m	GT /m	D /m	GT /m	D /m	GT /m
Fine sandstone	0.9	8	0.9	8	1.3	8	1.2	8	0.9	8
Sand mudstone	4.1	7.1	3.6	7.1	4	6.7	4	6.8	4.1	7.1
Coal	3	3	3.5	3.5	2.7	2.7	2.8	2.8	3	3

[1] STA-station; five monitoring stations were arranged along the haulage roadway at 20 m, 40 m, 60 m, 80 m, and 100 m away from the working face. [2] D-depth; the depth of each stratum according to the results of roof peeping. [3] GT-grand total, the depth of the count of overlying strata according to the results of roof peeping.

From Figure 6, it can be seen that the roof rock structure of the roadway changes little, the rock layer is fixed, and there is no obvious layer change. Roof bolts are all in the top coal seam, the end of the local roof bolt is close to the coal–rock interface, and there is a risk of failure. The end of the anchor cable is all in the sand mudstone in the upper part of the top coal, and the end of the anchor cable is far from the boundary between the sand mudstone and the top fine sandstone, which is in a stable layer. Combined with the results of borehole peeping, the strata with broken zones and fracture are located in the shallow part of the roadway top coal or the junction of the roadway top coal and sand mudstone. Therefore, under this geological condition and support strength, the damage depth of the roadway roof is small and the integrity of the roadway roof is better after mining.

3.3. Stress Characteristics of the Bolt in the Mining Roadway

In the range of 5~40 m in the advanced face of the haulage roadway, a measuring point is arranged every 5 m, the measurement is carried out along the advancing direction of the working face, and the stress of the bolts at each position is recorded. The stress of the three bolts in the auxiliary side of the same roadway section and the four bolts in the middle of the roof are measured. In order to facilitate the statistical results, the bolts in the side and the roof are respectively numbered clockwise. The position and number of the measured bolt are shown in Figure 2. The measurement results are shown in Figure 7.

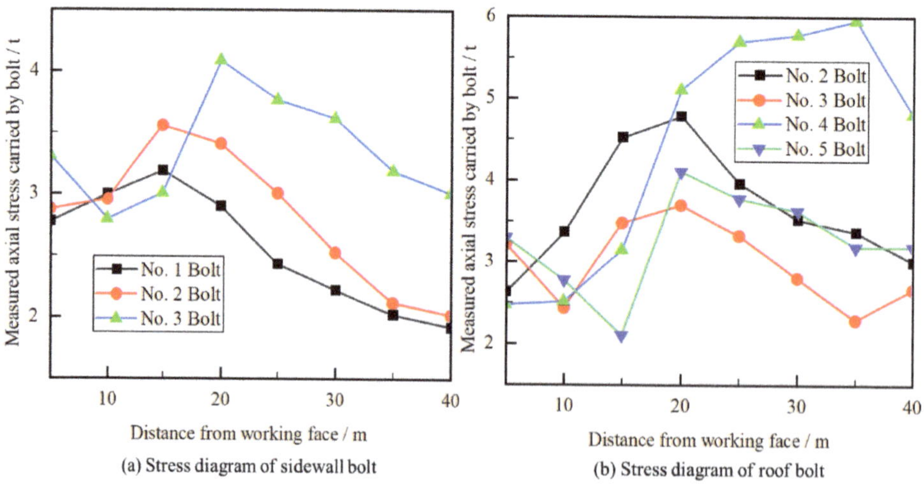

Figure 7. Stress diagram of bolts. (**a**) The stress diagram of a sidewall bolt; (**b**) The stress diagram of a roof bolt.

From Figure 7, it can be concluded that:

(1) It can be seen from the stress diagram of the sidewall bolt that the stress of the bolt in the top of the roadway side is larger than the other bolts, which reaches the maximum at 20 m from the working face, and the peak value is 4.1 t. Induced stress on the bolt falls as distance from the working face increases.

(2) According to the stress diagram of the roof bolt, the stress of the bolt (e.g., No. 4) in the middle of the roadway roof is greater than that on both sides (e.g., No. 2 and No. 5) of the roadway roof; see Figure 2. The stress reaches the maximum near 35 m from the mining face, and the peak value is 5.9 t. As the distance from the working face increases, the stress of some bolts (e.g., No. 2, No. 3, and No. 5) on the roof of the roadway tends to be stable at about 3 t.

(3) Generally speaking, in the monitoring area, the stress of the bolt at the top of the roadway is greater than that of the bolt at the side of the roadway, but the stress of the bolt at the roof and the side of the roadway is small, and there is a certain distance from the ultimate bearing capacity of the bolt. In addition, the working face roadway roof stability is good and less affected by working-face mining, with a high safety factor.

Combined with the end hanging roof structure, it can be found that the bolt in front of the working face is less stressed and in a stable state after being affected by the mining of the working face, far from reaching the breaking limit. After the mining of the working face, the roof of the roadway at the end of the working face has good integrity and stability under the suspension of the bolt and anchor cable. The roof of the roadway at a certain distance behind the working face is unstable and collapsed by the rotary pressure of the rock strata above the goaf.

3.4. Working Face End Hanging Structure Characteristics

Due to the limitation of field working conditions, the actual damage and stress of the end suspension structure cannot be directly measured, and only remote observation can be carried out. The end suspension structure mainly has the following characteristics:

(1) The hanging roof structure collapses periodically with the advance of the working face, and it collapses first at the end support edge and the corner of the coal pillar edge. The size of the caving range does not change much. The falling rock mainly comes from the top coal left by top coal caving and the sand mudstone above a small part of the top coal. The falling rock makes it difficult to fill the roadway behind the working face.

(2) In the hanging roof structure, the length of top coal is short, and most of the top coal without bolt support gradually collapses with the advance of the working face. The cantilever of the sand mudstone part above the top coal is longer. With the advancement of the working face, the sand mudstone cantilever will periodically collapse.

(3) With the advance of the working face, the shape of hanging roof structure is similar to arched-shape. In the axial direction of the roadway, the length of the arched-shape structure is about 10 m, and the arched-shape structure moves forward with the advance of the working face.

Combined with the previous measurement of the surrounding rock deformation of the haulage roadway in the working face, the peep of the roof rock structure, and the detection of the stress state of the roadway bolt, the analysis of the hanging roof characteristics of the working face end can be obtained:

Under the influence of working-face mining, the roadway roof in front of the working face has a maximum subsidence of 274 mm, and the maximum damage depth of the roof is about 2.7 m. It can be considered that, under this geological condition, after the roadway in front of the working face is affected by the mining of the working face, the integrity of the roof strata is good and there is no large-scale crushing and roof subsidence. After the working face is pushed over, it is easy to form a hanging roof structure near the end under the support of the end support and coal pillar.

4. Study on Working Face End Hanging Roof

4.1. Theoretical Calculation of Failure Depth and Breaking Distance

The length of the hanging roof is one of the key factors inducing roadway roof disasters [26,27]. It is generally believed that the supporting effect of bolt and anchor cable is mainly to suspend the broken rock strata in the roadway roof above the upper stable rock strata, as shown in Figure 2. In the supporting structure, the anchor cable plays a major role in suspension, so the condition for the collapse of the suspension structure is that the weight G of the rock stratum in the broken area is greater than the bearing capacity F of the anchor cable, that is, G > F.

The haulage roadway of the working face adopts the combined support of "anchor-net-cable". The anchor cable at the top of the roadway adopts the steel strand with the specification of $\varphi 17.8 \times 6250$ mm (the cross-sectional area is 193.45 mm^2 and the tensile strength is 1860 MPa), and the row spacing is 2000 × 1800 mm, 2-1-2 arrangement. According to the general calculation method, it can be considered that the number of anchor cables in 1 m roadway is 1.5, and the bearing capacity F provided by anchor cables in 1 m roadway is:

$$F = n \cdot P \cdot S \quad (1)$$

In the formula, n is the number of anchor cables in 1 m roadway, P is the tensile strength of anchor cables, and S is the cross-sectional area of anchor cables.

After calculation, the F value is 5.40×10^5 N.

The weight of broken coal-rock mass G_{coal} and G_{rock} in the plastic zone of 1 m roadway roof is:

$$G_{coal} = m_{coal} \cdot g = \rho_{coal} \cdot V_{coal} \cdot g \quad (2)$$

$$G_{rock} = m_{rock} \cdot g = \rho_{rock} \cdot V_{rock} \cdot g \quad (3)$$

In the formula, m_{coal} and m_{rock} are the weight of broken coal and rock mass, ρ_{coal} and ρ_{rock} are the density of broken coal and rock mass, V_{coal} and V_{rock} are the volume of broken coal and rock mass, and g is the gravity acceleration.

The height of the top coal is selected according to the average height of coal drawing, which is 3.23 m. The density of the coal body is 1390 kg/m^3, and the density of the rock mass is 2660 kg/m^3.

In addition, the G_{coal} value is 2.46×10^5 N, the G_{rock} value is $1.46\,h \times 10^5$ N, and h is the broken height of the sand mudstone roof.

It can be seen that, when the sand mudstone roof is complete, $F > G_{coal}$. At the same time, the bearing capacity of the anchor cable is greater than the weight of the broken surrounding rock mass, and the roof remains stable. From $F > G_{coal} + G_{rock}$, when h > 2.02 m, the bearing capacity of the anchor cable is insufficient to break, the roadway roof collapses, and the ultimate failure depth of the hanging roof structure is 5.25 m.

As the working face advances forward, the top rock strata of the working face roadway will form an arched-shape hanging roof structure under the support of the section protection coal pillar and the end support, as shown in Figure 8.

The relationship between the horizontal dimension of the curved hanging roof structure and its thickness is [28]:

$$H = 1.96\gamma \cdot \alpha^2 / 6_S \quad (4)$$

In the formula, H is the thickness of the arched-shape hanging roof, and the anchor cable and the roof is taken as a whole, taking 6 m. α is the horizontal size of the arched-shape hanging roof, 6_S is the unidirectional tensile strength of the coal body, taking 0.78 MPa, and γ is the bulk density of the top coal arched-shape hanging roof, taking 13 kN/m^3.

The α can be obtained as 13.55 m.

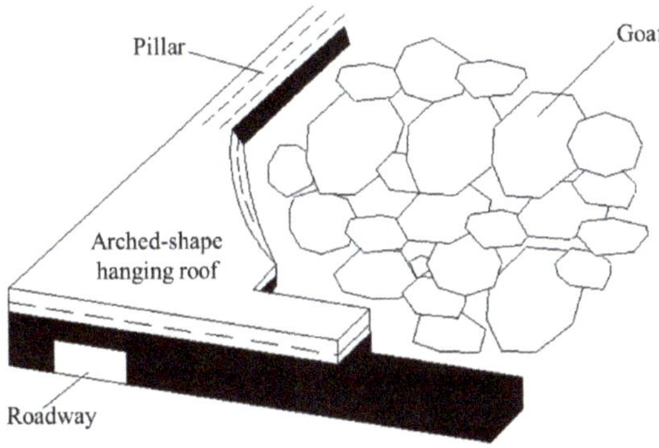

Figure 8. Arched-shape hanging roof at the end of mining roadway.

When the unidirectional tensile strength of the arched-shape hanging roof is constant, the greater the thickness of the arched-shape hanging roof, the greater the range of the arched-shape hanging roof. For the anchor-net support structure, as long as the anchor-net parameters are properly selected and the bolt (anchor cable) has sufficient preload, the low rock strata of the roadway roof are tightly connected by the "anchor-net-cable" structure to form a composite structure. At this time, the hanging roof structure can maintain a large strength, so it is easy to form an arched-shape hanging roof.

4.2. Numerical Simulation of Working Face End Hanging Structure

In order to analyze the influence of working-face excavation on the surrounding rock of a haulage roadway, especially the stress distribution of the surrounding rock of the haulage roadway and the characteristics of the plastic zone in front of and behind the working face under the action of mining stress, a FLAC3D numerical calculation model with a size of 500 m × 300 m × 80 m (length × width × height) is established according to the specific engineering geological conditions of the working face, and it is shown in Figure 9. We then apply displacement boundary conditions to the side and bottom of the model, and apply a vertical stress of 3 Mpa to the upper surface of the model.

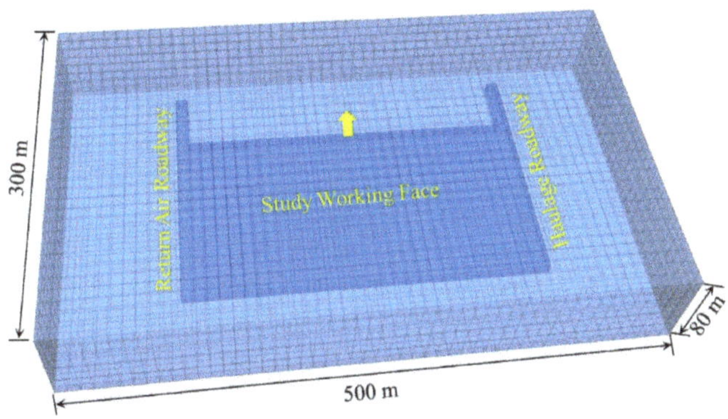

Figure 9. Numerical calculation model.

The numerical simulation of the roadway is full-section successive excavation, and the simulated excavation step is 1 m. The Mohr–Coulomb failure criterion has been adopted in this analysis. The physical and mechanical parameters of the model are shown in Table 2.

Table 2. Material parameters of each stratum in the model.

Surrounding Rock	Thickness/m	Elastic Modulus /GPa	Poisson's Ratio	Cohesion /MPa	Internal Friction Angle/°	Dilation Angle/°	Density /kg/m^3	Tensile Strength /MPa
Overlying strata	40	11.3	0.33	10.71	30.5	7.63	2550	2.8
Mudstone	4.5	11.5	0.17	8.60	32	8	2500	3.8
Fine sandstone	2	12.5	0.38	16.34	35.6	8.9	2950	7.52
Sand mudstone	4	12.18	0.18	9.15	33.49	8.37	2660	6.50
No. 2 Coal	6.5	2.37	0.26	2.21	34.14	8.54	1390	0.78
Sand mudstone	2	12.8	0.15	9.10	32.2	8.05	2650	4.18
No. 3 Coal	1	4.1	0.41	2.56	36.3	9.08	2750	1.38
Mudstone	1	11.5	0.17	8.60	32	8	2500	3.8
Sand mudstone	3	12.8	0.15	9.10	32.2	8.05	2650	4.18
Fine sandstone	5	12.5	0.38	16.34	35.6	8.9	2950	7.52
Sand mudstone	3	12.18	0.18	9.15	33.49	8.37	2660	6.50
Fine sandstone	8	12.5	0.38	16.34	35.6	8.9	2950	7.52

4.3. *Analysis of Numerical Simulation Results*

4.3.1. Plastic Zone and Stress Variation of the Roadway Surrounding the Rock in Front of the Working Face

The plastic zone and stress distribution of surrounding rock in front of the working face are shown in Figures 10 and 11.

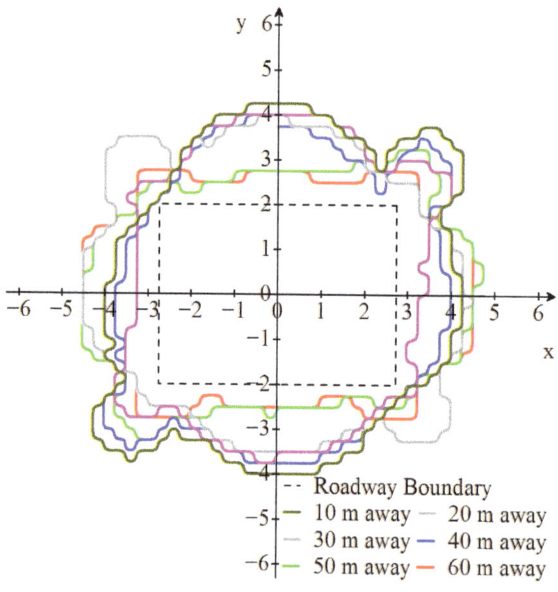

Figure 10. Distribution of the plastic zone of the surrounding rock of the roadway in front of the working face.

From Figure 10:

(1) As the distance from the working face gradually decreases, the range of the plastic zone of the roadway surrounding the rock gradually increases. When the range of the plastic zone around the rock mass reaches the maximum, the distance from the working face is 10 m and the maximum failure depth is 2.1 m. When the distance ahead of the working face is greater than 50 m, the range of the plastic zone of the roadway surrounding the rock tends to be stable, and the maximum failure depth is about 1 m.

(2) The failure process of the plastic zone of the roadway surrounding the rock shows obvious non-uniformity. When it is far away from the working face, the failure range of

the surrounding rock of the roadway side is larger than that of the roof and floor, and the failure is mainly concentrated in the center of the two sides. With the decrease of the distance from the working face, the rock layer at the shoulder of the roadway begins to be destroyed first, and the damage range and depth of the top gradually increase.

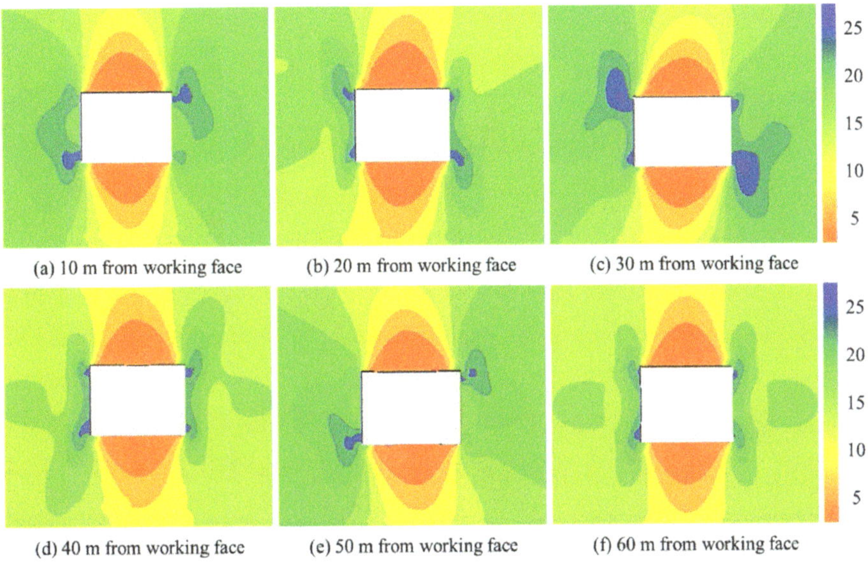

Figure 11. Vertical stress distribution of the surrounding rock of the roadway in front of the working face.

From Figure 11:

(1) Stress concentration occurs at the sharp corner of the roadway. As the distance from the working face decreases, the stress concentration phenomenon intensifies. The peak value of the stress concentration factor is 2.02, which is located near 10 m from the working face.

(2) The vertical stress of the roadway surrounding the rock shows obvious non-uniformity, and the stress concentration of the roadway sides and shoulders is more obvious than that of the roadway roof and floor. With the increase of the distance from the working face, the vertical stress of the roadway tends to be stable, and the stress concentration coefficient is stable at about 1.3.

In summary, in the range of the advancing working face, the plastic failure range of the roadway surrounding the rock gradually increases due to the influence of working-face mining. When the distance from the working face is 10 m, the failure range reaches the peak, and the maximum failure depth is 2.1 m, far from the ultimate failure depth of the hanging roof structure. Therefore, it is considered that, under the geological and supporting conditions, the disturbance of the working-face mining process has limited influence on the stability of the roadway.

4.3.2. Plastic Zone and Stress Change of the Roadway Surrounding the Rock behind the Working Face

The plastic zone and stress distribution of the surrounding rock of the roadway behind the working face are shown in Figures 12 and 13:

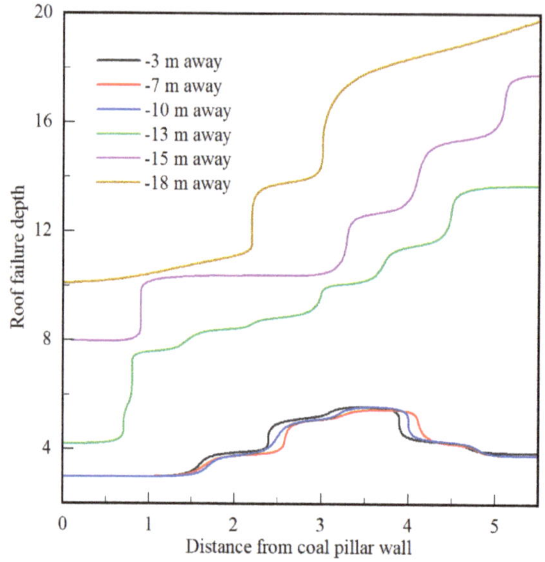

Figure 12. Depth of plastic zones around the roadway rock mass at various distances after the working face.

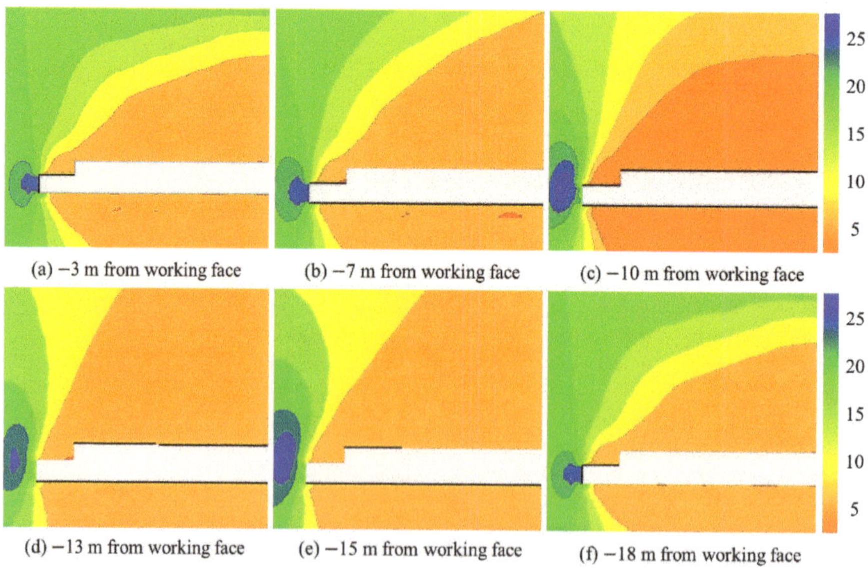

Figure 13. Vertical stress distribution of surrounding rock of the roadway behind the working face.

From Figure 12:

(1) As the working face advances forward, the failure depth of the rock strata at the top of the roadway increases slowly and the failure range changes near 13 m behind the working face. The failure depth increases from 2 m~3 m to 6 m~8 m. When the distance from the working face exceeds 15 m, the failure range of the roof in the goaf gradually stabilizes.

(2) In the range of 13 m behind the working face, it can be found that the maximum failure depth of the rock strata at the top of the roadway is 3 m, which does not reach

the calculated ultimate failure depth of the hanging roof structure. At the same time, the hanging roof distance of the roadway is 13 m, which is less than the caving length of the arched-shape hanging roof of 13.55 m, so the hanging roof structure is difficult to fall within this distance range.

(3) Outside the range of 13 m behind the working face, the failure depth of the roadway roof strata reaches more than 6 m, which has exceeded the ultimate failure depth of the hanging roof structure. At this time, the anchor cable breaks and the roof collapses. In addition, according to the calculation, when the span of the suspension structure exceeds 13.55 m, the suspension structure will break as a whole.

From Figure 13:

(1) The stress concentration phenomenon occurs in the protective coal pillar of the roadway behind the working face. As the working face advances forward, the stress concentration phenomenon intensifies, and the peak value of the stress concentration factor is 2.5, which is located 10 m behind the working face.

(2) When the distance from the working face exceeds 15 m, the vertical stress of the roadway surrounding the rock tends to be stable and the stress concentration factor is stable at about 1.8. At this time, the roof of the goaf behind the working face is completely collapsed, and the stress in the goaf is redistributed and gradually stabilized.

In summary, the numerical simulation results are consistent with the theoretical calculation results. In front of the working face, under the geological conditions and supporting strength, the rock surrounding the roadway remains stable. In the range of 13 m behind the working face, the hanging structure at the end of the working face remains stable. Outside the range of 13 m behind the working face, the anchor cable is not enough to bear the weight of the broken rock stratum of the roof, and the end hanging structure collapses and moves forward.

5. Discussion

5.1. Caving Condition Analysis of Working Face End Hanging Roof Structure

From the previous theoretical analysis, there are two forms of hanging roof structure collapse at the end of a coal mining face: one is that the roadway roof failure reaches the limit failure depth, the anchor cable breaks and fails, and the roadway roof falls; another case is that the length of the hanging roof structure of the roadway roof exceeds the limit span of the curved triangular plate. The numerical simulation results are analyzed, respectively.

(1) Failure depth analysis of roof

According to the theoretical analysis, when the destruction depth of the roadway roof exceeds 5.25 m, the roadway roof will fall under the weight of the rock mass in the broken area. Through numerical simulation analysis, the change of failure depth of the roadway surrounding the rock during mining can be obtained.

The maximum failure depth of the roadway roof strata in front of the mining face is 2.1 m. At this time, the broken area at the top of the roadway has not exceeded the anchoring range of the bolt. The roadway roof can still remain stable under the hanging of the bolt and anchor cable.

In the range of 13 m behind the working face, the damage depth of the roadway roof is less than 3 m. Outside the range of 13 m behind the working face, the damage depth of the roadway roof changes abruptly, reaching more than 6 m. At this time, the damage depth of the roadway roof has exceeded the anchoring range of the anchor cable, and the roof strata will fall due to the failure of the anchor cable. As the working face advances forward, the arched-shape hanging roof collapses periodically, and the new arched-shape structure gradually moves forward as the working face advances.

(2) Fracture analysis of arched-shape structure

According to the theoretical analysis, the two corners of the hanging arched-shape structure are on the two points behind the limit support or the coal pillar goaf. With the continuous advancement of the working face, the area of hanging roof structure increases, and the bending moment of hanging roof structure also increases. When the bending moment reaches a specific limit, it breaks at the two positions of the coal pillar and the end, the end and the support roof cutting line, and the end hanging; in this process, the working face continues to move forward and a new arched-shape hanging roof is formed at the front end.

Through calculation, when the fixed edge of the hanging structure exceeds 13.55 m, the hanging structure will collapse. In the work site, the first damage occurs at the top of the hanging structure of these two places, and a hanging structure length of about more than 10 m will break the overall subsidence. After they are destroyed, a new limit will appear on the principal stress line, thereby forming an arched-shape limit bending moment trace.

Because the influence of the rock stratum above the anchor cable anchoring rock stratum is neglected in the theoretical calculation, the rock stratum of the anchor cable anchoring stratum will produce certain pressure when it collapses to the goaf, so the theoretical calculation distance is larger than the actual working face roadway. The existence of a certain range of arched-shape hanging roof is conducive to the stability of the roadway end and ensures the safety of the mining face. However, when the hanging roof area is too large, there will be potential safety hazards such as gas accumulation and large-scale roof collapse.

5.2. *The Mechanism of Working Face End Hanging Roof*

Based on the above research, the formation mechanism of the end hanging roof of the working face can be preliminarily obtained. The main conclusions are as follows:

(1) The lithology of sand mudstone and fine sandstone strata above the top coal of the roadway is strong. Under the influence of working-face mining, the mine pressure of the roadway is not severe. After the working face is advanced, the sand mudstone and fine sand strata above the top coal of the roadway are less damaged and the integrity is better.

(2) In the rear of the working face, the top coal and immediate roof above the roadway will be under the pressure caused by the overlying strata rotating to the goaf. Because the sand mudstone and fine sandstone above the top coal have strong lithology and strong supporting ability, the mudstone above the fine sandstone is not easy to form a long cantilever structure in the process of rotating to the working face; therefore, the sand mudstone and fine sand strata above the top coal are not easy to collapse under the support of the working face end support and the roadway coal pillar, forming a hanging roof structure. The length of the hanging roof structure depends on the mechanical properties of the top coal, sand mudstone, fine sandstone, and mudstone, and the stability of the structure formed by their combination.

(3) Through theoretical calculation and numerical simulation, the caving conditions of the hanging roof structure are analyzed. Combined with the actual situation of the working face, the caving step distance of the curved hanging roof structure is 10 ~ 13.55 m, and the curved hanging roof structure moves forward with the advance of the working face.

6. Conclusions

(1) After the mining roadway is affected by working-face mining, the main performance of two-side bulging is more obvious, roof and floor deformation is not obvious, the whole is less affected by mining, and strata behavior is not obvious. Through the observation of the roof strata of the roadway, it is found that the strata of the roof of the roadway do not change much, and the integrity of the whole rock strata is better. In the shallow part of the rock, fracture development or crushing zone appears, while the deep rock is relatively complete.

(2) There are two cases of collapse of the end suspension structure: The first case is that the roadway roof breaking depth is too large, anchor cable failure occurs, and hanging roof structure collapses; the second case is that the rock stratum of the roadway hanging roof structure breaks as a whole and collapses together with the goaf.

(3) Combining theoretical analysis and numerical simulation, the caving step of the arched-shape hanging roof structure is 10~13.55 m, and the hanging roof structure moves forward with the advance of the working face.

(4) After the above research, it is believed that the formation of the arched-shape hanging roof structure is due to joint influence of many aspects, as follows: the lithology of the roadway top coal is strong and the hanging roof structure is less affected by the mining of the working face; the rotary pressure of the mudstone above the fine sandstone is insufficient; the arched-shape hanging roof structure is formed under the joint work of the coal pillar and the end support.

Next, we will study the treatment method of the arched-shape hanging roof. It is expected to provide a method to control the size of the arched-shape hanging roof structure to keep it within a reasonable range, so as to ensure the safe production of the working face.

Author Contributions: H.L.: writing, methodology. C.H.: data curation, writing—original draft. Z.H.: investigation. Q.L.: numerical simulation. H.W.: data curation. J.L.: methodology. D.Z.: numerical simulation. All authors have read and agreed to the published version of the manuscript.

Funding: This research was funded by the National Natural Science Foundation of China (Grant No. 51574224), the National Natural Science Foundation of China (Grant no. 52004289), and the Fundamental Research Funds for the Central Universities (Grant no. 2022XJNY01).

Institutional Review Board Statement: Not applicable.

Informed Consent Statement: Informed consent was obtained from all subjects involved in the study.

Data Availability Statement: The data that support the findings of this study are available from the corresponding author, upon reasonable request. The authors gratefully acknowledge the help and comments of the reviewers and editors.

Acknowledgments: The authors gratefully acknowledge the test site provided by the China National Coal Group Corporation. The authors gratefully acknowledge the help and comments of the reviewers and editors.

Conflicts of Interest: The authors declare no conflict of interest.

References

1. Qian, M.; Xu, J. Study on the "o-shape" circle distribution charactheristics of mining-induced fractures in the over-laying strata. *J. China Coal Soc.* **1998**, *05*, 20–23.
2. Huang, B.; Cheng, Q.; Zhao, X.; Xue, W.; Scoble, M. Using hydraulic fracturing to control caving of the hanging roof during the initial mining stages in a longwall coal mine: A case study. *Arab. J. Geosci.* **2018**, *11*, 603. [CrossRef]
3. Liu, J.; Liu, C.; Yao, Q.; Si, G. The position of hydraulic fracturing to initiate vertical fractures in hard hanging roof for stress relief. *Int. J. Rock Mech. Min. Sci.* **2020**, *132*, 104328. [CrossRef]
4. Brodny, J.; Tutak, M. Exposure to Harmful Dusts on Fully Powered Longwall Coal Mines in Poland. *Int. J. Environ. Res. Public Health* **2018**, *15*, 1846. [CrossRef] [PubMed]
5. Zang, C.W.; Zhuang, X.; Le Ma, C. Research of Large-Area Hard Roof Caving Prevention and Control System in Shortwall Mining Face. *Adv. Mater. Res.* **2013**, *734–737*, 802–808. [CrossRef]
6. Qian, M.; Shi, P. *Ground Pressure and Strata Control*; China University of Mining and Technology Press: Xuzhou, China, 2010.
7. Du, J.; Meng, X. *Mining Technology*; China University of Mining and Technology Press: Xuzhou, China, 2014.
8. Zhang, G. *Ventilation and Safety Technology*; China University of Mining and Technology Press: Xuzhou, China, 2011.
9. Li, C.; Li, S. Roof control technology for large unsupported roof at face end of coal mining face. *Coal Sci. Technol.* **2008**, *8*, 26–30.
10. Tan, Y.-L.; Zhang, L.-J. Gob-Side Entry with Limestone Roof Retained by Anchorage Cable and Individual Prop Support. Controlling Seismic Hazard and Sustainable Development of Deep Mines. In Proceedings of the 7th International Symposium on Rockburst and Seismicity in Mines (rasim7), Dalian, China, 21–23 August 2009; Rinton Press, Inc.: Princeton, NJ, USA, 2009; Volumes 1 and 2, pp. 143–148.
11. Huang, B.; Wang, Y. Roof Weakening of Hydraulic Fracturing for Control of Hanging Roof in the Face End of High Gassy Coal Longwall Mining: A Case Study. *Arch. Min. Sci.* **2016**, *61*, 601–615. [CrossRef]
12. Kai, W. Study of Reasonable Hanging Roof Length on Hard Roof. *Procedia Eng.* **2011**, *26*, 772–777. [CrossRef]
13. He, F.; Wang, L.; Yu, Y. Hanging roof rock beam model and its rheological analysis in goaf. *Mine Press. Roof Manag.* **2005**, *04*, 84–85+88. [CrossRef]
14. Wang, Z.; Wang, Y.; Li, J. Creep analysis of hanging roof beam model of empty region after excavation. *J. Liaoning Tech. Univ.* **2007**, *26*, 83–85.
15. Wang, J.; Gong, S.; Zhou, S. Analsis of structure of hanging triangle roof plate with curve side in fully-mechanized sublevel caving face. *China Coal Industry* **2012**, *38*, 36–40. [CrossRef]

16. Qin, L.; Zhang, L. Study on Structure and Stability of End Roof of Fully Mechanized Caving Face. *Mod. Coal Mines* **2000**, *06*, 14–16. [CrossRef]
17. Luan, H.; Lin, D.; Jia, C.; Sun, R. Mechanisms of Strata Behaviors Periodicity for Gob-side Entry Retaining. *J. Shandong Univ. Sci. Technol. Nat. Sci.* **2014**, *33*, 60–64. [CrossRef]
18. Yang, P.; Liu, C. Structure Forms of Basic Roof and Reasonable Supporting Parameters in Ends of Fully-Mechanized Top Caving Face. *J. Min. Saf. Eng.* **2012**, *29*, 26–32. [CrossRef]
19. Yang, J.; Liu, C.; Yu, B.; Wu, F. An analysis on strong strata behaviors and stress transfer of the roadway ap-proaching gob in triangle area of the face end. *J. Min. Saf. Eng.* **2016**, *33*, 88–95. [CrossRef]
20. Xu, Y. Structure coupling between hydraulic roof support and surrounding rock in extra-thick and hard coal seam with super large cutting height and longwall top coal caving operation. Ph.D. Thesis, China Coal Research Institute, Beijing, China, 2020. [CrossRef]
21. Pavlova, L.D.; Fryanov, V.N.; Keller, A.V.; Tsvetkov, A.B. Numerical study of the effect of console length of the main roof hanging on the geomechanical parameters of the mine face. *IOP Conf. Ser. Earth Environ. Sci.* **2019**, *377*, 012037. [CrossRef]
22. Xue, J.; Ma, Q.; Du, X.; Zhan, K.; Sun, B. Numerical simulation and control of a rockburst induced by main roof fracture in a deep coal seam. *Energy Sources Part A Recover. Util. Environ. Eff.* **2020**, 1–17. [CrossRef]
23. Dychkovskyi, R.; Shavarskyi, I.; Saik, P.; Lozynskyi, V.; Falshtynskyi, V.; Cabana, E. Research into stress-strain state of the rock mass condition in the process of the operation of double-unit longwalls. *Min. Miner. Depos.* **2020**, *14*, 85–94. [CrossRef]
24. Vu, T.T. Solutions to prevent face spall and roof falling in fully mechanized longwall at underground mines, Vietnam. *Min. Miner. Depos.* **2022**, *16*, 127–134. [CrossRef]
25. Chen, B.; Liu, C.; Wang, B. A case study of the periodic fracture control of a thick-hard roof based on deep-hole pre-splitting blasting. *Energy Explor. Exploit.* **2021**, *40*, 279–301. [CrossRef]
26. Liu, C.; Yang, Z.; Gong, P.; Wang, K.; Zhang, X.; Zhang, J.; Li, Y. Accident Analysis in Relation to Main Roof Structure When Longwall Face Advances toward a Roadway: A Case Study. *Adv. Civ. Eng.* **2018**, *2018*, 3810315. [CrossRef]
27. Li, P.; Zheng, X.; Guo, X.; Li, B.; Wang, C. Application of the Ultimate Hanging Arch Length of the Layered Roof with Anchors and the Gob-Side Entry Retaining. *Shock Vib.* **2021**, *2021*, 4338686. [CrossRef]
28. Dai, J.; Li, H.; Jiang, J. Elastic analysis hanging triangle roof plate with curve side on the end of sub-level caving mining face. *Chin. J. Rock Mech. Eng.* **2004**, *23(S2)*, 4757–4760. [CrossRef]

sustainability

MDPI

Article

Characteristics of Strata Behavior and Differentiated Control of Fully Mechanized Mining Working Face with Abnormal Roof

Qiang Fu [1,2,3,4], Ke Yang [1,2,3,4], Xiang He [2,3,4,*], Zhen Wei [1,2,3,4] and Qinggan Yang [1,3,4]

1 State Key Laboratory of Mining Response and Disaster Prevention and Control in Deep Coal Mines, Anhui University of Science and Technology, Huainan 232001, China
2 Institute of Energy, Hefei Comprehensive National Science Center, Hefei 230031, China
3 School of Mining Engineering, Anhui University of Science and Technology, Huainan 232001, China
4 Coal Mine Safety Mining Equipment Innovation Center of Anhui Province, Anhui University of Science and Technology, Huainan 232001, China
* Correspondence: xianghe_cumtb@126.com

Abstract: The roof control of a fully mechanized mining face has a significant impact on coal extraction. Therefore, information about strata behavior and differentiated control measures for abnormal roofs are needed. In the present research, we used the ground pressure theory to calculate the interval of the first and periodic weighting of working faces 11113, 11213, and 11313 in the Panbei mine. The maximum support working resistance required by the three working faces was calculated considering the influence of the hard roof, fault zone, and skip mining mode on the relationship between support–surrounding rock. In addition, we determined the hydraulic support for the three working faces taking into account the impact of roof lithology on ground pressure, as well as the layout and mining sequence of the working face. In order to ensure the safety of the working face mining, we implemented the laying mesh method, controlling the roof, providing side protection in the fault-affected area, and using the violent ground pressure control method for advanced deep hole pre-split blasting under hard roof conditions. The engineering practice showed that the effective control of roof ground pressure in abnormal areas was achieved using these control measures.

Keywords: ground pressure; numerical simulation; hard roof; tectonic influence; pre-split blasting

1. Introduction

The geological conditions of coal strata in China are complex and diverse. Even in the same mining area and the same coal seam, roof lithology and rock formation are different [1,2]. Engineering practices have shown that lithology and structural characteristics of coal seam roofs are the key factors affecting the design parameters of fully mechanized mining working face, equipment selection, and ground pressure control [3–5]. In most mining areas, the same set of fully mechanized mining equipment is used in multiple working faces with different roof lithologies. These differences complicate the performance of the equipment [6]. Since soft coal presents a diversity of roof conditions including soft roof and floor and given that thick and hard rock formation may result in faults and direct overload, a significant amount of research has been carried out to determine effective working face mining parameters and roof control designs.

With the aim of solving technical problems related to ground pressure control in "Three Soft" large mining heights, Tu et al. [7] optimized the working face parameters through the surrounding rock control theory and technology as well as a comprehensive gas treatment technology. Bahrani et al. [8] addressed issues associated with the simulation of jointed rock masses and ground support by employing both continuum and discontinuum numerical methods. Li et al. [9] proposed and validated an improved support method for roadways in broken rock mass under high geo-stress. These approaches were the basis for safe mining practices in the "Three Soft" coal seam with large mining height working faces.

Dychkovskyi et al. [10] developed a rendering algorithm of a 3D model of rock mass in terms of long-pillar mining of a coal seam using double-unit longwalls. Wang et al. [11] analyzed the mechanisms of spalling in very soft coal seam walls and proposed that the main technical ways to prevent and reduce the spalling of very soft coal seam walls were to reduce coal pressure and improve the shear strength of coal. Later, Vu proposed solutions to prevent face spalling and roof falling in fully mechanized longwalls in underground mines in Vietnam [12]. The effect of mining's advancing direction on the stability and control of the surrounding rock in the fault zone of the working face has also been studied. Babets et al. [13] developed a technique for rock strength assessment considering the rock mass disintegration and watering in the fault area. Li et al. [14] studied the mining stress in rock mass at the fault zone. These researchers used similar material simulation tests and determined the influence of the fault zone on stress transmission. Their results provided the basis and reference for reasonable reservation of the fault coal pillar and roadway layout. Additionally, the hinge balance theory of the main roof fault rock block was used by Li et al. [15] to study the load transmission law and difference of the main roof of work face goaf on both sides of the fault. Dai et al. [16] determined the first mining hanging wall and the first mining footwall of the fault and discussed the relationship between the retention of the fault coal pillar and the advancing direction. Ju et al. [17] focused on large overhang spans, severe weighting, and difficulties in controlling thick and hard roof working faces. These researchers applied the key layer theory to determine the mechanism of interaction between key layer breaking and ground pressure. Fan et al. [18,19] determined disaster characteristics and proposed technical measures for preventing hard roof impact ground pressure. Cui et al. [20] investigated the load distribution, energy accumulation, and release, as well as the weighting strength characteristics of the first and periodic weighting in the process of hard roof fracture. Guo et al. [21] discussed the technique of controlled caving of hard roofs and proposed a method to calculate the reasonable suspended roof length. He et al. [22,23] used numerical simulation to analyze the mechanism of deep hole pre-splitting blasting and hydraulic pre-splitting roof weakening. Sun et al. [24] proposed a model based on the traditional CCM and the classical Carranza's model to analyze support-surrounding rock interaction considering pre-reinforcement. He et al. [25–27] applied deep hole pre-splitting blasting and hydraulic directional fracturing method to perform industrial experiments on hard roof weakening with good results. Yang et al. [28] analyzed the relationship between support-surrounding rock under hard roofs of thin bedrock conditions. These researchers calculated the support working resistance, and proposed methods for controlling fully mechanized mining working faces. In summary, fully mechanized mining technology has been developed and applied to complex and diverse mining coal seams. However, most of these studies have considered the ground pressure and control of working faces under the same coal seam roof lithology. Little research has been performed to determine the pressure and roof control of adjacent working faces with different coal seam roof lithologies.

Hence, based on experimental analysis and theoretical calculations, we analyzed the characteristics of ground pressure in the fully mechanized mining working faces of the zone with an abnormal roof and the mechanism of roof lithology changes on ground pressure. With this information, we proposed the mechanisms for selecting proper fully mechanized mining equipment and ground pressure control methods. This information is of great significance for safe and efficient mining.

2. Case Study

The A3 coal was obtained from group A coal roof in layers of the Panbei Mine in the Huainan mining area, China. The first mining working face was 11113, followed by 11313, and finally working face 11213. The working face parameters are shown in Figure 1. The comprehensive analysis of the borehole columnar diagram and field borehole coring data indicated that, along the direction of the coal seam, the immediate roof thickness gradually decreased and tended to be directly covered by a thick and hard roof, as shown in Figure 2.

In addition, the coal seam angle gradually increased, and the mining area displayed dense faults with a density of 25 faults/km^2. These issues complicated the design and safe mining of working faces. The main physical and mechanical parameters of rocks and coal of the roof and floor of this mining area were obtained in the laboratory using the coal and rock physical parameters test.

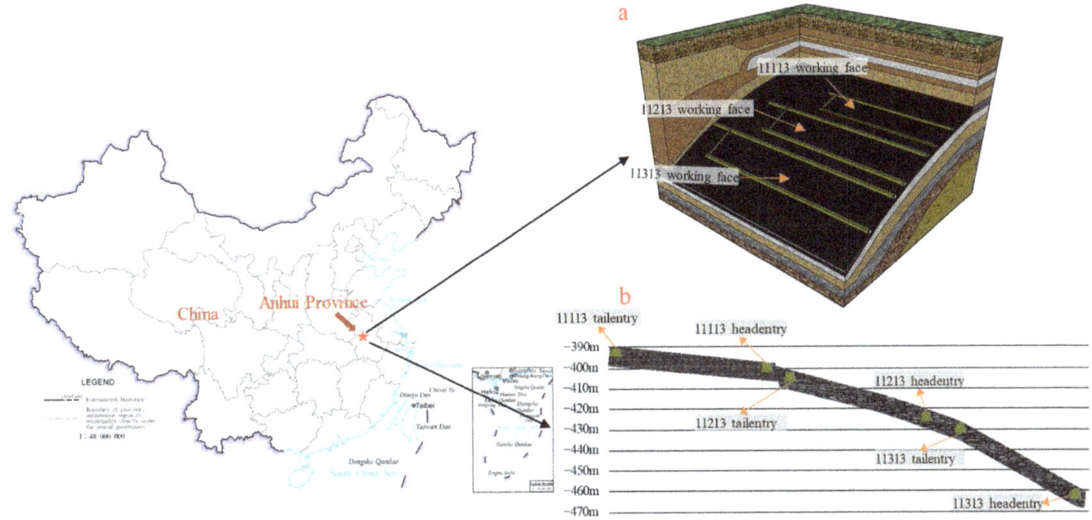

Figure 1. Layout and section of mining area. (**a**) Plane layout of mining area, working face of group A coal. (**b**) Section of group A coal, mining area along the uphill direction.

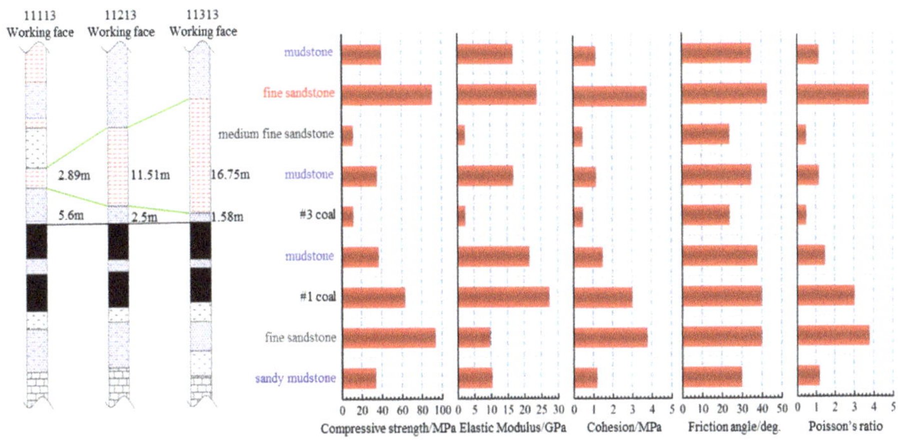

Figure 2. Comprehensive histogram and physical and mechanical parameters of different working faces.

According to the main roof classification method, the N ratio of immediate roof thickness h_i to coal seam mining height h_m is expressed as $N = h_i / h_m$. Description of the three basic types of working face roof rock formations were:

1. Working face 11113, $N > 5$, belonged to class I roof condition. The thicknesses of overlying rock formations were small, and the roof rock contained fractures.
2. Working face 11213, $2 < N < 5$, belonged to class II roof condition. There was an immediate roof of size 2.0~9.0 m between the main roof and the coal seam.

3. Working face 11313, $N < 2$, belonged to class III roof condition. The lithology of the main roof is medium-fine sandstone. A small amount of 1~2 m mudstone roof was contained between the main roof and the coal seam and almost no immediate roof was present.

3. Characteristics of Strata Behavior with Abnormal Roof

3.1. Caving Characteristics of Abnormal Roof

In the present research, we considered the engineering conditions and used the calculation method of periodic weighting interval previously reported in [11,14]. The following assumptions were made: in the overlying load of the main roof calculation, the rock formation thickness was considered a real value; in the case of stress and fracture interval of rock mass, rock formation thickness was taken as the equivalent rock formation thickness described by the rock quality designation (RQD) index.

The main roof's first weighting interval L was calculated using Equation (1):

$$L = h\sqrt{\frac{\sigma_t}{q}} \tag{1}$$

where L is the fracture interval when the main roof was calculated as beam (m); h corresponds to the thickness of the main roof rock formation (m); σ_t represents the main roof rock formation tensile strength (MPa); q is the load bear by main roof rock formation (MPa).

The overlying load rock formation q on the main roof [11] was calculated using Equation (2):

$$(q_n)_1 = \frac{E_1 h_1^3(\gamma_1 h_1 + \gamma_2 h_2 + \cdots + \gamma_n h_n)}{E_1 h_1^3 + E_2 h_2^3 + \cdots + E_3 h_n^3} \tag{2}$$

where $(q_n)_1$ represents the load acting on the main roof rock formation in the nth layer (kPa); E_1, h_1 are the elastic modulus (GPa) and thickness (m) of main roof rock formation; E_n, h_n indicate the elastic modulus (GPa) and thickness (m) of rock formations above the main roof; γ_1, γ_n correspond to bulk density of the main roof and above the rock formation (kN/m^3).

When $(q_{n+1})_1 < (q_n)_1$, q in the first weighting interval becomes $(q_n)_1$.

According to the coring test, the RQD value of medium-fine sandstone in the main roof was 0.60. The main roof's first weighting interval and periodic weighting interval were calculated for the three working faces considering the mechanical parameters of the rock in the borehole coring. Results are shown in Table 1.

Table 1. Main roof weighting intervals of different types of roofs.

Working Face	Roof Category	q_1 (kPa)	$(q_2)_1$ (kPa)	$(q_3)_1$ (kPa)	First Weighting Interval (L/m)	Periodic Weighting Interval (m)
Working face 11113	I	74.0	33.6	/	29.6	12.1
Working face 11213	II	309.2	417.4	95.2	38.0	15.5
Working face 11313	III	449.6	638.7	108.7	44.7	18.2

3.2. Weighting Characteristic of Abnormal Roof

3.2.1. Working Face 11113

In the working face 11113 (Class I roof), no main roof rock formation was observed in the collapse zone. It is likely that the immediate roof rock formation fell during the mining of the working face, and that the covered 4.2 m-thick fine sandstone (20 m from the coal seam) entered the fracture zone. Thus, hydraulic support carried the load of rock mass in the collapse zone within the roof control zone, as shown in Figure 3. The first weighting of the main roof was not significant.

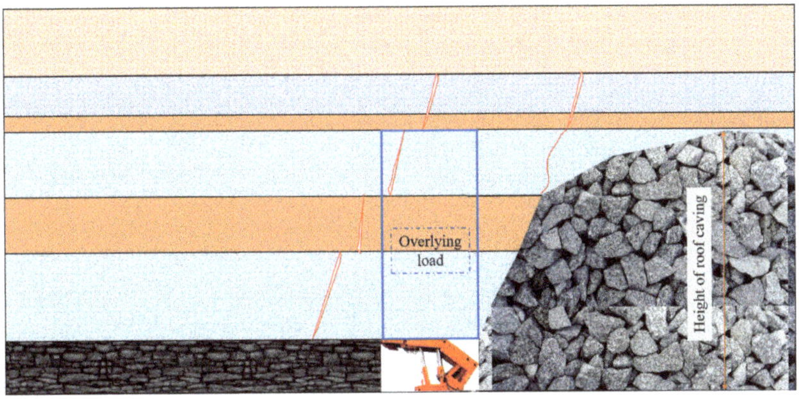

Figure 3. Loading on hydraulic support in the longwall panel of working face 11113.

The support stress was calculated using Equation (3) [11]:

$$P = \frac{L_a \cdot \gamma h L_z' \cdot \cos\alpha \cdot K}{\eta'} \tag{3}$$

where L_a indicates the length of the working face controlled by each support (m), which was set as 1.75 m; γ corresponds to the bulk density of immediate roof rock formation in the collapse zone (kN/m^3), 25 kN/m^3; h is the immediate roof thickness in the collapse zone (m), 9~15 m; L_z' represents the immediate roof length of rock beam (m), $L_z' = L_d + L_h + L_z'$ (L_d is end distance; L_h is the sum of top beam and front beam length (m); $L_d + L_h$ indicates the maximum top control distance (m); L_{zx} represents the limit overhang length of immediate roof rock formation after support (m), L_z' = 6 + 1 = 7.0 m; α-coal seam dip angle (°), 10; K is the safety factor, 1.2; η' is the coefficient considering the waste dump on the shield beam and that the top beam was not perpendicular to the column, 0.85.

The support working resistance required for working face 11113 was 3832~6386 kN.

3.2.2. Working Face 11213

According to the sequence of the working face 11113-11313-11213, working face 11213 corresponded to an island working face. F11213-1 fault exposed at 31 and 43 m away from the setup entry of the working face in tailentry and headentry, respectively, also affected the first weighting of the working face. The mining influence and overlying strata are shown in Figure 4.

The two-layer hard rock formations in the roof of working face 11213 were located within the elevation of the goaf fracture zone (45 m). The roof of working face 11213 was affected by the mining of working faces 11113 and 11313 and their corresponding faults. For this reason, a large fracture was observed as a result of the damage in the inclined direction. As shown in Figure 4, working face mining may cause gradual collapse along the slope, forming a simple support on three sides and a fixed support on one side. The support stress was calculated using Equation (4) [11], which considered the periodic fracture of the main roof.

$$P = \frac{L_a(\gamma h L_z' + \sum_{i=1}^{n} \gamma_{li} h_{li} L_{lki}) \cos\alpha \cdot K}{\eta'} \tag{4}$$

where γ_{l1} corresponds to the average bulk density of the first main roof and its additional rock formation in the collapse zone (kN/m^3), 26 kN/m^3; h_{l1} is the thickness of the first main roof and its additional rock formation in the collapse zone (m), 15 m; L_{lk1} indicates the rock block length of the first main roof in the collapse zone (m), 7.7 m (half of the main roof periodic weighting interval); values of other parameters were: L_a = 1.75 m, γ = 25 kN/m^3, h = 2.5 m, L'_z = 7.0 m, α = 18°, K = 1.2, η' = 0.85.

Later, we substituted the corresponding data into Equation (4). Our calculations indicated that the working resistance required to support working face 11213 was 8084 kN.

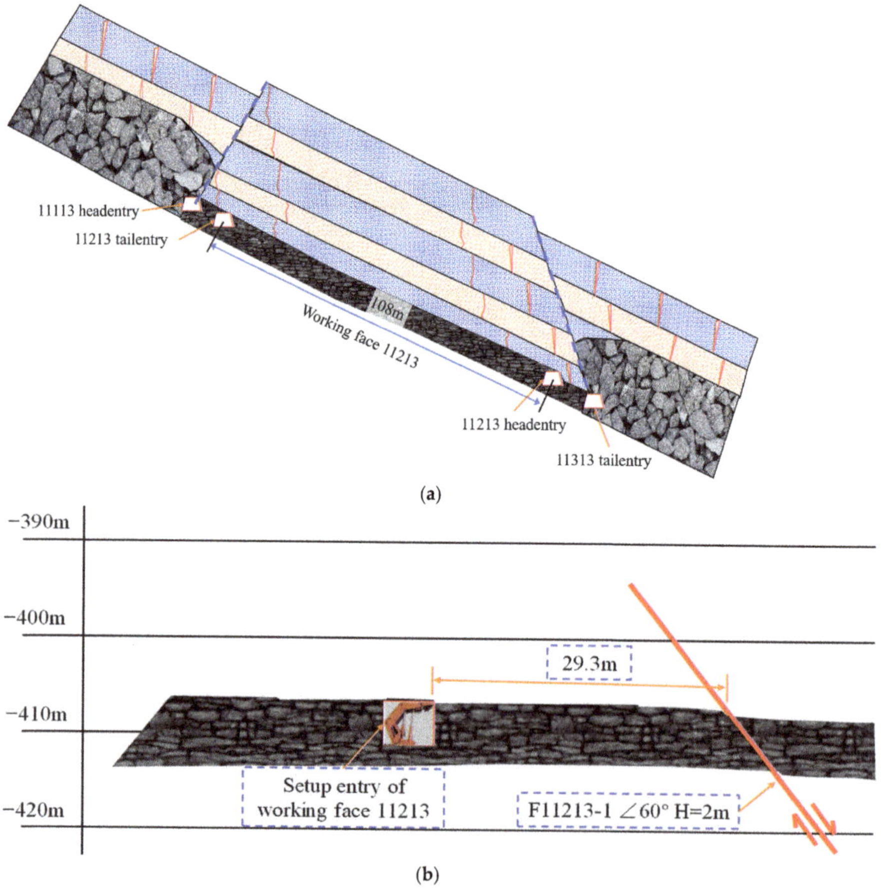

Figure 4. Structure of overlying strata near the setup entry in the longwall panel of working face 11213. (**a**) Rock structure in the inclined direction. (**b**) Fault influence (along the tailentry section).

3.2.3. Working Face 11313

Working face 11313 corresponds to class (III) with a thin immediate roof where the local main roof directly covered the working face. As shown in Figure 5, according to the section along the headentry direction near the setup entry, the lower part of the working face was affected by FX1 and F11313-9 faults. The tailentry and headentry exposed F11313-9 faults at 110 m and 26.3 m away from the setup entry, respectively. The upper part of the working face was not affected by the fault within the first weighting range.

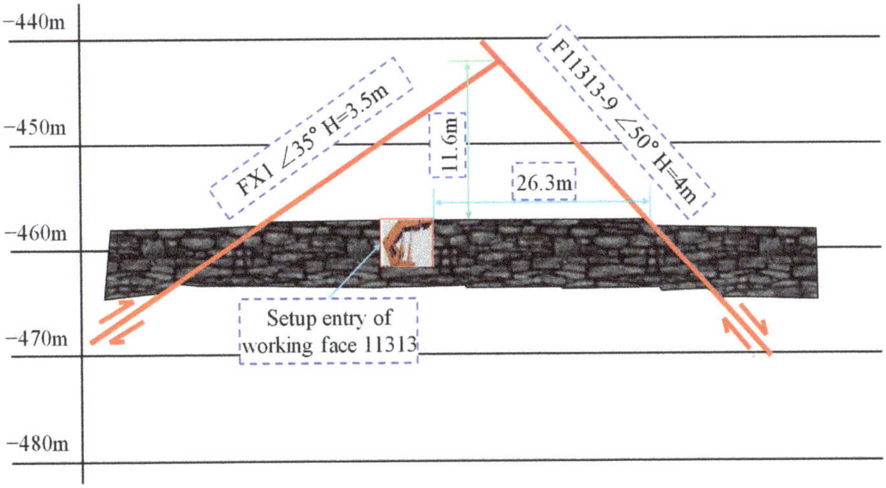

Figure 5. Section along the longwall panel near the setup entry in working face 11313.

(1) Assuming that the first layer of the 16.75 m thick medium-fine sandstone did not separate during the collapse of the whole layer, the 8.3 m thick sandy mudstone, which displayed certain stability, entered the fracture zone. In addition, the first weighting interval was 44.7 m. According to Equation (4), the support working resistance required for the first weighting main roof was as high as 239 kN.

(2) The on-site coring near the setup entry of the working face indicated that the rock formation in the roof developed fissures and presented poor integrity. Data indicated that the average uniaxial compressive strength of medium-fine sandstone was relatively low, with a value of 63.5 MPa and a statistical average RQD of 0.60. According to this, the product of the actual thickness of medium-fine sandstone and RQD value (defined as equivalent thickness in this case) was 16.75 × 0.6 = 10 m. Thus, the support working resistance required for working face 11313 for the first weighting was 14,132 kN.

In summary, the hydraulic support working resistance required during the first weighting of the three working faces varied greatly (3832~14,132 kN). This probably occurred because of differences in roof lithology in the mining area. However, the support working resistance required for periodic weighting was 6135~9842 kN, which affected the unified selection of mining support and fully mechanized mining equipment.

4. Numerical Simulation

4.1. Model Establishment

According to the geological characteristics of the working face 11313, a numerical simulation model of the thick hard roof was obtained using FIAC3D. The size of the model was selected as length × width × height = 250 m × 250 m × 200 m. In addition, the elastic-plastic constitutive was selected, the Mohr–Coulomb criterion was used for the failure criterion, and the goaf was filled with weakened material. Displacement boundaries were used at the bottom and on both sides of the model, and stress boundaries are used at the upper part of the model, As shown in Figure 6. In order to simulate the action of the overlying strata, an equivalent uniform load of 9.3 MPa was applied to the top of the model, the lateral pressure coefficient was 1.16, and the horizontal stress of the model was 10.8 MPa. According to Zhao et al. [29–31], the impact factor D of blasting excavation is 0.7 and the details are shown in Table 2.

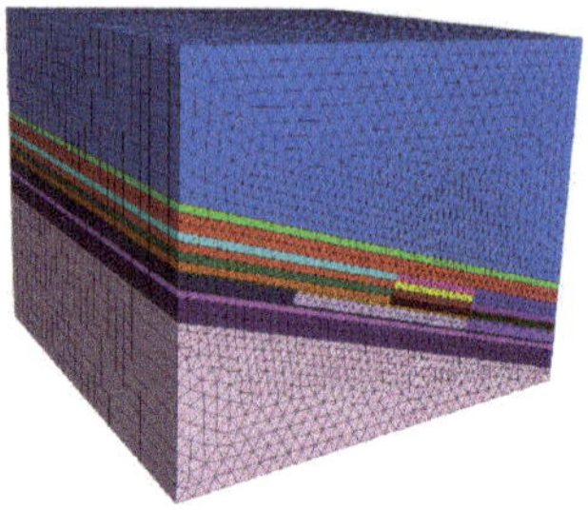

Figure 6. Numerical model of 11313 working face.

Table 2. Value of damage factor after pre-split blasting.

Rock Mass Characteristics	Damage Factor D
The blasting effect is poor, and the hard rock roadway is partially damaged	D = 0.8
Small and medium-scale blasting caused blasting of rock mass	D = 0.7 Good blasting effect D = 1.0 Poor blasting effect
Mass production blasting caused significant disturbances	D = 1.0 Good blasting effect D = 0.7 Poor blasting effect

The relationship between the deformation parameters of the rock mass before and after the damage is shown in Equation (5). The strength parameters, cohesion, and internal friction angle after blasting were 0.5 times [32,33] those before blasting. The basic physical and mechanical parameters of the roof after blasting are shown in Table 3.

$$\begin{aligned} E_1 &= (1-D)E \\ \mu_1 &= \mu \end{aligned} \tag{5}$$

where E, G, μ are the elastic modulus, shear modulus, and Poisson's ratio of the undamaged rock, respectively; E_1, G_1, μ_1 are the corresponding quantities of rock blasting damage, respectively.

Table 3. Physical and mechanical parameters of the basic roof after blasting.

Lithology	Bulk Modulus/Gpa	Shear Modulus/GPa	Cohesion/MPa	Internal Friction Angle/°
Medium-fine sandstone	3.6	3.9	1.5	20

4.2. Analysis of Results

The stress of the surrounding rock in the stope is shown in Figure 7. The stress arch rotates to the blasting side, and the stress arch height increases with the blasting depth, from spoon shape to oval shape. When the blasting thickness is less than 9 m, the stress arch height increases gradually from 30 m to 40.2 m with the increase in blasting thickness. The inclination angle of the stress arch remained unchanged, the side inclination angle of the working face was 60°, and the inclination angle of the stress arch on the lower side of the working face was about 20°. When the blasting thickness was more than 12 m, the stress arch height increased to 54.2 m, the inclination angle of the stress arch on the side of the working face was 60°, and the inclination angle of the stress arch on the lower side of the working face suddenly changed to 57°. The side abutment pressure of the headentry was significantly released, which benefits the support.

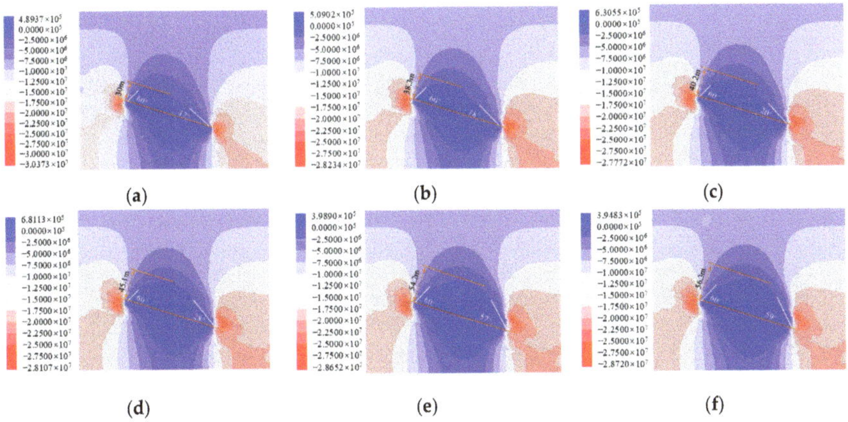

Figure 7. Stress of surrounding rock along working face tendency after blasting. (**a**) The thickness of the main roof pre-split blasting is 0 m; (**b**) The thickness of the main roof pre-split blasting is 3 m; (**c**) The thickness of the main roof pre-split blasting is 6 m; (**d**) The thickness of the main roof pre-split blasting is 9 m; (**e**) The thickness of the main roof pre-split blasting is 12 m; (**f**) The thickness of the main roof pre-split blasting is 15 m.

Figure 8 shows the distribution of the abutment pressure when the thickness of the main roof pre-split blasting was 12 m. In the range beyond 80 m in front of the coal wall, the front abutment pressure also remained constant, and the coal body entered a state of in-situ stress. In the range of 25~80 m in front of the coal wall, the coal body was affected by the mining disturbance. As the distance from the coal wall decreased, the abutment pressure above the coal body also gradually increased. When the distance from the coal wall was 20 m~25 m, the abutment pressure reached a peak value of 23.4 MPa, and the stress concentration factor reached 1.67 (the in-situ stress is 13.7 MPa). Compared with the tailentry, the influence of the front abutment pressure on the blasting zone and the distance of the peak abutment pressure from the working face increased. The peak value of the abutment pressure also decreased to 16.2 MPa, and the stress concentration factor was reduced to 1.3. The degree of stress concentration on the coal body in front of the working face can be significantly reduced after the roof blasting, and the stress environment can be effectively improved.

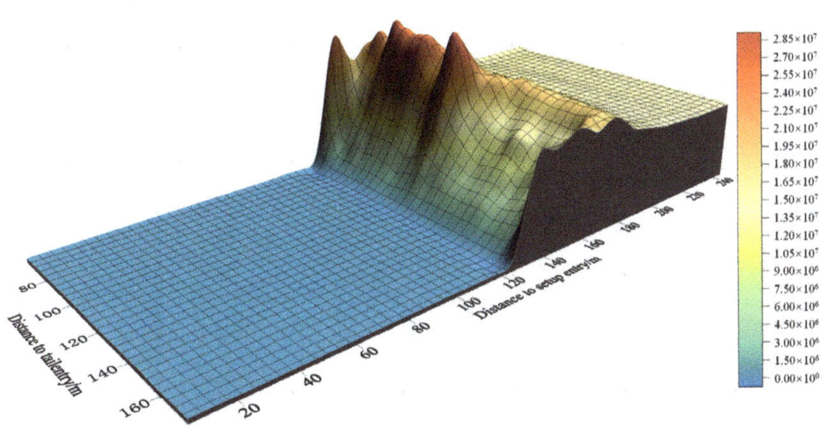

Figure 8. Distribution of abutment pressure.

5. Ground Pressure Control

5.1. Support Selection of Working Face

Since working face 11113 was the first mining face of A3 coal in the mine, and considering the previous experience of the Huainan mining area with fully mechanized mining, it was proposed that mining should be performed in two layers, with an average mining height of the top layer of 2.8 m. With respect to working face 11313, the theoretical analysis showed that the rated working resistance of 9200 kN support did not satisfy the strength requirements of first weighting, and the support working resistance required for working face periodic weighting was also in danger of exceeding the rated working resistance. In order to unify mining area technologies and equipment, and to reduce the production cost of the working face, the deep hole pre-splitting roof technology was selected in order to reduce the strength of working face 11313 during the first weighting and local regional periodic weighting. Thus, after performing theoretical calculations for working face 11113 ground pressure, the hydraulic support models of the three working faces were ZZP7200/18/38, ZZ9200/24/50, and ZZ9200/24/50, respectively.

5.2. Scheme of Deep Hole Pre-Split Blasting

5.2.1. Initial Caving

The lower roof of working face 11313 was affected by the fault, and the integrity of the main roof was damaged. Hence, a hole in working face 11313 5 m away from the setup entry was drilled along the slot and directed to the roof. In addition, blasting was carried out. The blast hole layout scheme is shown in Figure 9, and the blasting technical parameters are shown in Table 4.

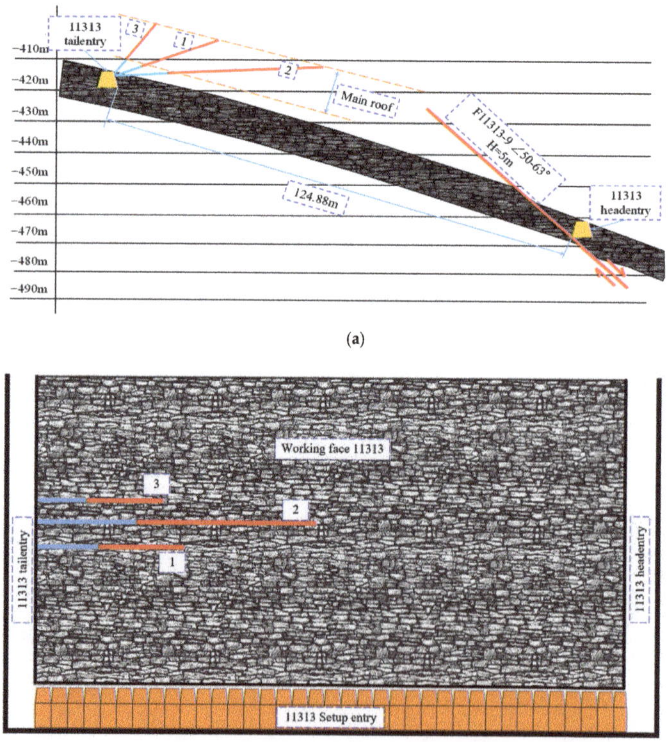

Figure 9. Layout of roof blast hole near 11313 setup entry. (**a**) Blast hole layout in the inclination direction; (**b**) Blast hole layout in the strike direction.

Table 4. Technical parameters of deep hole blasting near setup entry.

Roadway Name	Blast Hole Serial Number	Blast Hole Depth/m	Dip Angle/°	Blast Hole Diameter/mm	Charge Length/m	Blocking Length/m
11313 tailentry	1#	19	34	94	14	5
	2#	55	3	94	45	10
	3#	17	70	94	16	4

5.2.2. Periodic Weighting Caving

In some areas of working face 11313, the main roof thickness was greater than 16.75 m. It was also observed that the main roof directly covered the working face, which displayed good integrity, with the roof hanging over 15 m. The advanced deep hole blasting technology was applied to these areas with frame pressing risk, as shown in Figure 10. According to the engineering geological analysis, the deep hole blasting pre-splitting roof technology should be selected when the fault structure is in use, details are shown in Table 5.

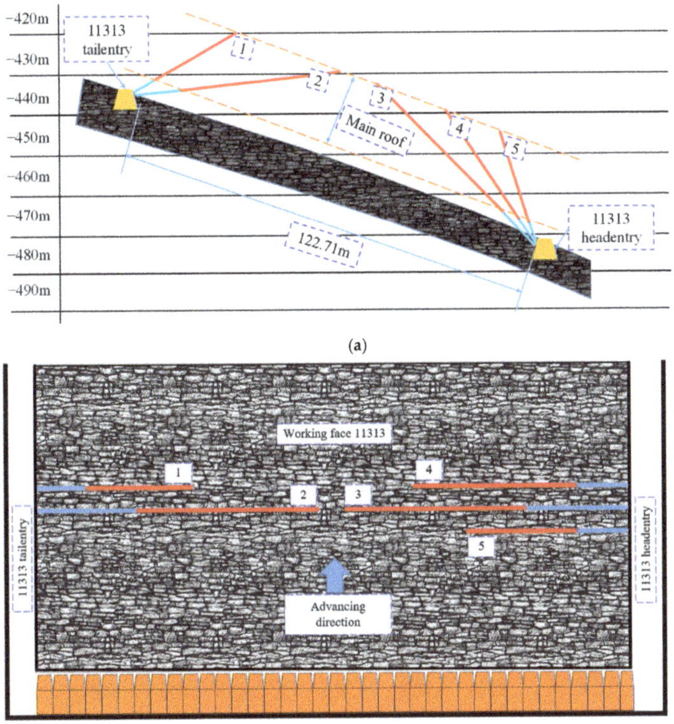

Figure 10. Layout of roof blast hole during advancing mining of working face 11313. (**a**) Blast hole layout horizon. (**b**) Blast hole parameters.

Table 5. Technical parameters of long suspended roof section deep hole blasting.

Roadway Name	Blast Hole Serial Number	Blast Hole Depth/m	Dip Angle/°	Blast Hole Diameter/mm	Charge Length/m	Blocking Length/m
11313 tailentry	1#	27	34	94	21	6
	2#	65	9	94	55	10
11313 headentry	3#	65	45	94	55	10
	4#	37	65	94	30	7
	5#	30	85	94	24	6

5.3. Roof Control Scheme in Fault Zone

Since the fault zone on the roof was relatively broken and difficult to control and considering that a layered mesh laying under the layered mining was needed, the mesh laying method was improved. Thus, we proposed the roof mesh laying and hanging method with advanced hanging mesh protection, as shown in Figure 11. This method was useful in controlling the roof fall and rib spalling in the fault zone and broken roof area.

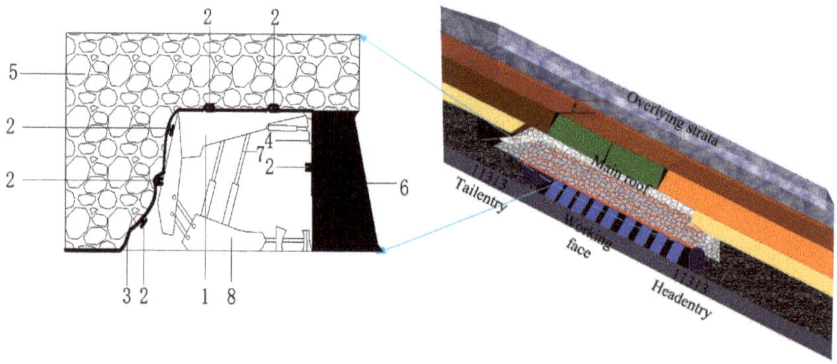

Figure 11. Support cross-section of rib spalling and roof fall protection for working face (1-hydraulic support top beam; 2-I steel; 3-anchor chain mesh; 4-side protecting board; 5-workface broken roof; 6-workface coal wall; 7-hydraulic prop; 8-support base).

5.4. Field Measurement of Ground Pressure

According to field measurements, the first weighting interval of working face 11113 was 33 m and the weighting strength was 34 MPa. Thus, the support met the mining demand. During weighting, the average dynamic loading efficiency was 1.3, and the weighting was relatively stable. The broken roof mesh laying technology had a good control effect on the roof, the weighting interval basically conformed to the theoretical value, and the hydraulic model met the mining demand.

When mining the 27 m in working face 11213, the main roof broke and collapsed. The working face weighting strength was reduced by increasing the initial force and reducing the mining height. The length of the periodic weighting step was 15 m, which was basically consistent with the theoretical weighting interval, and the weighting strength was within the bearing range of support.

When the deep hole blasting technology was applied to working face 11313, the large weighting interval and weighting strength were avoided. After advancing about 15 m into the main roof rock formation, layer separation and caving occurred. As shown in Figure 12, during periodic weighting, the support value reached the support resistance required by the working face. In the subsequent mining process, the advanced deep hole pre-splitting technology effectively prevented the risk of frame pressing in the working face.

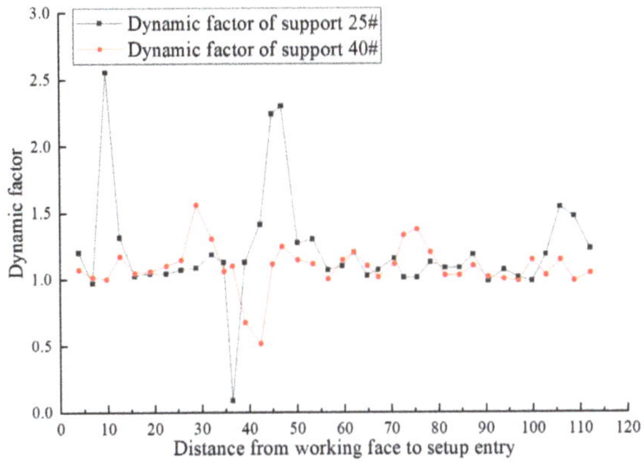

Figure 12. Dynamic load coefficient near blasting zone.

6. Discussion

Deep hole pre-split blasting has been proven to be an effective technical measure to control the ground pressure within a certain limit [22–25]. An effective method based on deep hole pre-split blasting to control roof collapse for the adjacent working faces with different roof conditions has been proposed in this paper. In order to determine the working face parameters and roof control design with different roof conditions in the same mining area, coal seam characteristics and roof differences were considered. The geo-mechanical evaluation of the surrounding rock and the overall prediction of the ground pressure within the mining area were the basis for the selection of fully mechanized mining equipment and ground pressure control.

First of all, it is necessary to obtain detailed information on the roof strata of adjacent working faces and conduct theoretical calculations to find the hard strata [34]. Secondly, according to the results of theoretical calculations, it is necessary to identify the range of pressure variation that the working face may face during the process of mining. Finally, in the process of mining, the characteristics of roof pressure, numerical simulation, and field monitoring results must be used to develop the design of the weakened roof scheme for different working faces.

Although the method has been successfully applied to the working face 11313, it is still necessary to carry out a detailed study to determine the scope of application and propose a suitable model for the control of the stopping roof of the adjacent working face.

7. Conclusions

Based on theoretical analysis, numerical simulation, and field measurements of ground pressure induced by working face 11113, 11213, and 11313 mining, the main conclusions are as follows:

(1) As the deep hole pre-split blasting thickness of the main roof increases, the stress arch rotates to the blasting side, and the stress arch height increases with the blasting depth from spoon shape to oval shape. The stress arch height increased to 54.2 m, with an inclination angle of the stress arch on the side of the working face of 60°.

(2) Herein, a two-deep hole pre-splitting blasting scheme to set up the entry section and long suspended roof section was proposed. This helped reduce the first weighting strength of working face 11313 below 34 Mpa and the roof suspension distance within 15 m, which effectively weakened the impact of severe ground pressure.

(3) The roof mesh laying and hanging method with advanced hanging mesh protection was proposed. This method effectively controlled roof fall and rib spalling.

Author Contributions: Conceptualization, Q.F.; Data curation, Z.W.; Formal analysis, X.H.; Investigation, K.Y.; Resources, Q.Y.; Writing—original draft, Q.F.; Writing—review & editing, K.Y., X.H. and Z.W. All authors have read and agreed to the published version of the manuscript.

Funding: This study was supported by the Institute of Energy, Hefei Comprehensive National Science Center under Grant No. 21KZS217 and No. 21KZS215, Major special projects of science and technology in Shanxi Province (No. 20191101016), Open Research Grant of Joint National-Local Engineering Research Centre for Safe and Precise Coal Mining (Grant No. EC2021014).

Data Availability Statement: The data used for conducting classifications are available from the corresponding author upon request.

Acknowledgments: All authors contributed to this paper. Qiang Fu prepared and edited the manuscript. Xiang He, Ke Yang and Zhen Wei substantially contributed to the data analysis and revised the article. Qinggan Yang reviewed the manuscript and processed the investigation during the research process. And we would like to express our thanks to Yanqun Yang from Shanxi Coking Coal Group Co., Ltd. for his support and assistance with the manuscript. This study was supported by the Institute of Energy, Hefei Comprehensive National Science Center under Grant No. 21KZS217 and No. 21KZS215, Major special projects of science and technology in Shanxi Province (No. 20191101016), Open Research Grant of Joint National-Local Engineering Research Centre for Safe and Precise Coal Mining (Grant No. EC2021014).

Conflicts of Interest: The authors declared no potential conflict of interest concerning this article's research, authorship, and/or publication.

References

1. Zhu, S.; Jiang, Z.; Zhou, K.; Peng, G.; Yang, C. The characteristics of deformation and failure of coal seam floor due to mining in Xinmi coal field in China. *Bull. Eng. Geol. Environ.* **2014**, *73*, 1151–1163. [CrossRef]
2. Wang, J. Development and prospect on fully mechanized mining in Chinese coal mines. *Int. J. Coal Sci. Technol.* **2014**, *1*, 253–260. [CrossRef]
3. Zhang, C.; Tu, S.; Zhao, Y. Compaction characteristics of the caving zone in a longwall goaf: A review. *Environ. Earth Sci.* **2019**, *78*, 27–42. [CrossRef]
4. Gao, R.; Kuang, T.; Zhang, Y.; Zhang, W.; Quan, C. Controlling mine pressure by subjecting high-level hard rock strata to ground fracturing. *Int. Coal Sci. Technol.* **2021**, *8*, 1336–1350. [CrossRef]
5. Wang, J.; Yu, B.; Kang, H.; Wang, G.; Mao, D.; Liang, Y.; Jiang, P. Key technologies and equipment for a fully mechanized top-coal caving operation with a large mining height at ultra-thick coal seams. *Int. Coal Sci. Technol.* **2015**, *2*, 97–161. [CrossRef]
6. Patyk, M.; Bodziony, P.; Krysa, Z. A Multiple Criteria Decision Making Method to Weight the Sustainability Criteria of System Selection for Surface Mining. *Energies* **2021**, *14*, 3066. [CrossRef]
7. Tu, H.; Tu, S.; Yuan, Y.; Wang, F.; Bai, Q. Present situation of fully mechanized mining technology for steeply inclined coal seams in China. *Arab. Geosci.* **2015**, *8*, 4485–4494. [CrossRef]
8. Bahrani, N.; Hadjigeorgiou, J. Influence of stope excavation on drift convergence and support behavior: Insights from 3D continuum and discontinuum models. *Rock Mech. Rock Eng.* **2018**, *51*, 2395–2413. [CrossRef]
9. Li, G.; Ma, F.; Guo, J.; Zhao, H.; Liu, G. Study on deformation failure mechanism and support technology of deep soft rock roadway. *Eng. Geol.* **2020**, *264*, 105262. [CrossRef]
10. Dychkovskyi, R.; Shavarskyi, I.; Saik, P.; Lozynskyi, V.; Falshtynskyi, V.; Cabana, E. Research into stress-strain state of the rock mass condition in the process of the operation of double-unit longwalls. *Min. Miner. Depos.* **2020**, *14*, 85–94. [CrossRef]
11. Jia-Chen, W. Mechanism of the rib spalling and the controlling in the very soft coal seam. *J. China Coal Soc.* **2007**, *32*, 785–788.
12. Vu, T.T. Solutions to prevent face spall and roof falling in fully mechanized longwall at underground mines, Vietnam. *Min. Miner. Depos.* **2022**, *16*, 127–134. [CrossRef]
13. Babets, D.; Sdvyzhkova, O.; Shashenko, O.; Kravchenko, K.; Cabana, E.C. Implementation of probabilistic approach to rock mass strength estimation while excavating through fault zones. *Min. Miner. Depos.* **2019**, *13*, 72–83. [CrossRef]
14. Li, T.; Mu, Z.; Liu, G.; Du, J.; Lu, H. Stress spatial evolution law and rockburst danger induced by coal mining in fault zone. *Int. J. Min. Sci. Technol.* **2016**, *26*, 409–415. [CrossRef]
15. Li, Z.; Wang, C.; Shan, R.; Yuan, H.; Zhao, Y.; Wei, Y. Study on the influence of the fault dip angle on the stress evolution and slip risk of normal faults in mining. *Bull. Eng. Geol. Environ.* **2021**, *80*, 3537–3551. [CrossRef]
16. Dai, J.; Jiang, J. Influence of mining sequence of hanging wall and foot wall on mining-induced stress of fault coal pillar. *J. Min. Saf. Eng.* **2016**, *33*, 35–41.
17. Ju, J.; Xu, J. Structural characteristics of key strata and strata behaviour of a fully mechanized longwall face with 7.0 m height chocks. *Int. Rock Mech. Min. Sci.* **2013**, *58*, 46–54. [CrossRef]
18. Fan, J.; Dou, L.; He, H.; Du, T.; Zhang, S.; Gui, B.; Sun, X. Directional hydraulic fracturing to control hard-roof rockburst in coal mines. *Int. Min. Sci. Technol.* **2012**, *22*, 177–181. [CrossRef]

19. Dou, L.; Mu, Z.; Li, Z.; Cao, A.-Y.; Gong, S.-Y. Research progress of monitoring, forecasting, and prevention of rockburst in underground coal mining in China. *Int. Coal Sci. Technol.* **2014**, *1*, 278–288. [CrossRef]
20. Cui, F.; Dong, S.; Lai, X.; Chen, J.; Cao, J.; Shan, P. Study on rule of overburden failure and rock burst hazard under repeated mining in fully mechanized top-coal caving face with hard roof. *Energies* **2019**, *12*, 4780. [CrossRef]
21. Guo, W.; Wang, H.; Dong, G.; Li, L.; Huang, Y.-G. A case study of effective support working resistance and roof support technology in thick seam fully-mechanized face mining with hard roof conditions. *Sustainability* **2017**, *9*, 935. [CrossRef]
22. He, J.; Dou, L.; Mu, Z.; Cao, A.-Y.; Gong, S.-Y. Numerical simulation study on hard-thick roof inducing rock burst in coal mine. *J. Cent. South Univ.* **2016**, *23*, 2314–2320. [CrossRef]
23. Lu, A.; Dou, L.; Bai, J.; Chai, Y.; Zhou, K.; Kan, J.; Cao, J.; Song, S. Mechanism of Hard-Roof Rock Burst Control by the Deep-Hole Blasting, Numerical Study Based on Particle Flow. *Shock. Vib.* **2021**, *2021*, 9527956. [CrossRef]
24. Sun, Z.; Zhang, D.; Fang, Q.; Dui, G.; Tai, Q.; Sun, F. Analysis of the interaction between tunnel support and surrounding rock considering pre-reinforcement. *Tunn. Undergr. Space Technol.* **2021**, *115*, 104074. [CrossRef]
25. He, H.; Dou, L.; Fan, J.; Du, T.; Sun, X. Deep-hole directional fracturing of thick hard roof for rockburst prevention. *Tunn. Undergr. Space Technol.* **2012**, *32*, 34–43. [CrossRef]
26. Wang, F.; Tu, S.; Yuan, Y.; Feng, Y.; Chen, F.; Tu, H. Deep-hole pre-split blasting mechanism and its application for controlled roof caving in shallow depth seams. *Int. Rock Mech. Min. Sci.* **2013**, *64*, 112–121. [CrossRef]
27. Huang, B.; Liu, J.; Quan, Z. The reasonable breaking location of overhanging hard roof for directional hydraulic fracturing to control strong strata behaviors of gob-side entry. *Int. Rock Mech. Min. Sci.* **2018**, *103*, 1–11. [CrossRef]
28. Yang, J.; Liu, C.; Yu, B. Application of confined blasting in water-filled deep holes to control strong rock pressure in hard rock mines. *Energies* **2017**, *10*, 1874. [CrossRef]
29. Zhao, Y.X.; Wang, T.; Jiang, Y.D.; Pan, L.; Zhang, K.X. Application of Hoek-Brown criterion in numerical simulation of ground pressure features in multi-seam longwall mining. *J. China Coal Soc.* **2013**, *38*, 970–976.
30. Asheghi, R.; Shahri, A.A.; Khorsand, Z.M. Prediction of uniaxial compressive strength of different quarried rocks using metaheuristic algorithm. *Arab. J. Sci. Eng.* **2019**, *44*, 8645–8659. [CrossRef]
31. Zhao, Y.; Yang, T.; Zhang, P.; Zhou, J.; Yu, Q.; Deng, W. The analysis of rock damage process based on the microseismic monitoring and numerical simulations. *Tunn. Undergr. Space Technol.* **2017**, *69*, 1–17. [CrossRef]
32. Hoek, E.; Kaiser, P.K.; Bawden, W.F. *Support of Underground Excavations in Hard Rock*; Balkema: Rotterdarn, The Netherlands, 1995.
33. Cheng, S.; Ma, Z.; Gong, P.; Li, K.; Li, N.; Wang, T. Controlling the deformation of a small coal pillar retaining roadway by non-penetrating directional pre-splitting blasting with a deep hole, A case study in Wangzhuang coal mine. *Energies* **2020**, *13*, 3084. [CrossRef]
34. Wang, X.; Zhu, W.; Xu, J.; Han, H.; Fu, X. Mechanism of overlying strata structure instability during mining below unconsolidated confined aquifer and disaster prevention. *Appl. Sci.* **2021**, *11*, 1778. [CrossRef]

sustainability

MDPI

Article

Research on the Influence of Backfilling Mining in an Iron Mine with Complex Mining Conditions on the Stability of Surface Buildings

Huaibin Li [1,2,*], Zhenpeng Guo [1], Xinzhu Hua [3], Bibo Dai [1], Xuemin Zeng [1] and Yuemao Zhao [4]

1 State Key Laboratory of Safety and Health for Metal Mines, Maanshan 243000, China
2 School of Safety Science and Engineering, Anhui University of Science and Technology, Huainan 232001, China
3 School of Mining Engineering, Anhui University of Science and Technology, Huainan 232001, China; xzhhua@aust.edu.cn
4 College of Energy and Mining Engineering, Shandong University of Science and Technology, Qingdao 266590, China
* Correspondence: lhb_aust@126.com or 2021068@aust.edu.cn

Abstract: The safe and efficient extraction of ore bodies under buildings contributes to the continued exploitation and utilization of mineral resources. To effectively reduce the mining-induced disturbance caused to surface buildings, the Vertical Crater Retreat (VCR) delayed backfilling method was applied at the Wuji iron mine according to mining theory and engineering experience, and the surface deformation under buildings was investigated through numerical simulation and on-site monitoring. The results show that stress concentration occurs at the top of the ore body and in surrounding rocks between ore veins, and the direction of the principal stress coincides with that of the main bearing capacity of the rock mass. With the increasing distance between the surrounding rock and the ore body, the stress disturbance in the surrounding rock attenuates gradually. The surface subsidence and deformation caused by ore mining exhibits the same northwest trend as the strike of the ore body, and the surface deformation in the X direction is greater than that in the Y direction. The surface deformation of the mining area near the 1# air shaft, Wuji village, steam autoclave workshop of the brick factory, and 1# substation is relatively large, while the surface subsidence, horizontal deformation, inclination, and curvature are all smaller than the critical values allowed for buildings and structures. This research provides a theoretical basis and engineering guidance for the prevention and control of surface deformation during mining under buildings and structures.

Keywords: mining under buildings; Wuji iron mine; surface deformation; mining methods; numerical simulation

1. Introduction

In China, there exist ample mineral resources located under buildings (structures), water bodies, and railways, and a significant proportion of ore bodies are located under buildings [1–3]. Previous studies have shown that the iron ore bodies under structures, water bodies, and railways total 3.26 billion tons in China [4]. Mining under structures, water bodies, and railways can further realize the utilization of underground resources, improve economic benefits from mining, and reduce damage to the geological and natural environment. However, surface deformation and subsidence can be caused during mining under these conditions. Therefore, research on safe mining under structures, water bodies, and railways is of great significance for the development and utilization of mineral resources in China.

At present, research on and application of coal mining under buildings has been widely conducted, and regulations for coal mining under structures, water bodies, and railways have been formulated in China. Nonetheless, a comprehensive theory and corresponding regulations for metal mining in the same situation have not been fully developed [5]. The characteristics of ore rocks in metal mines differ from those in coal mines, which are characterized by heterogeneous lithology and multiple structures. As a result, theory and methods developed for coal mining under structures, water bodies, and railways cannot be used directly for metal mines [6–8].

In recent years, studies on metal mining technology under buildings have been conducted by several research institutions in China. When mining the ore body under buildings, appropriate mining methods should be employed based on the occurrence conditions of the ore body, so that the deformation of surface buildings caused by underground mining can be controlled within the allowable range. Based on the mining practice under structures and waterbodies in the South Tannery, Jing Wenyu et al. [9] put forward a two-step backfilling mining method for gently inclined ore bodies, illustrated the mechanism of mining pressure in the stope, and proposed a mining pressure control scheme (that is, mining from the middle to the two wings, multiple unloadings in different zones, and stable settlement) to fully recover underground resources and extend the service life of the mine. Through numerical simulation, Wang Shanfei et al. [10] simulated different mining sequence schemes for the mining of the Sanshandao gold mine under buildings and waterbodies, and concluded that alternating room-and-pillar mining is beneficial for controlling surface deformation. Based on existing experience and the occurrence conditions of the Panlong lead–zinc ore body, Guo Mingming et al. [11] used a mechanized upward horizontal layered backfilling mining method, and verified that this method will not cause damage to surface highways. Wu Zhenyu et al. [12] simulated the use of a sublevel open stoping with delayed backfilling method and the pillar-retaining method on the shallow ore body in an iron ore mine. The results showed that the surface deformation is relatively small, and this mining method can ensure the safety of surface buildings. To prevent surface deformation during iron mining under buildings and highways, Guo Qi et al. [13] adopted sublevel open stoping with the delayed backfilling method and the pillar-retaining mining method, which can ensure the stability of surface buildings and roads. Although retaining safety pillars can mitigate the issue of significant surface deformation, it also entails a loss of mineral resources.

Surface deformation can be used to measure the safety of mining under structures, water bodies, and railways. Researchers have also conducted extensive investigations of surface deformation around mining areas after comprehensively analyzing and adopting appropriate mining methods under these conditions [14–18]. By using an electronic dumpy level, Zhao Yunchang [19] monitored the surface deformation and studied the subsidence mechanism around a mining area, and obtained a theoretical and scientific basis for mining subsidence, compensation for collapsed land, and comprehensive utilization. Xia Kaizong et al. [20–22] used GPS and a Leica DNA03 digital level to monitor surface deformation around the Chengchao iron mine. The results showed that, with the expansion of mining in depth and plane, the horizontal stress in the mining area releases continuously and a failure zone can be formed. Through the use of statistics and correlation analysis, Feng Yi [23] summarized the relationship between mining subsidence control and drilling and grouting, and proposed grouting and subsidence control techniques in the mine, which alleviate the mining subsidence caused by the exploitation of mineral resources under the three bodies. Based on ground deformation GPS data, dumpy level, and 3D laser scanning, Song Xugen et al. [24] studied the ground deformation above and around the mining site and goaf, established the mechanical model of surrounding rock failure around the goaf, and described the rock movement and mechanism of deformation at the measuring points. Huaizhan Li et al. [25] analyzed the influence of strip super-high water filling mining on surface buildings by means of theoretical analysis, numerical simulation and field monitoring. In Zhiqiang Yang et al.'s study [26], according to the site-specific geological conditions, the stability of the backfilling body and the characteristics of surface subsidence due to mining of −450 m level were analyzed using physical modeling. Furthermore, the surface subsidence and the stability of the stope were achieved using FLAC3D. Binglei Li et al. [27] studied the NMR-based damage characterization of backfill material in host rocks under dynamic loading using a true-triaxial testing machine. Due to the complexity of the occurrence of metal deposits, control measures to solve the surface deformation caused by underground mining are not unified, and the methods for analyzing surface deformation are relatively limited. Therefore, a combination of multiple analysis methods is needed to conduct in-depth research on the impact of mining of ore bodies under buildings on surface deformation.

In the early mining stage, the safety pillar-retaining method was used in the Wuji iron mine to protect surface buildings and structures, resulting in a large accumulation of mineral resources in the shallow area. To safely mine the remaining shallow and deep ore bodies, the occurrence conditions of ore bodies were analyzed, various mining methods were comprehensively compared and analyzed, and the most appropriate mining method was determined in this study. Additionally, numerical simulations and on-site monitoring were used to obtain the surface deformation caused by mining under buildings, such as surface subsidence, horizontal displacement, inclination, and curvature. Based on the Chinese standard of the Code for Design of Nonferrous Metal Mining, the influence of backfilling mining in iron mines with complex mining conditions on the stability of surface buildings was illustrated.

2. Selection of Mining Methods

2.1. Geological Conditions of the Mining Area

Wuji iron mine is located in Huoqiu County, Anhui Province. The deposit is located on the south bank of the middle and upper reaches of the Huaihe River. It belongs to the Huaihe River Class II terrace, with relatively flat terrain and an elevation of +40~+57 m. The deposits are covered by the quaternary system, with a thickness ranging from 32 m to 157 m. The hanging wall of Wuji iron mine is mainly biotite plagioclase gneiss and the footwall is mainly magnetite hornblende plagioclase gneiss. A YT-28 rock drill is used for drilling, the depth of the blast holes is 1.8 m and the diameter of blast holes is 38~40 mm. The falling height of each layer is controlled within 2.0 m. The blasting ore is transported to the ore pass by a 2 m 3 scraper, and then transferred to the main transport level.

Wuji iron mine is located in the east wing of the Zhouji inverted steering area. The trend of the ore body in the southern area is 30–45° NW; the direction of the ore body in the northern area is north–south, and the ore body is inclined SW–W. In the shallow part of exploration line 4, the ore body is inclined to the northeast, and the dip angle of the ore body is generally 35–60°, with certain local sections exceeding 70°. The ore body of Wuji iron mine exhibits a gradual slope in the southern area and a steep incline in the northern area, while, vertically, it gradually descends into deeper levels and becomes steeper. The ore body is a monoclinal structure with an inverted skew wing at the north and south ends. Due to several tectonic changes, secondary folds have developed. The structural stress of exploration lines 0–8 is more prominent, showing an S-shaped form. The deposit is approximately 4000 m long from north to south and 150–600 m wide from east to west, with an average length of about 300 m. The burial depth of the ore body is about 45–1000 m, with an elevation of around 0–950 m. The planned mining elevation is +54-–650 m.

Within the boundary of the surface movement of the Wuji iron mine, there is a large number of mine-owned buildings and villages. Figure 1 shows the overall distribution of buildings and villages. The numbers in Figure 1 represent the exploration line. If the traditional method of retaining safety pillars is used to protect surface structures, the long-term development of the mine will be hindered. As the mining activities gradually extend deeper, the retaining scope of the safety pillars becomes larger, until almost all deep ore bodies in the mine are within this scope of safety pillars. According to satellite images, there are nearly one thousand private houses in Wuji village alone. If the scheme of relocating surface villages and industrial buildings is adopted, it will lead to huge labor costs and substantial economic and time expenditures, which will directly affect the normal production of the mine. To improve the effective utilization of mineral resources, appropriate mining methods should be selected to reduce the impact of mining on surface deformation.

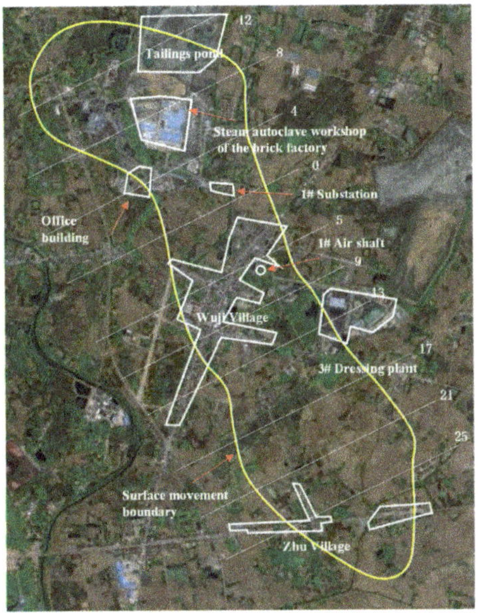

Figure 1. Distribution of surface buildings in Wuji iron mine.

2.2. Selection of Mining Methods

In the Wuji iron mine, the ore body is mainly inclined (an inclination angle of about 50°), thick, and large (an average thickness of about 20 m). Therefore, this type of ore body was selected for research. Based on the engineering geological conditions and experience with mining, three schemes were preliminarily selected: sublevel open stoping with the delayed backfilling method (Scheme I), the VCR delayed backfilling method (Scheme II), and the lateral caving stage open stope delayed backfilling method (Scheme III).

To obtain the appropriate method for mining in Wuji, the advantages and disadvantages of the three mining methods were analyzed. In Scheme I, medium deep hole blasting was adopted; it is a simple, safe process with high production capacity. However, this scheme requires sublevel rock drilling engineering, which requires a large amount of mining preparation work. Additionally, it is difficult to manually charge up deep holes and control the boundary of the mining site, which may lead to over-mining and over-blasting. In Schemes II and III, down hole drilling, large hole blasting, and scraper mining were used. This method is characterized by highly technical equipment, high production efficiency, a simple stope structure, a small amount of development and mining preparation work, and a large production capacity of the stope. However, a large drilling depth and complex blasting process are required, which is expensive and may cause the punching phenomenon induced by blasting, and inability to support the upper wall roof. Scheme III will inevitably cause a large amount of blasting through lateral collapse blasting and a high rate of large blocks, which cause significant disturbance to adjacent mining areas or backfilling bodies.

In addition, a comprehensive evaluation was conducted on the three mining schemes from the perspectives of mining techniques, mining safety, and economic benefits. Table 1 shows the relative evaluation of the comprehensive qualitative indicators.

Table 1. Qualitative relative evaluation of mining methods.

No.	Content	Scheme I	Scheme II	Scheme III
1	Mining techniques	Simple	Simple	Simple
2	Backfilling techniques	Complex	Complex	Complex
3	Adaptability to ore body	Poor	Poor	Poor
4	Worker proficiency	Skilled	Requires training	Skilled
5	Production efficiency	General	High	High
6	Safe working conditions	Excellent	General	General
7	Mining equipment investment	High	High	High
8	Bulk fraction	High	High	High
9	Mining costs	66 CNY/t	64 CNY/t	63 CNY/t
10	Recycling rate	High	Medium	Low

The above schemes were compared qualitatively and quantitatively. Although the recovery rate of Scheme I is relatively high, the difficulty in controlling the boundary of the upward fan-shaped deep hole stope may lead to over-blasting and over-mining, and the cutting ratio is large and the cost is high. Thus, Scheme I was ruled out. During the application of Scheme III in inclined or steeply inclined ore bodies, mining from the hanging wall to the footwall can lead to significant depletion. Additionally, a large blasting volume is required in this scheme, and it can have a significant impact on blasting vibration. Considering the characteristics of deep mining at Wuji iron mine (southern area), Scheme III is not recommended. After analysis and comparison, Scheme II was selected for mining, as shown in Figure 2. When the VCR delayed backfilling method is used for mining, it is necessary to arrange a rock drilling chamber horizontally on the upper part of the ore block. The rock drilling truck is used to drill down the vertical deep hole to the bottom level of the lower part of the ore body, and the ball charge funnel blasting is adopted. The bottom structure of the ore extraction forms a trench, and the two mining sites share a common ore extraction roadway.

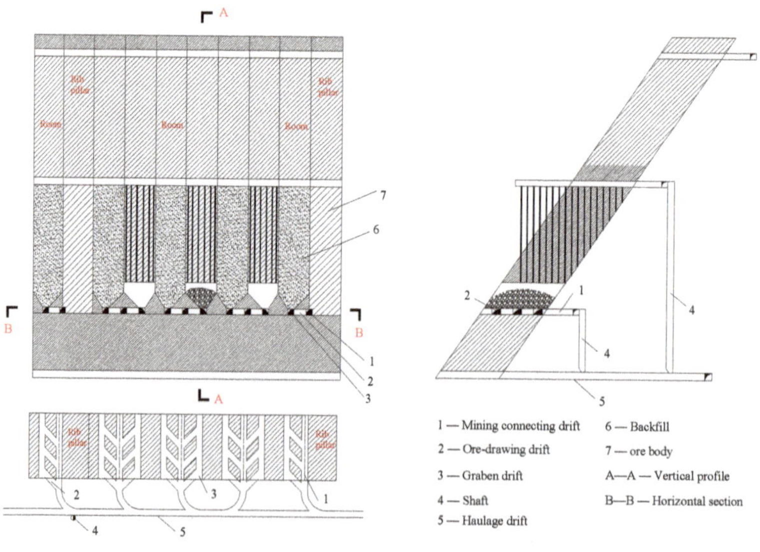

Figure 2. VCR delayed backfilling method.

3. Estimation of Mechanical Parameters of Rock Mass

3.1. Rock Mechanical Experiments

According to the international experimental standards of rock mechanics, the rock was processed into φ 50 × 100 mm and φ 50 × 50 mm cylinders with coring machines, cutting machines, and stone grinding machines in the laboratory for uniaxial compression, Brazilian splitting tests, and shear tests, respectively. Mechanical tests were conducted on samples from the surrounding rock, ore, weathering zone, and quaternary of Wuji iron mine using a testing machine. Additionally, cemented tailings were used as the backfill material, and the backfill was composed of tailings and cement in a ratio of 8:1. The physical and mechanical parameters of different types of rocks were obtained, as shown in Table 2.

Table 2. Physical and mechanical parameters of rock samples.

Category	Density (g/cm^3)	Compressive Strength (MPa)	Tensile Strength (MPa)	Elastic Modulus (GPa)	Poisson Ratio μ	Cohesion (MPa)	Internal Friction Angle (°)
Surrounding rock	2.84	124.56	7.62	60.23	0.23	16.60	43.85
Ore body	3.49	163.59	15.01	62.34	0.20	53.25	42.43
Weathering zone	2.68	50.47	5.55	34.97	0.26	11.08	44.36
Quaternary	1.92	1.65	0.17	0.23	0.35	0.03	25.00
Backfilling body	1.83	4.35	0.31	0.57	0.25	0.56	60.00

3.2. Hoek–Brown Parameters

Based on the Griffith strength theory, Hoek and Brown proposed the Hoek–Brown strength criterion, and later put forward a modified version based on the geological strength index (GSI). The expression is shown below.

$$\sigma_1 = \sigma_3 + \sigma_{ci}\left(m_b\frac{\sigma_3}{\sigma_{ci}} + s\right)^a \tag{1}$$

where σ_1 is the maximum principal stress; σ_3 is the minimum principal stress; σ_{ci} is the uniaxial compressive strength of intact rock; m_b, s, and a are Hoek–Brown constants of the rock mass, which can all be expressed as functions of the GSI.

The Hoek–Brown failure criterion was used to determine the rock mass properties. Hoek et al. [28] proposed the following equations for calculating the constants (m_b, s, and a) of the rock mass:

$$m_b = m_i \exp^{\left(\frac{GSI-100}{28-14D}\right)} \tag{2}$$

$$s = \exp^{\left(\frac{GSI-100}{9-3D}\right)} \tag{3}$$

$$a = \frac{1}{2} + \frac{1}{6}\left(e^{-GSI/15} - e^{-20/3}\right) \tag{4}$$

where m_i refers to the complete rock material constant. The ore body and surrounding rock are mainly migmatite. As shown in Table 3, m_i = 29 [28,29]. The rock mass in the weathered zone is mainly composed of gneiss, schist, migmatite, migmatite granite, and the iron ore body of the neoarchean Wuji Formation. The lower limit of m_i is 12. The m_b refers to the material constant of the rock mass; s and a are the characteristic constants of the rock mass; D is the disturbance coefficient, which is related to the degree of damage to the rock mass during excavation. Most rock masses in the deep part of the ore body are less disturbed by blasting, and, as shown in Table 3, D = 0.5 [28,29]. The calculated Hoek–Brown constants are listed in Table 3.

Table 3. Hoek–Brown parameter values.

Rock Formation	GSI Value	m_i	m_b	s	a
Ore body	65	29	5.477	0.0094	0.502
Surrounding rock	45	29	2.113	0.0007	0.508
Weathering zone	20	12	0.689	0.0001	0.544

3.3. Mechanical Parameters of Rock Mass

According to the mechanical parameters of rock mass obtained from laboratory tests and rock mass quality evaluation, the Hoek–Brown strength criterion was used to estimate the uniaxial compressive strength, tensile strength, deformation modulus, Poisson's ratio, shear strength, and other mechanical parameters of different rock masses.

Uniaxial compressive strength of rock masses can be obtained as follows:

$$\sigma_{cmass} = \sigma_{ci} s^a \tag{5}$$

Uniaxial tensile strength of rock masses can be obtained as follows:

$$\sigma_{tmass} = -s\sigma_{ci}/m_b \tag{6}$$

The expression for the deformation modulus E_m of the rock mass is:

$$E_m = \begin{cases} \left(1 - \frac{D}{2}\right)\sqrt{\frac{\sigma_c}{100}}10^{\left(\frac{GSI-10}{40}\right)}, (\sigma_c \le 100MPa) \\ \left(1 - \frac{D}{2}\right)10^{\left(\frac{GSI-10}{40}\right)}, (\sigma_c > 100MPa) \end{cases} \tag{7}$$

The expressions for shear strength of rock mass (c and φ values) are:

$$c' = \frac{\sigma_{ci}[(1+2a)s + (1-a)m_b\sigma_{3n}](s + m_b\sigma_{3n})^{a-1}}{(1+a)(2+a)\sqrt{1 + (6am_b(s + m_b\sigma_{3n})^{a-1})/((1+a)(2+a))}} \tag{8}$$

$$\varphi' = \sin^{-1}\left[\frac{6am_b(s + m_b\sigma_{3n})^{a-1}}{2(1+a)(2+a) + 6am_b(s + m_b\sigma_{3n})^{a-1}}\right] \tag{9}$$

where $\sigma_{3n} = \sigma_{3max}/\sigma_c$, and σ_{3max} is the upper limit value of the limit stress in the relationship between the Hoek–Brown criterion and Mohr–Coulomb criterion; σ_{ci} is the uniaxial compressive strength of the rock mass.

The relationship between σ_{3max} and the strength σ_{cmass} of rock mass is:

$$\frac{\sigma_{3max}}{\sigma_{cmass}} = 0.47\left(\frac{\sigma_{cmass}}{\gamma H}\right)^{-0.94} \tag{10}$$

where σ_{cmass} is the uniaxial compressive strength of the rock mass, MPa; γ is the bulk density of the rock mass, MN/m^3; and H is the burial depth of the study area, m. When the horizontal stress is greater than the vertical stress, γH is replaced by horizontal stress.

The above equations are used to calculate the mechanical parameters of different rock masses, as shown in Table 4.

Table 4. Physical and mechanical parameters of rock masses.

Category	Compressive Strength (MPa)	Tensile Strength (MPa)	Deformation Modulus (GPa)	Poisson Ratio μ	Cohesion (MPa)	Internal Friction Angle (°)
Surrounding rock	31.76	0.75	21.14	0.23	1.86	42.38
Ore body	45.92	2.76	25.12	0.20	3.72	41.76
Weathering zone	12.63	0.53	11.26	0.26	1.53	43.12
Quaternary	0.41	0.04	0.12	0.35	0.01	18.00
Backfilling body	4.35	0.31	0.57	0.25	0.56	20.00

4. Numerical Simulation Analysis of Surface Deformation

4.1. Establishment of Numerical Simulation

The mining range of the ore body from the planned mining elevation of +54 m to −650 m was simulated, the surface deformation characteristics during the mining process were studied, and the impact on the stability of surface buildings after complete mining of the ore body was obtained.

The three-dimensional numerical calculation of the mining and backfilling process of the ore body was carried out using FLAC3D software. First, a mining geological model was established by connecting the ore bodies of each section with the geological boundary in the vertical direction by the 3D mine software. Second, Midas software was imported, and a 3D solid model of the mine suitable for meshing was established through Boolean operations. The ore body grid was encrypted. After meshing and grouping the model, a 3D numerical mesh model was finally established (see Figure 3). The range of the three-dimensional numerical grid model was 4000 m in the EW direction and 5300 m in the NS direction. Tetrahedral grids were used, with a total of 1.35 million grids and 300,000 nodes established.

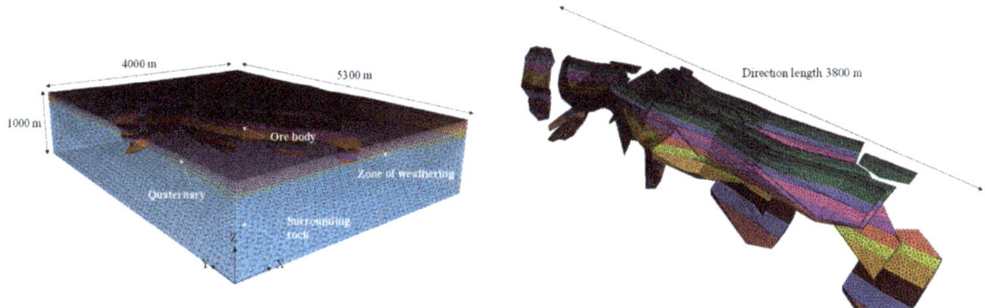

(**a**) The grid model of the mine (**b**) The grid model of the ore body

Figure 3. 3D Numerical grid model of Wuji iron mine.

The Mohr–Coulomb strength criterion was used to describe the strength characteristics and failure criteria of rock mass in the simulation. The self-weight stress was set as the initial stress field, and normal constraints were applied around the model and on the ground. The top surface of the model was a free surface without any constraints. According to the physical and mechanical parameters of rock mass in Table 4, the material parameters of different rock masses in the model were assigned. After the establishment of the three-dimensional numerical model of the mine, the initial stress balance was carried out. The maximum principal stress at the bottom of the numerical model was about 24.5 MPa, which was consistent with the actual principal stress in the project.

4.2. Stress and Displacement Analysis of Ore Body Surrounding Rock

The stability of the mining stope is an important factor affecting the surface displacement of mines. The cumulative displacement of surrounding rocks in each stope results in the final displacement of the mining surface. Therefore, analyzing the changes in stress and displacement of the stope-surrounding rock during the mining process is of great significance for studying surface deformation.

Figures 4 and 5 show the stress vector and displacement of the Wuji iron mine, respectively. It can be seen that reduction in the principal stress occurs in the surrounding rock of the hanging wall of the ore body, and the stress concentration occurs in the top of the ore body and some surrounding rocks between the ore veins. The direction of principal stress in the stress concentration area at the top of the ore body is nearly horizontal, and that of the rock mass between the ore veins is in the inclined direction of the ore body. The stress concentration at the top of the ore body is mainly caused by the rock mass in ore veins supporting the surrounding rock at the top, indicating that the rock mass in the ore veins bears a certain amount of upper rock load. The direction of the principal stress is the direction of the main bearing capacity of the rock mass, indicating that the surrounding rock in the ore veins plays an important supporting and protective role in the mining process. Therefore, the surrounding rock of the hanging wall and the footwall of the ore vein should be protected to reduce the damage caused by blasting to the surrounding rock during the mining process.

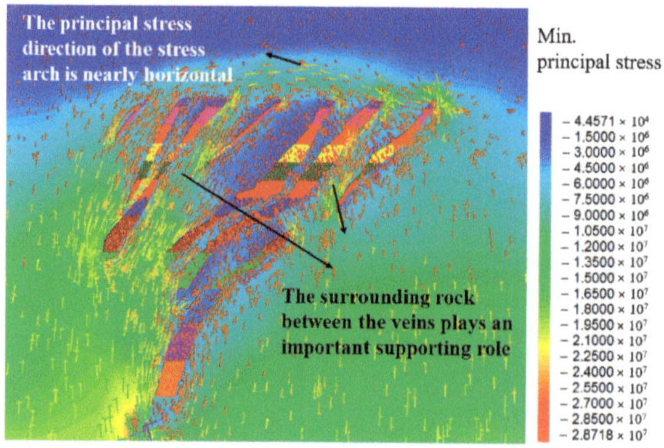

Figure 4. Stress vector in the surrounding rock.

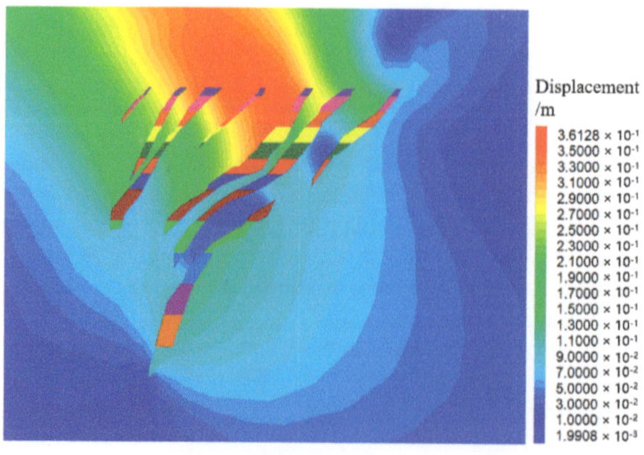

Figure 5. Displacement cloud diagram.

The mining-induced displacement of the surrounding rock of the ore body mainly occurs on the hanging wall of the ore body and gradually extends to the surface. Among them, the maximum settlement displacement of the surface located at the hanging wall of the ore body is relatively large, about 33 cm. The vertical displacement near the rock pillar in the ore veins significantly decreases. This is because the rock pillar bears the upper stress and slows down the settlement of the hanging wall of the ore body. It can be seen that with the gradual accumulation of the surrounding rock displacement at each stope, the magnitude of surface displacement will be ultimately affected. Therefore, it is necessary to strengthen the support of the surrounding rock in the hanging wall of the mining area during the mining process.

4.3. Numerical Simulation Analysis of Surface Deformation

4.3.1. Calculation Principles and Limits of Surface Deformation

The analysis of the stability and damage done to surface buildings and structures within the range of influence of underground mining is of great significance for safety and maintaining the normal life of residents. The main patterns of surface deformation caused by underground mining include settlement, horizontal deformation, inclination, and curvature, and their calculation functions are shown in Equations (11)–(14). Figure 6 shows the calculation diagrams of various parameters of surface deformation.

$$W_{\max} = qm\cos\alpha \tag{11}$$

$$\varepsilon = \frac{u_E - u_F}{EF} = \frac{\Delta u}{l_2^0} \tag{12}$$

$$i_{AB} = \frac{w_B - w_A}{l_{AB}^0} = \frac{\Delta w}{l_{AB}^0} \tag{13}$$

$$k_B = \frac{i_{BC} - i_{AB}}{0.5(BC + AB)} = \frac{\Delta i}{0.5\left(l_2^0 + l_1^0\right)} \times 10^{-3} \tag{14}$$

where $W_{\max}$, ε, i_{AB}, and k_B represent surface subsidence, horizontal deformation, inclination, and curvature, respectively; q is the subsidence coefficient, and q = 0.15 is taken in the backfilling mining method; m is the mining thickness.

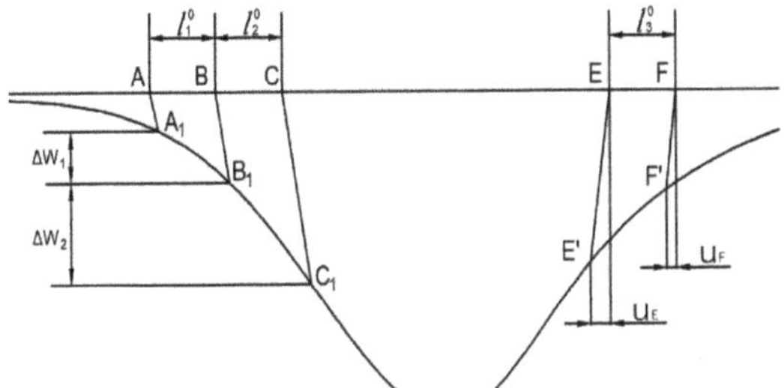

Figure 6. Schematic diagram for calculation of surface inclination, curvature, and horizontal deformation.

The Code for Design of Nonferrous Metal Mining (GB-50771-2012) provides a more detailed classification of the levels of buildings and structures that need to be protected on the surface. Different critical deformation values are used for buildings and structures with different protection levels, as shown in Table 5. Based on the theory of surface deformation and the relevant regulations, the allowable critical values for the displacement and deformation of surface buildings in Wuji iron mine are as follows: the surface inclination i = 3 mm/m, the surface curvature k = 0.2 × 10^{-3}/m, and the surface horizontal deformation ε = 2 mm/m. Based on the simulated surface deformation data, the impact of underground mining on surface structures in the Wuji iron mine is evaluated below.

Table 5. Allowable values for surface displacement and deformation of buildings and structures.

Buildings and Structures Protection Level	Inclination (mm × mm)	Curvature k (10^{-3}/m)	Horizontal Deformation ε (mm × mm)
I	±3	±0.2	±2
II	±6	±0.4	±4
III	±10	±0.6	±6
IV	±10	±0.6	±6

4.3.2. Surface Subsidence

The mining of ore bodies causes the displacement and deformation of surrounding rock, and when the displacement of surrounding rock develops and reaches the surface, subsidence and deformation can be caused. To scientifically analyze the impact of underground mining on the surface buildings and structures at the Wuji iron mine, a surface subsidence cloud map of the Wuji iron mine was simulated, as shown in Figure 7. The relevant displacement data were extracted, and Tecplot, Sufer, and other software types were applied for data processing. Finally, the three-dimensional diagram of land surface subsidence was obtained, as shown in Figure 8.

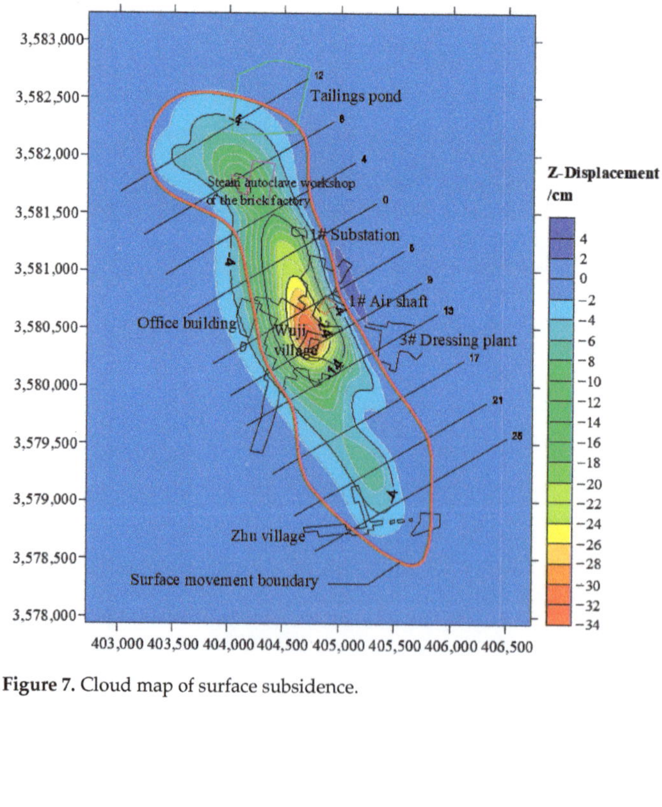

Figure 7. Cloud map of surface subsidence.

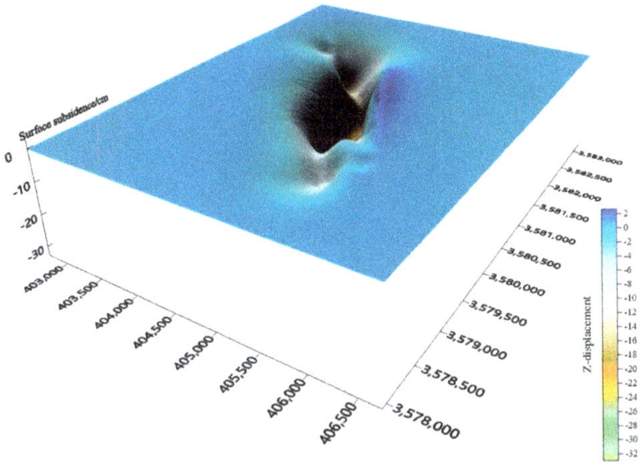

Figure 8. 3D schematic diagram of surface subsidence.

It can be seen that the surface subsidence range of the mining area is distributed in a narrow strip, and the overall distribution of surface subsidence is located above the ore body in a northwestern direction, which is consistent with the trend of the ore body. The surface subsidence is distributed in three regions: north, central, and south. The area of maximum subsidence occurs in the center of the ore body, followed by the northern area, with the smallest amount of subsidence occurring in the southernmost end of the ore body.

To further observe the changes in surface subsidence, cross-sections of subsidence were made along lines 8, 5–9, and 21, and longitudinal sections were made along the surface subsidence area, as shown in Figures 9 and 10. There are three extreme points of surface subsidence in space, which are located near the steam autoclave workshop of the brick factory (line 8, 14.8 cm), the center of Wuji village in the central area (lines 5–9, 33 cm), and in the southern area (line 21, 7.9 cm). The amount of surface subsidence is the greatest at lines 5–9, and the monitoring of surface subsidence should be strengthened at these positions. The surface subsidence increases in line 8 and line 21, which means that further detection and monitoring are required at this location. Additionally, the reasons for the increase in subsidence should be analyzed to ensure safe mining under buildings.

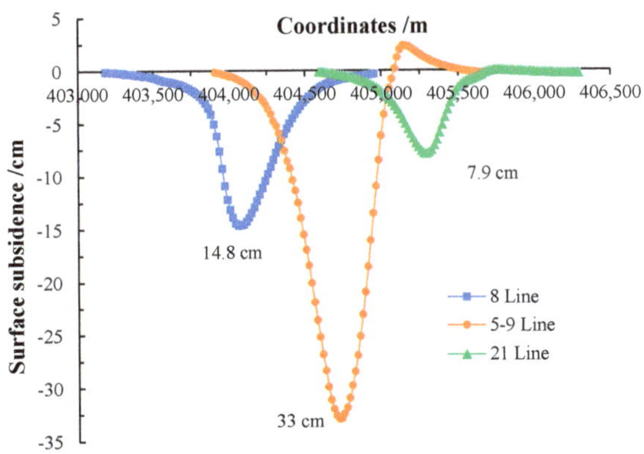

Figure 9. Cross-section of surface subsidence along the exploration line.

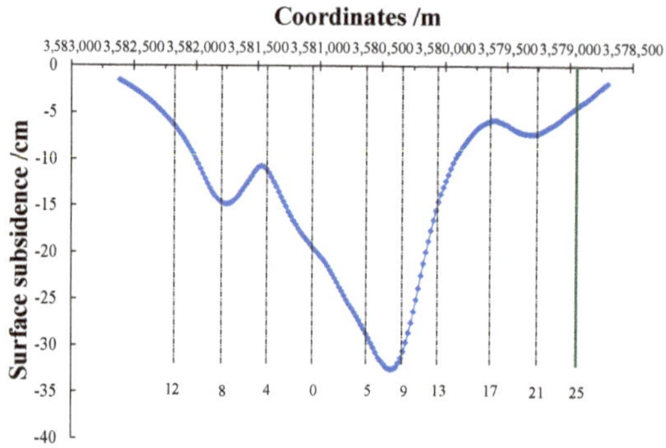

Figure 10. Longitudinal section of surface subsidence in the mining area.

4.3.3. Horizontal Deformation of the Surface

The horizontal deformation of the surface has a significant destructive effect on buildings, especially the impact of tensile deformation. Figure 11 shows the cloud map of surface horizontal displacement in the X and Y directions, while Figure 12 shows the cloud map of surface horizontal deformation in the X and Y directions. It can be seen that there are two extreme points for horizontal displacement and deformation in the X direction, located near the steam autoclave workshop of the brick factory (line 8) and in the center of Wuji village (lines 5–9). The extreme points of horizontal displacement and deformation in the Y direction are located near the filling station (line 8), near the 1# air shaft (line 5), and in the center of Wuji village (line 9). Although there are fewer extreme points for horizontal deformation in the X direction, the deformation value is greater than that in the Y direction. The maximum horizontal deformation values in the X and Y directions near the steam autoclave workshop are 1.05 mm/m and 0.47 mm/m, respectively. This is mainly because the distribution pattern of horizontal deformation on the surface is closely related to the occurrence status of ore bodies, and the extreme point of deformation mainly appears in the upper wall of the inclined ore body and the edge of the horizontal ore body. The ore body in the Wuji iron mine ranges from tilted to steeply inclined, and has a SW–W tendency. Horizontal surface displacement frequently occurs on the hanging wall of the ore body, mainly in the SW–W direction. Therefore, the surface displacement in the X direction is relatively large, while the displacement in the Y direction is relatively small. According to the relevant provisions of the Code for Design of Nonferrous Metal Mining, the above surface deformation values are all lower than the allowable critical deformation value of buildings and structures by 2 mm/m.

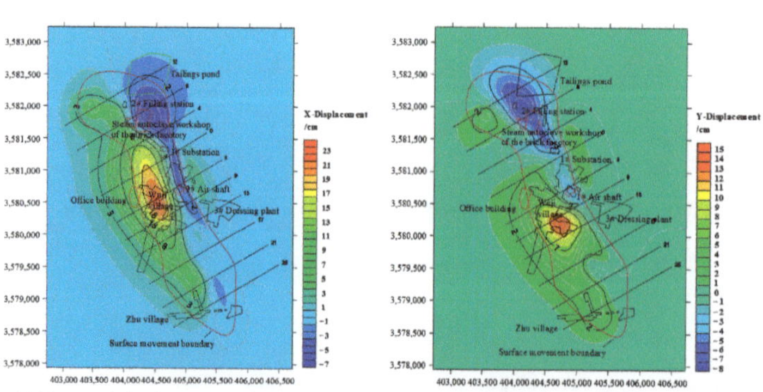

(**a**) Horizontal displacement in the X direction (**b**) Horizontal displacement in the Y direction

Figure 11. Cloud map of surface horizontal displacement.

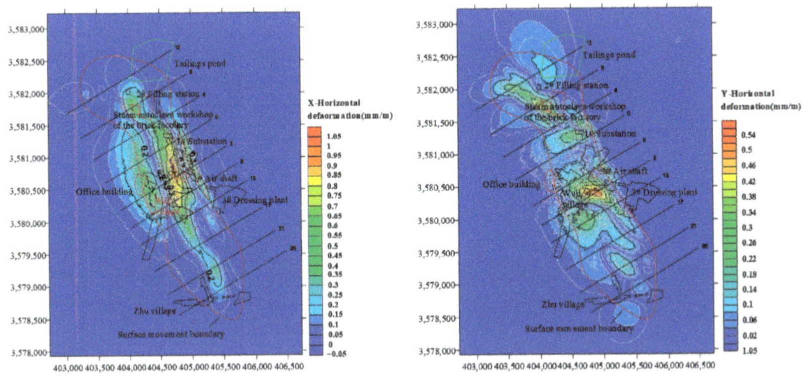

(**a**) Horizontal deformation in X direction (**b**) Horizontal deformation in Y direction

Figure 12. Cloud map of horizontal surface deformation.

4.3.4. Surface Inclination Characteristics

The ground inclination can cause the center of gravity of buildings to shift. This not only affects the stability of the building, but also causes additional stress inside the building and the redistribution of the bearing capacity of the foundation. Figure 13 shows a cloud map of the surface inclination. It can be seen that the distribution of surface inclination is related to the ore body. The surface inclination in the X direction is distributed in a strip along the direction of the ore body, while that in the Y direction displays a block-like distribution. There are two extreme points in the surface inclination of the X direction, which are all located near line 5. One is from the 1# air shaft to Wuji village (1.24 mm/m), and the other is in the central area of Wuji village (0.87 mm/m). There are four extreme points in the surface inclination of Y located near the steam autoclave workshop of the brick factory (lines 4–8, 0.71 mm/m), the 1# substation (line 0, 0.76 mm/m), the 1# air shaft (line 5, 0.66 mm/m), and the southeast of Wuji village (lines 9–13, 0.69 mm/m). Although there are fewer extreme points of surface inclination in the X direction than in the Y direction, the extreme values in the X direction are all greater than those in the Y direction. According to the relevant provisions of the Code for Design of Nonferrous Metal Mining, the above surface deformation values are all lower than the allowable critical deformation value for buildings and structures by 3 mm/m.

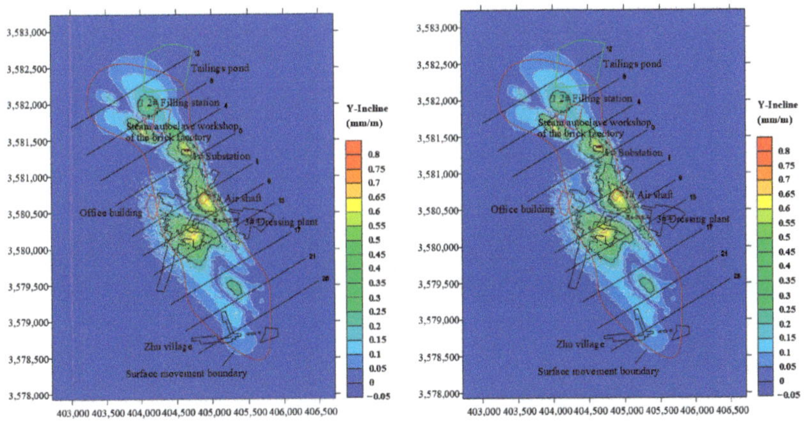

(**a**) X direction (**b**) Y direction

Figure 13. Cloud map of surface inclination.

4.3.5. Surface Curvature Characteristics

Surface curvature represents the degree of change in surface inclination. The existence of surface curvature makes the surface change from its original plane to a curved surface, and then causes the redistribution of foundation reactions in buildings and structures, changing from the original uniform foundation reaction and forming loading and unloading zones. Figure 14 shows the cloud map of surface curvature in this mining area. It can be seen that the distribution range of surface curvature in the mining area is relatively large, but there are no exact extreme points of the surface curvature. The maximum curvature of the mining area in the X direction is $0.02 \times 10^{-3}/\mathrm{m}$, and that in Y direction is $0.017 \times 10^{-3}/\mathrm{m}$. According to the relevant provisions of the Code for Design of Nonferrous Metal Mining, the above surface deformation values are all lower than the allowable critical deformation value of $0.2 \times 10^{-3}/\mathrm{m}$ for buildings and structures. Therefore, the damage caused to buildings by the underground mining activities would be relatively small.

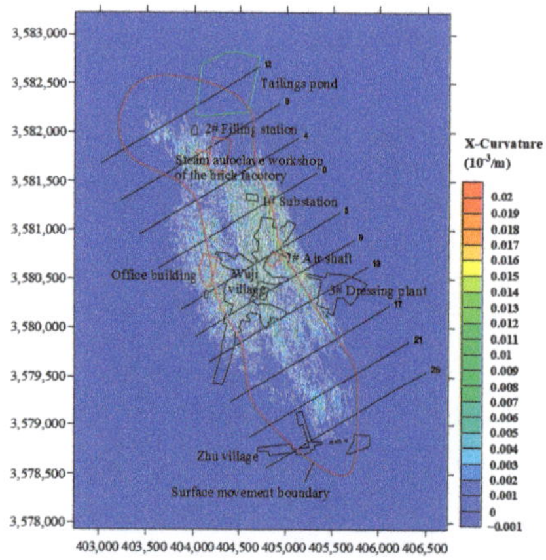

(**a**) X direction

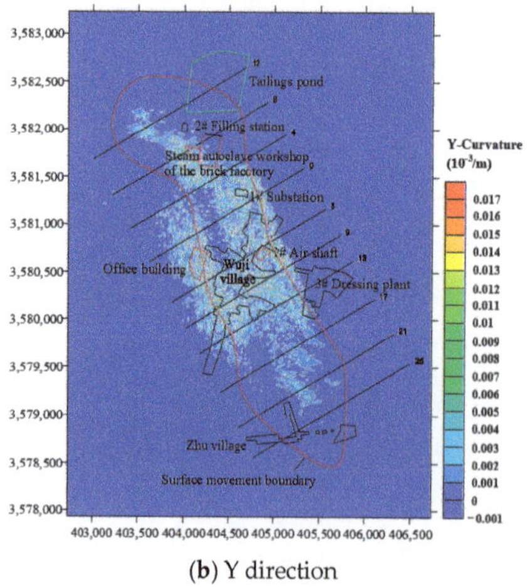

(**b**) Y direction

Figure 14. Cloud map of surface curvature.

5. On-Site Monitoring

To protect surface buildings and structures in the mine, shallow ore bodies are not mined and roof pillars are left in. According to the distribution of surface buildings and structures, the maximum mining depth in the area north of line 0 is −95 m, while that in the area south of line 0 is −130 m. As of June 2021, the mining area has reached the middle section of −310 m.

To accurately study the surface deformation law caused by the underground mining of the Wuji iron mine, 156 displacement monitoring points were arranged on the surface of the mining area in 2013. The RTK and electronic dumpy level were used for displacement monitoring. Figure 15 shows the location distribution of surface displacement monitoring points.

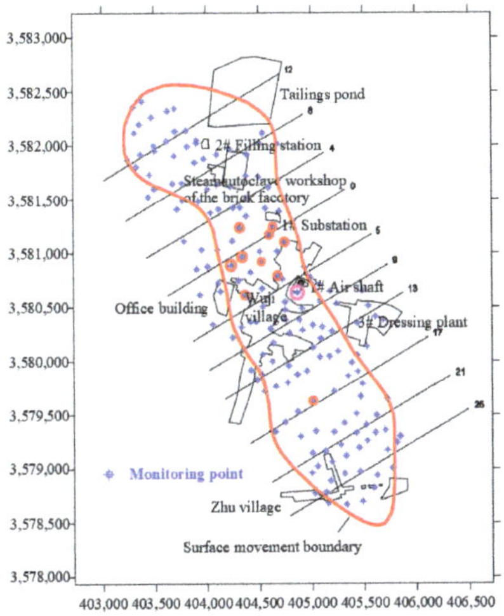

Figure 15. Distribution of surface monitoring points in the mining area.

According to the surface displacement monitoring data of the mine from June 2021, the settlement and horizontal displacement of each monitoring point were plotted, as shown in Figures 16 and 17. Due to the large number of monitoring points, some monitoring points were selected longitudinally along the ore body. The maximum subsidence value is 11 cm with local fluctuation, and the maximum fluctuation value is 13 cm. The maximum surface displacement in the X direction of the mining area is 8 cm, and that in the Y direction is 12 cm.

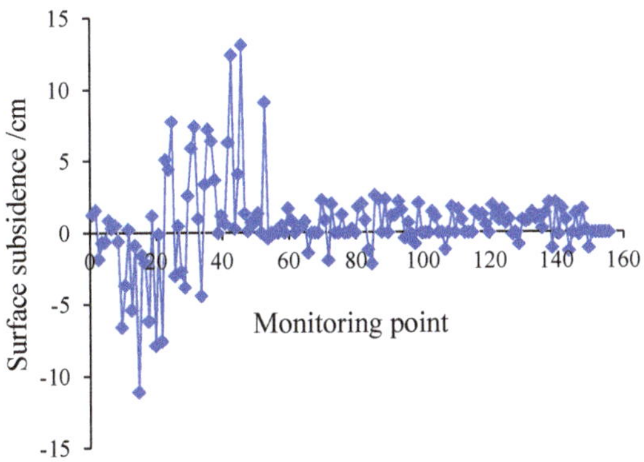

Figure 16. Subsidence values of surface monitoring points in the mining area.

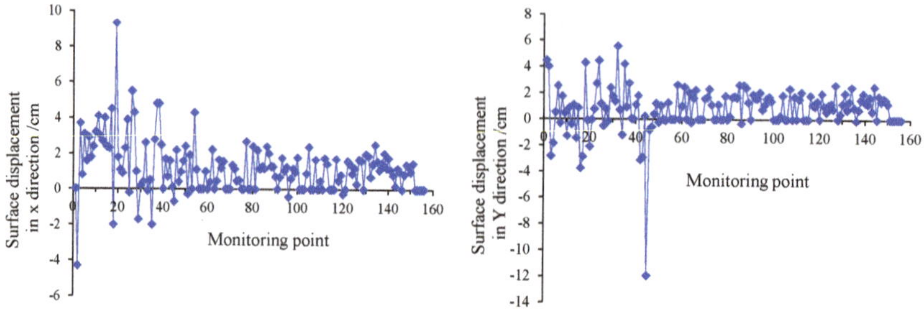

(**a**) X direction (**b**) Y direction

Figure 17. Horizontal displacement of surface monitoring points in the mining area.

To study the change in surface deformation at different locations, the distribution of surface subsidence and horizontal displacement was obtained, as shown in Figures 18 and 19. It can be seen that the surface deformation of the mining area is consistent with the mining direction of the ore body. The larger values of surface subsidence and horizontal displacement in the mining area occur in the area north of line 0. This is mainly because the depth of the highest mining middle section in the mining area north of line 0 is shallower than the highest mining depth in the mining area south of line 0. Moreover, there are more ore bodies extracted in the mining area north of line 0, and the surface is severely affected by mining. The distribution of surface displacement and subsidence in the mining area is consistent with the current mining situation.

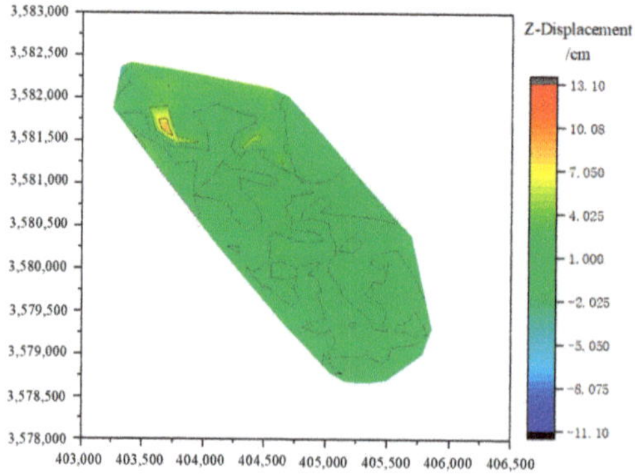

Figure 18. Subsidence distribution map of surface monitoring points in the mining area.

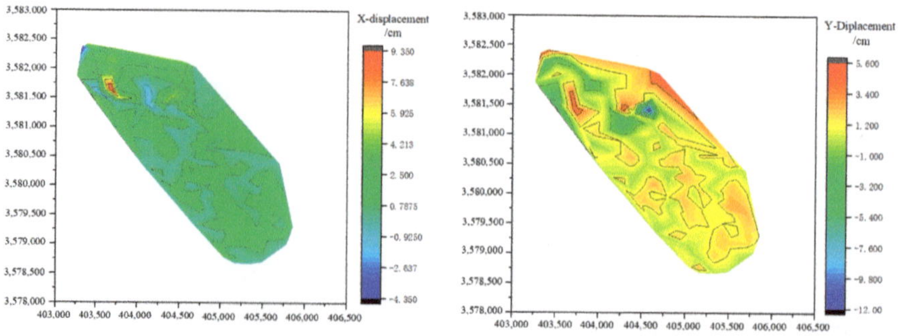

(**a**) X direction (**b**) Y direction

Figure 19. Horizontal displacement distribution of surface monitoring points in the mining area.

6. Conclusions

Based on underground mining theory and engineering practice, the mining method of the mining area was determined, a detailed study was conducted on the surface deformation under this mining method, and the movement of ground under buildings during complex iron ore mining was described.

(1) Based on the engineering geological conditions of the Wuji iron mine, a comprehensive evaluation was conducted of the mining scheme in terms of mining technology, mining safety, economic benefits, and other aspects, and the VCR delayed backfilling method was finally selected for mining.
(2) The stress concentration occurs at the top of the ore body and in some surrounding rocks between the ore veins. This is mainly because the rock mass between the ore veins supports the top surrounding rock, and the direction of the principal stress is consistent with that of the main bearing capacity of the rock mass. Therefore, the surrounding rocks of the hanging walls and footwalls of ore veins should be protected to reduce damage to the surrounding rock caused by blasting during the mining process.
(3) The distribution pattern of surface subsidence obtained from numerical simulation is the same as that obtained by on-site monitoring, which is consistent with that of the ore body (NW trend). After the mining of the ore body is completed, the maximum subsidence area appears in the middle zone of the ore body, followed by the northern zone, and the stress concentration is the smallest in the southernmost zone of the ore body.
(4) The gradual accumulation of deformation in the surrounding rock is the main cause of mining-induced rock deformation in the mining area, ultimately affecting surface deformation. The maximum value of surface settlement in this mining area is 33 cm, which is less than the specified critical value of surface subsidence.
(5) The surface deformation is related to the strike of the ore body, and the surface deformation in the X direction is greater than that in the Y direction of the mining area. The maximum horizontal deformation in the X and Y directions are 1.05 mm/m and 0.47 mm/m, respectively, the maximum curvature in the X and Y directions are 0.02×10^{-3}/m and 0.017×10^{-3}/m, respectively, and the maximum inclination in the X and Y directions are 1.24 mm/m and 0.76 mm/m, respectively. The above surface deformation values are all smaller than the allowable critical deformation values of buildings and structures specified in the Code for Design of Nonferrous Metal Mining.
(6) This research provides a theoretical basis and engineering guidance for the prevention and control of surface deformation during mining under buildings and structures.

Author Contributions: All the authors contributed to publishing this paper. H.L. conceived and designed the research. Z.G. performed the field investigations and experiments, presented the numerical simulation. X.H. and H.L. wrote the original manuscript. B.D., X.Z. and Y.Z. participated in the data analysis and manuscript modification. All authors have read and agreed to the published version of the manuscript.

Funding: The State Key Laboratory of Safety and Health for Metal Mines (2021-JSKSSYS-02), Key project of Natural Science research in Universities of Anhui Province (KJ2021A0455), and Research Start-up Fund for introduction of talents of Anhui University of Science and Technology (2021yjrc01).

Institutional Review Board Statement: Not applicable.

Informed Consent Statement: Not applicable.

Data Availability Statement: Data are available upon request from the authors.

Acknowledgments: Great appreciation goes to the editorial board and the reviewers of this paper.

Conflicts of Interest: No potential conflict of interests were reported by the authors.

References

1. Lian, D.; Wang, Y. Comprehensive evaluation system of "three under" coal mining. *J. China Univ. Min. Technol.* **2005**, *1*, 100–104. (In Chinese)
2. Cheng, L.; Li, X.; Ye, Z.; Sun, X. Surface stability analysis of "three under" mining within subsidence basin in a gold mine. *Met. Mine* **2019**, *3*, 56–61. (In Chinese)
3. Guo, W.; Ma, Z.; Bai, E. Current status and prospect of coal mining technology under buildings, water bodies and railways, and above confined water in China. *Coal Sci. Technol.* **2020**, *48*, 16–26. (In Chinese)
4. Xu, M. *Mining Induced Failure Characteristics and Deformation Prediction Method of High-Grade Highway*; China University of Mining and Technology: Beijing, China, 2017.
5. Yi, Z. *Research on Safety Technology of "Three Under" Resources Exploitation of Xinqiao Mining*; Central South University: Changsha, China, 2008.
6. Villegas, T.; Nordlund, E.; Dahnér-Lindqvist, C. Hangingwall surface subsidence at the Kiirunavaara Mine, Sweden. *Eng. Geol.* **2011**, *121*, 18–27. [CrossRef]
7. Villegas Barba, T.F.; Nordlund, E. Numerical analyses of the hangingwall failure due to sublevel caving: Study case. *Int. J. Min. Miner. Eng.* **2013**, *4*, 201–223. [CrossRef]
8. Marschalko, M.; Yilmaz, I.; Křístková, V.; Fuka, M.; Bednarik, M.; Kubečka, K. Determination of actual limit angles to the surface and their comparison with the empirical values in the Upper Silesian Basin (Czech Republic). *Eng. Geol.* **2012**, *124*, 130–138. [CrossRef]
9. Jing, W.; Zhang, G. Study on "Three-under" mining techniques in South area of the Xikuangshan mine. *Hunan Nonferrous Met.* **2003**, *6*, 1–4. (In Chinese)
10. Wang, S.; Wang, K.; Ma, F.; Lu, R. Optimization of three-underground mining technology and disaster prevention in Sanshandao gold mine. *Gold Sci. Technol.* **2020**, *28*, 734–742. (In Chinese)
11. Guo, M.; Li, J.; Li, H.; Dongxu, C. Safety impact analysis of "three under" mining for Panlong lead-zinc mine. *Nonferrous Met. Eng.* **2020**, *10*, 107–112. (In Chinese)
12. Wu, Z.; Wang, J.; Liu, P.; Li, Z.; Chen, H. Analysis on the influence of security pillar mining on surface structures of an iron mine in shandong province. *Mod. Min.* **2022**, *38*, 207–211. (In Chinese)
13. Guo, Q.; Ma, C.; Chen, Y. Optimization on shaft safety pillar and numerical simulation on surface deformation in "three under" mining. *Min. Res. Dev.* **2020**, *40*, 117. (In Chinese)
14. Guo, W.; Ma, Z.; Jiao, Y.; Chen, X.; Bai, E. Coal mining under television signal tower and its dynamic tilt adjustment protection technology. *J. China Coal Soc.* **2023**, *48*, 15–27. (In Chinese)
15. Ma, F.; Zhao, H.; Zhang, Y.; Guo, J.; Wei, A.; Wu, Z.; Zhang, Y. GPS monitoring and analysis of ground movement and deformation induced by transition from open-pit to underground mining. *J. Rock Mech. Geotech. Eng.* **2012**, *4*, 82–87. [CrossRef]
16. Ma, F.; Zhao, H.; Yuan, R.; Guo, J. Ground movement resulting from underground backfill mining in a nickel mine (Gansu Province, China). *Nat. Hazards* **2015**, *77*, 1475–1490. [CrossRef]
17. Zhao, H.; Ma, F.; Zhang, Y.; Guo, J. Monitoring and analysis of the mining-induced ground movement in the Longshou Mine, China. *Rock Mech. Rock Eng.* **2013**, *46*, 207–211. [CrossRef]
18. Cai, L.; Liu, Y.; Meng, W.; Xu, Z.; Wang, S. High-voltage power lines pylons foundation subsidence dynamic prediction and monitoring method in mining area. *Metal Mine* **2016**, *3*, 168–171. (In Chinese)
19. Zhao, Y. Mechanism analysis of surface subsidence based on the application of "three under" coal mining technology. *Bull. Surv. Mapp.* **2020**, *6*, 156–157+162. (In Chinese)
20. Xia, K.Z.; Liu, X.M.; Chen, C.X.; Fu, H.; Zhang, H.; Zhang, J.; Wu, Y. Surface deformation in west area of Chengchao iron mine induced by underground mining. *Chin. J. Rock Mech. Eng.* **2014**, *33*, 1572–1588. (In Chinese)
21. Long, L.; Chen, C.; Xia, K.; Deng, Y.Y. Research on the surface subsidence laws during underground mining in the west of Chengchao iron mine. *Min. Res. Dev.* **2016**, *36*, 47–52. (In Chinese)
22. Deng, Y.; Chen, C.; Xia, K. Cause analysis of surface collapse in western area of Chengchao iron mine. *Rock Soil Mech.* **2019**, *40*, 743–758. (In Chinese)
23. Feng, Y. Mining Subsidence Control and Drilling Grouting Engineering Example Analysis. *Coal Geol. China* **2022**, *34*, 33–38. (In Chinese)
24. Song, X.; Chen, C.; Pang, H.; Kai-Zong, X.; Shan, C.; Kuo-Yu, Y.; Chao-Yi, S. Study of relationship between underground mining and surface deformation in metal mines. *Rock Soil Mech.* **2018**, *39*, 425–436.
25. Li, H.; Guo, G.; Zhai, S.C. Mining scheme design for super-high water backfill strip mining under buildings: A Chinese case study. *Environ. Earth Sci.* **2016**, *75*, 1–12. [CrossRef]
26. Yang, Z.; Zhai, S.; Gao, Q.; Li, M. Stability analysis of large-scale stope using stage subsequent filling mining method in Sijiaying iron mine. *J. Rock Mech. Geotech. Eng.* **2015**, *7*, 87–94. [CrossRef]
27. Li, B.; Lan, J.; Si, G.; Lin, G.; Hu, L. NMR-based damage characterisation of backfill material in host rock under dynamic loading. *Int. J. Min. Sci. Technol.* **2020**, *30*, 329–335. [CrossRef]

28. Hoek, E.; Carranza-Torres, C.; Corkum, B. Hoek-Brown failure criterion-2002 Edition. In Proceedings of the North American Rock Mechanics Society, Toronto, QC, Canada, 7–10 July 2002.
29. Hoek, E.; Brown, E.T. Practical estimates of rock mass strength. *Int. J. Rock Mech. Min. Sci.* **1997**, *34*, 1165–1186. [CrossRef]

sustainability

MDPI

Article

Difference in Surface Damage between Deep and Shallow Mining of Underground Coal Resources in China

Weitao Yan [1,2], Junting Guo [1,*] and Shaoge Yan [2]

1 State Key Laboratory of Groundwater Protection and Utilization by Coal Mining, Beijing 102009, China; yanweitao@hpu.edu.cn
2 State Collaborative Innovation Center of Coal Work Safety and Clean-Efficiency Utilization, Henan Polytechnic University, Jiaozuo 454003, China; yanshaoge@hpu.edu.cn
* Correspondence: guojunting@nicenergy.com

Abstract: The mining of underground coal resources often results in extensive damage to the ground surface, particularly in China, which has a large amount of coal resources. However, the laws of surface damage caused by the mining of deep and shallow underground coal resources are relatively different. This study analyzes the difference in surface damage induced by deep and shallow mining and its mechanism by field measurement and similar material simulation experiments. Surface damage is mainly manifested in the form of cracks, which can be categorized as permanent and dynamic cracks. Permanent cracks occur above the mining boundary of shallow and deep coal mines. Dynamic cracks (including dynamic stretching cracks and stepped cracks) only appear above the goaf in shallow mining. This disparity is due to the fact that strata movement in deep mining occurs in a "three zones" mode, with the failure height of strata increasing with the mining degree in an "S" shape. However, rock strata movement in shallow mining follows a "two zones" mode, with the failure height of rock strata increasing exponentially with the mining degree. Thus, the rock strata are prone to slide and become unstable in the form of benched rock beams, producing dynamic cracks, such as stepped cracks on the surface. This research improves the existing mining subsidence theory and provides technical support for relevant mines to take targeted treatment measures.

Keywords: surface damage; formation mechanism; deep mining; shallow mining

1. Introduction

China has the largest coal production in the world, with a production of 4.5 billion tons in 2022, accounting for 54% of the world's total output [1]. The extraction of a large amount of underground coal resources leads to a large range of surface damage, including surface water accumulation [2], cracks [3,4], underground pipeline damage, and vegetation damage [5,6]. It has a serious impact on the local environment and undermines the sustainable development of the local economy, society, and ecology (Figure 1).

Coal seams can be classified into shallow, middle-deep, and deep, based on their depth and relevant mining science knowledge. A coal seam buried more than 400 m deep is considered a deep coal seam, a coal seam buried less than 200 m deep is a shallow coal seam, and a coal seam buried between 200 and 400 m deep is a middle-deep coal seam (Figure 1) [7,8].

Mining of underground coal seams with different burial depths results in varying degrees of surface damage. The impact of underground mining spreads through the overburden and ultimately affects the surface, causing surface damage [9]. The deeper the coal seam is buried, the longer the transmission path required for the mining impact to spread to the surface, and the greater the mining impact reduction, the greater the surface damage range, and the smaller the surface damage degree [10].

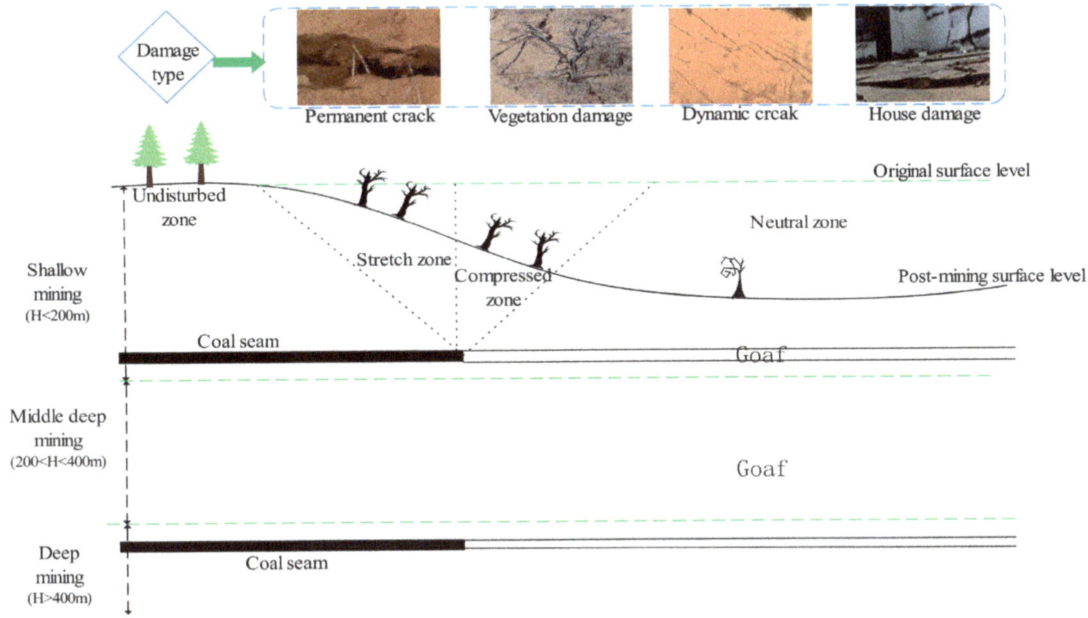

Figure 1. Surface damage induced by mining.

At present, many scholars have conducted extensive research on shallow and deep mining, leading to various research results. Scholars have integrated a variety of monitoring technologies to conduct multisource, integrated surface damage monitoring [11–15]. The law of mining surface damage is clarified on the basis of multisource fusion data [16,17]. The damage mechanism of overburden and surface is revealed by various means [18–21], and the prediction model of surface subsidence is built by combining various algorithms [22–25].

However, most of the results only analyze the surface damage caused by either shallow or deep coal seams. What are the similarities and differences between them? Why do these differences exist? What are the different formation mechanisms? Understanding these factors is essential to develop effective mitigation measures in the later stage. However, there is currently limited research on the formation mechanism of this difference. Therefore, this study conducted in-depth research on the formation mechanism of differences through actual measurements and similar model experiments.

2. Study Area and Its Surface Failure Characteristics

2.1. Working Face for Shallow Mining

Working face 22407 is located in Ordos, Inner Mongolia Autonomous Region. The mining depth of the working face is 130 m, with 75 m of bedrock and 55 m of loose layer. The dip angle of the coal seam is 1°, and the size of the working face is 3224 m × 284 m. The working face is mined by using a comprehensive mechanized mining method, and the roof is managed by using the full caving method. The advancing speed of the working face is 15 m per day. The mined 2-2 coal seam has a simple structure and stable endowment, which belongs to the stable coal seam. The overlying rock formation is dominated by sandstone and interspersed with mudstone and sandy mudstone. The working face is mostly thick siltstone and fine-grained sandstone with good stability. The lithology of the overlying rock layer is medium-hard (Figure 2) [25].

Geological time	Numbers of rock layers	Names of rocks	Thickness (m)	Lithology
Quaternary System	20	Aeolian sand	15.66	
	19	Loess	26.34	
	18	Gritstone	13.47	
Zhiluo Formation	17	Sandy mudstone	4.89	
	16	Siltstone	5.35	
	15	Medium sandstone	6.47	
	14	Fine sandstone	4.57	
	13	Siltstone	4.88	
	12	Medium sandstone	13.98	
Yan'an Formation	11	Siltstone	2.02	
	10	Fine sandstone	6.87	
	9	Fine sandstone	3.64	
	8	Medium sandstone	4.42	
	7	Fine sandstone	5.66	
	6	Sandy mudstone	3.29	
	5	Medium sandstone	2.84	
	4	Siltstone	6.68	
	3	2-2 coal seam	5.39	
	2	Siltstone	5.80	
	1	Fine sandstone	4.41	

Figure 2. Borehole log of working face 22407.

The ratio of the mining depth to the mining thickness of working face 22407 is small, and surface cracks are fully developed. The cracks can be divided into permanent cracks and dynamic cracks. The permanently stretching crack appears above the boundary of the working face, and the width of the crack increases with the advancement of the working face until it reaches the maximum width (Figure 3).

(a) (b) (c)

Figure 3. Surface discontinuous subsidence. (**a**) Permanently stretching crack; (**b**) Dynamic stretching crack; (**c**) Stepped crack.

Dynamic cracks appear directly above the goaf and can be subdivided into dynamic stretching cracks and stepped cracks. The width of dynamic cracks first increases with the advance of the working face, then decreases after reaching the maximum value, and finally closes.

2.2. Working Face for Deep Mining

Working face 11071 is located in Yuzhou City, Henan Province. The average dip angle of the coal seam is 4°, the average mining thickness of the coal seam is 4 m, and the average mining depth is 508.5 m. The mining method is full-thickness comprehensive mechanized mining with a strike long wall retreating type, and the advancing speed of the working face is 55 m per month.

The direct roof of working face 11071 consists of sandy mudstone that is dark gray in color, with developed horizontal bedding. The layer can be seen with small white mica fragments and is rich in plant fossil fragments and oriental comb teeth. It has a thickness of about 1.0 m. The main roof is made up of medium-coarse-grained sandstone, also dark gray in color, and is mainly composed of quartz and feldspar with a small amount of argillaceous inclusions. The layer mostly contains muscovite and siderite particles with poor sorting, and the overall thickness is about 30 m. The direct bottom is an interbedding of mudstone and siltstone, with gray siltstone containing a small amount of plant fossil fragments and a large number of muscovite slices. The siltstone and mudstone form a strip with a thickness of about 3.3 m. The old bottom is the interbedding of sandy mudstone or fine-grained sandstone, with developed microwave-like bedding and a thickness of about 13 m. The overlying strata of the working face are medium-hard to soft strata.

During the mining of working face 11,071, the surface subsidence basin occurs (Figure 4a). Obvious cracks appear on the surface road above the mining boundary of the working face (Figure 4b), and they extend forward with the advance of the working face. These cracks are permanent; however, no dynamic cracks are observed on the surface above the working face.

(a) (b)

Figure 4. Surface subsidence situation.

3. Methods and Results

This study analyzes the law of overburden movement under shallow and deep mining by using a similar material simulation method to reveal the mechanism of the difference in surface damage between shallow and deep mining.

3.1. Similar Material Simulation in Shallow Mining

3.1.1. Design and Production of Model

The working face 22407 served as the sample for the model test, with a plane model frame measuring 3000 mm × 1600 mm × 250 mm selected. The model experiment design diagram is shown in Figure 5.

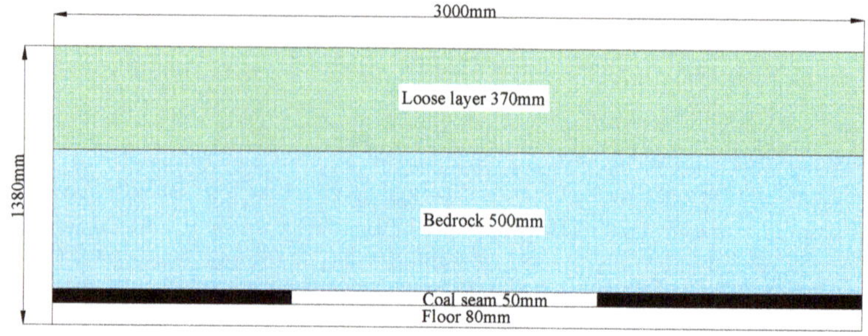

Figure 5. Design diagram of model test.

In accordance with the geological and mining conditions of the simulated working face, 45° was selected as the boundary angle. The geometric dimension relationship between the simulated working face and the model frame was compared and analyzed, and the geometric dimension ratio of 1:150 was finally selected. The unit weight ratio is 0.6, based on the characteristics of overburden and simulated materials. The stress ratio and time ratio can be obtained by using the following formula:

$$\alpha_\sigma = \alpha_{\text{fl}} \times \alpha_1 = 0.6 \times (1/150) = 0.004 \quad (1)$$

$$\alpha_t = \sqrt{\alpha_1} = \sqrt{1/150} = 0.082 \quad (2)$$

where: α_σ is the stress ratio; α_{fl} is the unit weight ratio; α_1 is the geometric dimension ratio; and α_t is the time ratio.

In accordance with the experience and the simulated lithology of each rock stratum, the bedrock is mainly composed of sand, with calcium carbonate and gypsum used as cementation auxiliary materials. The loose layer consists mainly of loose sand, supplemented by sawdust. The mica powder is evenly spread as the joint surface between the rock layers after each layer is paved. The material ratio of each layer of rock is shown in Table 1.

Table 1. Proportioning table of model one.

Number	Name	Thickness (cm)	Weight/kg	Material Consumption				
				Sand/kg	Calcium Carbonate/kg	Gypsum/kg	Water/kg	Sawdust/kg
1	Aeolian sand	36.98	301.46	262.01				39.45
2	Sandy mudstone	3.26	36.08	32.07	2.81	1.20	2.16	
3	Siltstone	3.57	39.01	34.14	1.46	3.41	2.34	
4	Medium-grained sandstone	4.32	49.99	39.99	3.00	7.00	3.00	
5	Fine-grained sandstone	3.05	34.03	29.17	1.46	3.40	2.04	

Table 1. *Cont.*

Number	Name	Thickness (cm)	Weight/kg	Material Consumption				
				Sand/kg	Calcium Carbonate/kg	Gypsum/kg	Water/kg	Sawdust/kg
6	Siltstone	3.25	35.52	31.08	1.33	3.11	2.13	
7	Medium-grained sandstone	9.28	107.38	85.91	6.44	15.03	6.44	
8	Siltstone	1.35	14.75	12.91	0.55	1.29	0.89	
9	Fine-grained sandstone	4.58	51.10	43.80	2.19	5.11	3.07	
10	Fine-grained sandstone	2.43	27.11	23.24	1.16	2.71	1.63	
11	Medium-grained sandstone	2.95	34.14	27.31	2.05	4.78	2.05	
12	Fine-grained sandstone	3.77	42.07	36.06	1.80	4.21	2.52	
13	Sandy mudstone	2.19	24.24	21.54	1.88	0.81	1.45	
14	Medium-grained sandstone	1.89	21.87	17.50	1.31	3.06	1.31	
15	Siltstone	2.23	24.37	21.32	0.91	2.13	1.46	
16	Coal seam	5	42.86	38.57	3.43	0.86	2.57	
17	Siltstone	10	115.71	92.57	6.94	16.20	6.94	

3.1.2. Overburden and Surface Damage Law

The model was dried for one week and then excavated 15 cm at a time. The next stage of excavation could only be conducted after the model's movement became stable. This experiment assumed that the movement of the model rock stratum could reach a stable state 2 h after each excavation. Therefore, the time interval between the two excavations was about 2 h. The method of full height and section excavation at one time was adopted during the simulated excavation.

In the process of gradual excavation, the overburden failure at each stage was as follows (Figure 6).

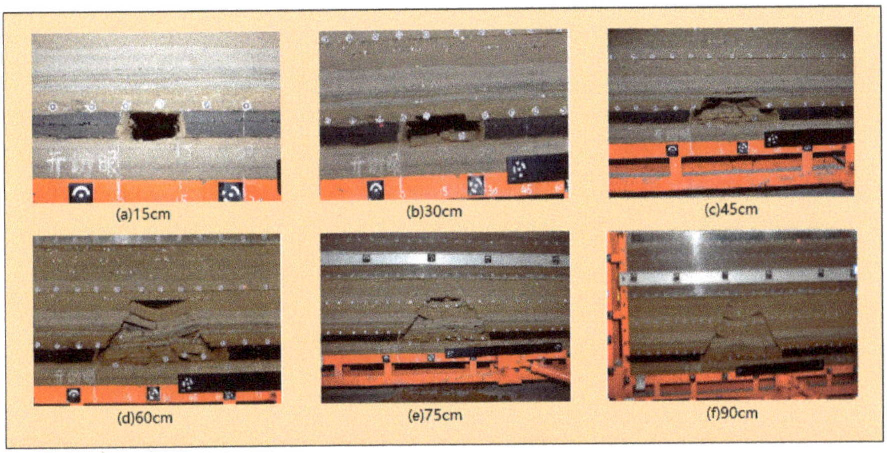

Figure 6. Overburden failure at each stage.

(1) When the working face was excavated to a depth of 15 cm, the roof above the coal seam was not damaged, and no falling phenomenon occurred.

(2) When the working face was excavated to a depth of 30 cm, the false roof above the coal seam was broken. Broken rock blocks accumulated in blocks in the goaf.

(3) When the working face was excavated to a depth of 45 cm, the tensile stress of the direct roof above the coal seam exceeded its tensile strength, and a tensile break occurred in the middle of the direct roof.

(4) When the working face was excavated to a depth of 60 cm, the old roof above the coal seam was broken. With the forward excavation of the working face, the roof of the coal seam continued to collapse, and the collapse height continued to increase, forming an approximate trapezoidal collapse space above the collapsed rock block.

(5) When the working face was excavated to a depth of 90 cm, the overlying bedrock of the coal seam was completely broken. Figure 7a,b show the overburden and surface damage, respectively. The damage to the overlying rock and surface can be described as shown in Figure 7c. The bedrock and loose layer above the goaf were all broken, and the overburden movement in the shallow coal seam mining followed a "two zones" mode, where only caved and fractured zones existed. The overburden was damaged directly up to the surface. Two types of dynamic crack were found above the goaf, which were stepped crack and dynamic stretching crack. The surface above the mining boundary exhibited a permanently stretching crack due to horizontal tension. This condition was consistent with the observed surface damage pattern in shallow mining.

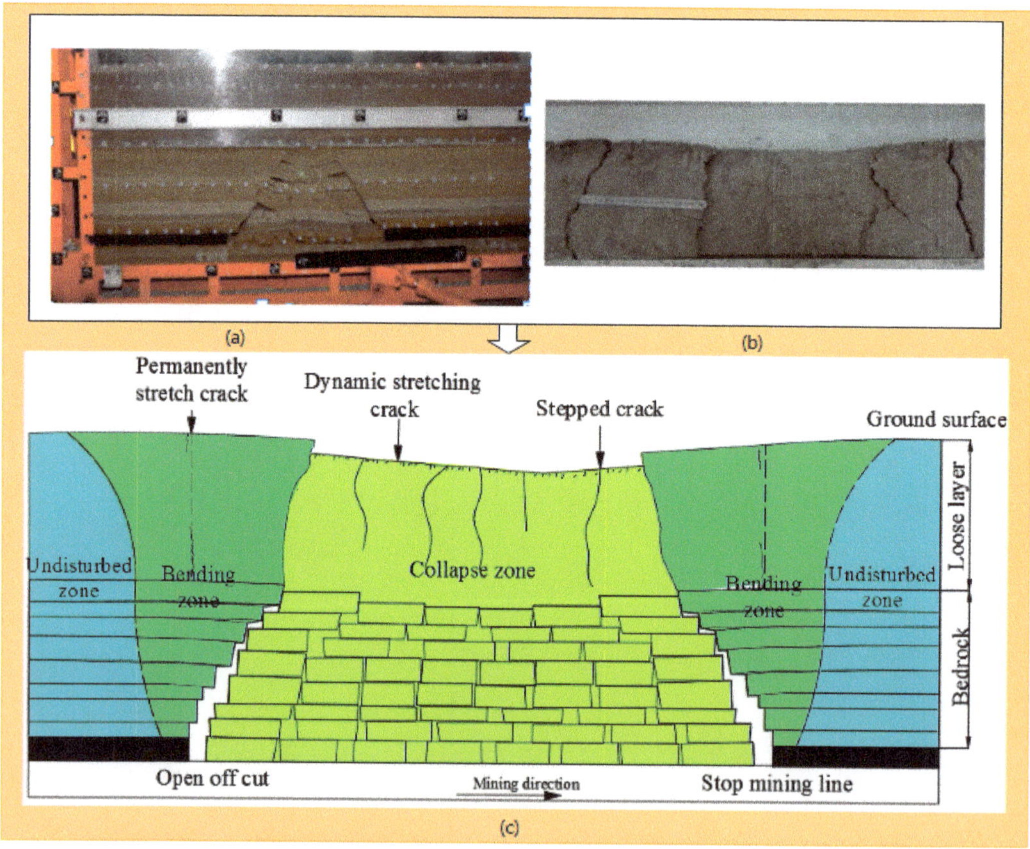

Figure 7. Overburden and surface damage induced by shallow mining. (**a**) Overburden damage; (**b**) Surface damage; (**c**) Schematic.

3.2. Similar Material Simulation in Deep Mining

3.2.1. Design and Production of Model

The working face 11071 was selected for the similar material simulation experiment in deep mining. In combination with the geological and mining conditions and experimental conditions of the mining area, a scale of 1:300 was selected, with a model size of 3.0 m × 0.25 m × 1.66 m. The simulated mining depth was 480.00 m, and the simulated coal thickness was 6 m, as shown in Figure 8.

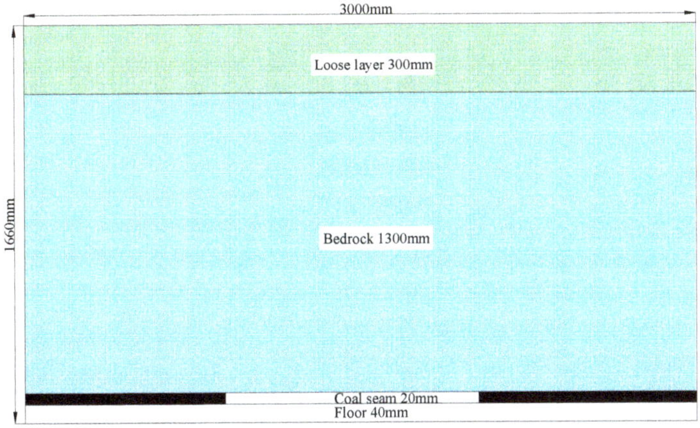

Figure 8. Design diagram of deep mining model test.

On the basis of the geological and mining conditions of the mining area and the available literature, the similarity coefficients selected for the experiment were as follows:

(1) Geometric ratio

$$\alpha_L = L_m / L_p = 1/300 \tag{3}$$

(2) Time ratio

$$\alpha_t = t_m / t_p = \sqrt{a_L} = \sqrt{1/300} = 17.32 \tag{4}$$

(3) Unit weight similarity coefficient

$$\alpha_r = \mathrm{fl}_m / \mathrm{fl}_p = 0.6 \tag{5}$$

(4) Strength ratio

$$\alpha_\sigma = \frac{\sigma_m}{\sigma_p} = \frac{\mathrm{fl}_m \times L_m}{\mathrm{fl}_p \times L_p} = \alpha_{\mathrm{fl}} \times \alpha_L = 0.002 \tag{6}$$

The material ratios are shown in Table 2.

Table 2. Material selection and matching for the similar material simulation experiment.

Number	Name	Thickness (cm)	Weight/kg	Material Consumption				
				Sand/kg	Calcium Carbonate/kg	Gypsum/kg	Water/kg	Sodium Borate/g
1	Loose layer	30	375.5	331.8	17.7	25.9	41.6	417.2
2	Medium-grained sandstone	2.79	47.38	40.61	3.38	3.38	5.26	52.64
3	Mudstone	10.84	180.7	161.91	10.08	8.71	20.09	200.75
4	Siltstone	0.93	15.62	13.89	0.87	0.87	1.74	17.36

Table 2. *Cont.*

Number	Name	Thickness (cm)	Weight/kg	Material Consumption				
				Sand/kg	Calcium Carbonate/kg	Gypsum/kg	Water/kg	Sodium Borate/g
5	Medium-grained sandstone	1.66	28.19	24.16	2.01	2.01	3.13	31.32
6	Mudstone	2.98	49.15	43.27	4.4	1.48	5.47	54.61
7	Siltstone	6.44	108.3	96.28	6.03	6.03	12.03	120.3
8	Medium-grained sandstone	6.24	106.18	91	7.58	7.58	11.8	117.98
9	Siltstone	4.03	67.29	59.8	3.73	3.73	7.48	74.77
10	Sandy mudstone	2.29	37.73	33.01	3.77	0.94	4.19	41.92
11	Siltstone	5.95	99.96	88.81	5.57	5.57	11.09	111.05
12	Medium-grained sandstone	1.72	29.21	25.03	2.09	2.09	3.25	32.45
13	Sandy mudstone	1.74	28.62	25.05	2.86	0.72	3.18	31.81
14	Medium-grained sandstone	6.79	115	98.6	8.23	8.23	12.78	127.87
15	Siltstone	2.3	37.16	33.03	2.06	2.06	4.13	41.29
16	Medium-grained sandstone	5.3	89.88	77.05	6.42	6.42	9.99	99.87
17	Sandy mudstone	40.72	677.3	594.46	54.89	27.85	75.27	752.58
18	Siltstone	4.57	76.85	67.4	4.72	4.72	8.54	85.39
19	Medium-grained sandstone	2.56	43.47	37.26	3.1	3.1	4.83	48.3
20	Sandy mudstone	7.7	127.01	111.11	12.71	3.18	14.12	141.11
21	Siltstone	2.07	34.6	30.75	1.92	1.92	3.84	38.44
22	Medium-grained sandstone	10.38	181.6	155.63	12.98	12.98	20.16	201.76
23	Coal seam	2	23.36	20.77	1.82	0.78	2.6	25.96
24	Siltstone	1.6	14.28	12.7	0.79	0.79	1.59	15.87
25	Sandy mudstone	2.4	39.48	34.54	3.94	0.98	4.38	43.86

3.2.2. Overburden and Surface Damage Law

The model was dried for one week and then excavated for 10 cm at a time. The next stage of excavation could only be conducted after the model's movement became stable. This experiment assumed that the movement of the rock stratum could reach a stable state 2 h after each excavation. Therefore, the time interval between the two excavations was about 2 h.

The method of full height at one time was adopted during the simulated excavation. In the process of gradual excavation, the overburden failure at each stage was as follows (Figure 9).

As the excavation proceeded, the direct roof continued to collapse, until it stabilized at a height of 200 cm. The fractured zone extended 116 cm from the coal seam, and the maximum height of the caved zone was 3.2 cm, as shown in Figure 10. Figure 10a shows the overburden damage. Figure 10b illustrates the damage of the overlying rock. The overburden movement in deep coal seam mining was characterized by a "three zones" mode, with the caved zone, fractured zone, and sagging zone existing simultaneously.

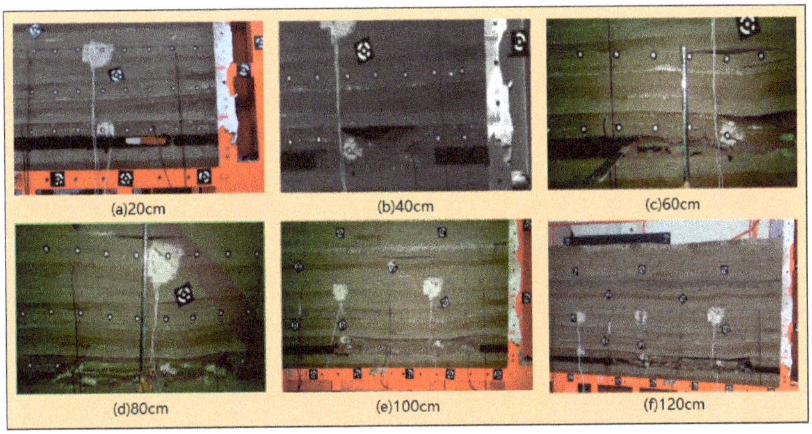

Figure 9. Overburden failure induced by deep mining at each stage.

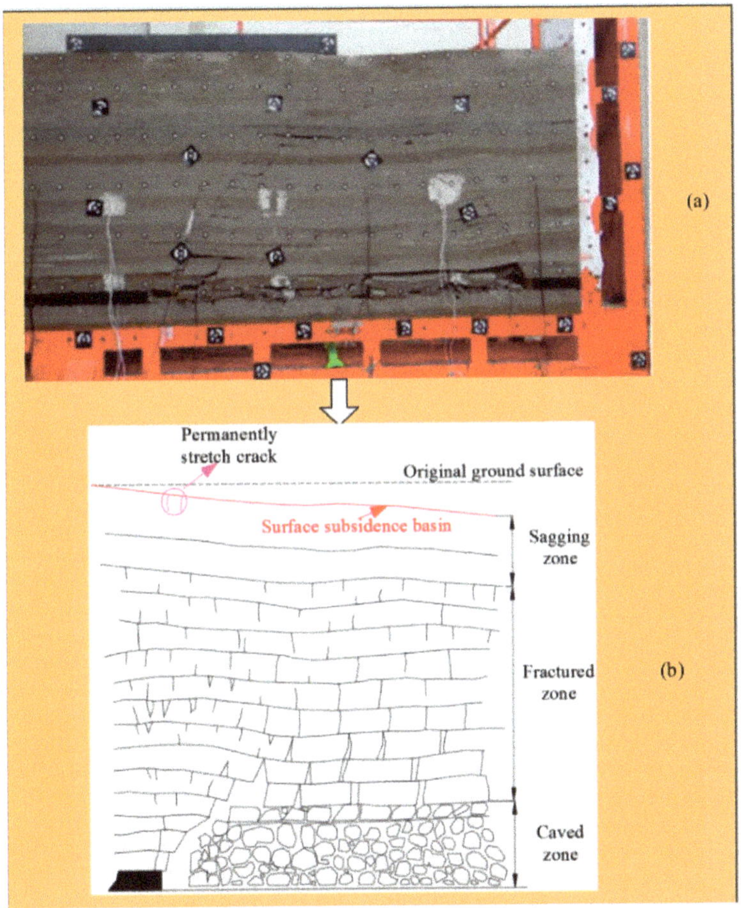

Figure 10. Overburden damage induced by deep mining. (**a**) Overburden damage; (**b**) Schematic.

Permanently stretched tensile cracks appeared on the surface above the mining boundary due to horizontal tension, and the surface subsidence occurred in the continuous form of a surface subsidence basin above the goaf. This condition was consistent with the observed surface damage law of deep mining.

4. Discussion

4.1. Formation Mechanism of Permanently Stretch Cracks

Overlying strata move toward the goaf after an underground coal seam is mined. The strata above the mined coal seam are broken layer by layer, and the broken rock blocks fill the goaf. In the vicinity of the mining boundary, the broken rock strata are in the form of cantilever beams. From the mined coal seam upward, the length of the overhanging rock block increases, as shown in Figure 11a. The overhanging rock block and its overlying loose layer can be simplified as wedge block B in Figure 11b. Before mining, wedge block B is subject to the thrust of its left rock stratum, the thrust of its right rock stratum, and the supporting force of the underlying rock stratum. The balance of mechanics is maintained under the combined action of these forces.

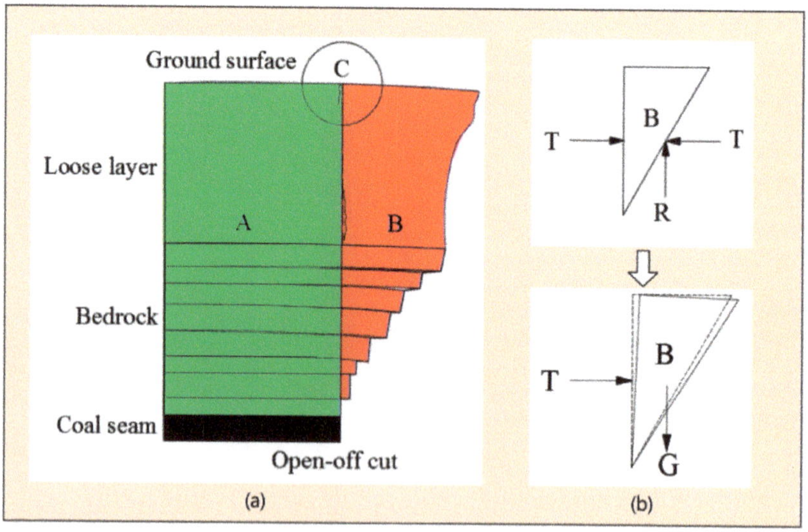

Figure 11. Formation mechanism of permanently stretched crack. (**a**) Wedge; (**b**) Force analysis.

After mining, the thrust T and supporting force R of the right and underlying strata on B decrease because of the movement of the right and lower pumice strata toward the goaf, resulting in rock block B tilting to the lower right under the combined action of the thrust of the left rock strata on it and its own self-weight force G. Thus, B and A rock blocks stretch at surface point C, and tensile cracks appear. Rock block B cannot be restored after it tilts, so the crack will remain permanent after it appears, which is called a permanently stretched crack.

The above processes occur in shallow and deep coal seams, making permanently stretched cracks easily occur on the surface above the mining boundary.

4.2. Formation Mechanism of Dynamic Crack

The impact of underground mining on the ground surface is mainly reflected in two aspects: the subsidence range and maximum subsidence value. According to the mining subsidence theory, when the working face is advanced to a certain distance, the maximum subsidence value of the surface will reach the maximum, and it will not increase in the future. To effectively represent the subsidence effects of each mining stage, the concept of the mining degree was introduced. The degree of mining is calculated as follows:

$$D = \frac{L}{H} \tag{7}$$

where D is the mining degree, L is the mining width, and H is the mining depth.

When the mining degree D reaches 1.2–1.4, the surface subsidence value reaches the maximum. Subsequently, the working face is excavated, and the maximum subsidence value of the surface will no longer increase, but only increase the subsidence range. According to the results of the two similar models, the change of the fracture height of the rock formation with the mining degree is shown in Figure 12.

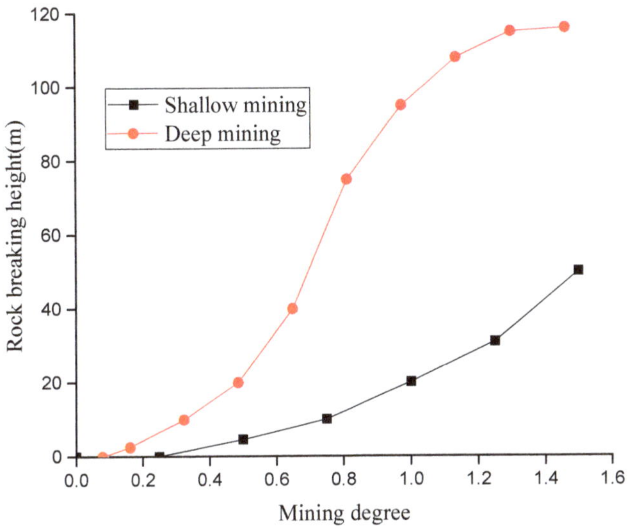

Figure 12. Development law of rock damage height.

Under the condition of deep burial, the damage height of rock strata in the model test increases with the mining degree, showing an S-shaped development trend.

(1) The inflection point of the rock damage height development curve is about 0.6–0.8. Before the inflection point, with the increase of the mining degree, the overburden failure height increased rapidly, and the increase rate gradually increased.

(2) At the inflection point (mining degree 0.6–0.8), the overburden failure height is still increasing, and the increase rate is the maximum.

(3) After exceeding the inflection point, with the increase of mining degree, the overburden failure height still continues to increase, but the increase rate gradually decreases.

(4) When the mining degree reaches 1.2–1.4, the overburden failure height reaches the maximum value. The damage height of overlying rock does not increase further during subsequent mining.

In accordance with the theory of "three zones," the rock strata above the maximum damage height up to the surface form the sagging zone, where the overlying rock moves continuously without experiencing discontinuous deformation, such as cracks. Therefore, no dynamic crack occurs on the surface above the working face under the condition of deep mining. The overburden movement in deep coal seam mining is "three zones" mode.

However, the development law of rock damage height in shallow coal mining is obviously different from that in deep mining. The damage height of rock strata in shallow thick coal seam mining increases exponentially with the mining degree.

(1) With the increase of mining degree, the overburden failure height increases rapidly, and the increase rate gradually increases.

(2) The larger the mining degree is, the larger the rock damage height caused by the same, until the damage extends to the bedrock surface.

In this case, the overburden is broken directly to the surface of the bedrock, and there is no sagging zone in the overburden, only a collapse zone and a fracture zone, that is, the overburden movement in shallow coal seam mining is in a "two zones" mode.

According to literature [25], the rock strata migrate in the form of bench rock beams under shallow mining conditions. The stability of the step rock beam mainly depends on the relationship between the friction and shear force at the overlap of rock blocks. If the shear force is greater than the friction force, then the step rock beam structure will slide and lose stability.

Rock block N_2 falls completely on the collapsed rocks and presents a compacted state, so R_2 is equal to P_2.

According to the moment equilibrium equations $\sum M_A = 0$ and $\sum M_B = 0$, Equations (8) and (9) can be derived.

$$Q_A + Q_B = P_1 \tag{8}$$

$$T = \frac{lP_1}{2(h-a-w)} = \frac{P_1}{i - 2\sin\theta_{max} + \sin\theta} \tag{9}$$

where i is the rock fragmental size, $i = l/h$.

The shear force of point B can be calculated in accordance with Equation (10) as follows:

$$Q_B = T\sin\theta_2 \tag{10}$$

θ_2 is very small; hence, Q_B can be ignored, and $Q_A = P_1$ can be concluded.

According to literature [25], the relationship between the friction force and the shear force at the overlap of the broken rock block of the bench rock beam under the mining of shallow coal seams is as follows:

$$T\tan\varphi = \frac{P_1}{i - 2\sin\theta_{max} + \sin\theta}\tan\varphi < P_1 = Q_A \tag{11}$$

where T is the horizontal thrust. Q_A is the shear forces of points A. P_1 is the suffered load of the rock N_1 i is the rock fragmental size. θ is the rock rotation angle.

In accordance with the above formula, the shear force is greater than the friction force, and the joint of the broken rock block is prone to slip and instability. The sliding and instability of the roof cause the overlying strata to cut off and create step subsidence.

The loose layer relative to the bedrock is similar to a random medium, and its antideformation ability is mostly negligible when faced with the step subsidence transmitted from the bedrock. Thus, the step subsidence is transmitted to the ground surface, which is represented by a stepped crack.

As shown in Figure 13b, rock blocks N_1 and N_2 are broken during mining. N_2 is subject to sliding instability, and step subsidence occurs between the N_1 and N_2 rock blocks. As the working face continues to advance, sliding instability also occurs between N_1 and the left rock block. The step subsidence between the N_1 and N_2 rock blocks disappears. Similarly, the stepped crack and dynamic tensile crack on the surface above the working face appear first and then disappear, with an automatic closing function.

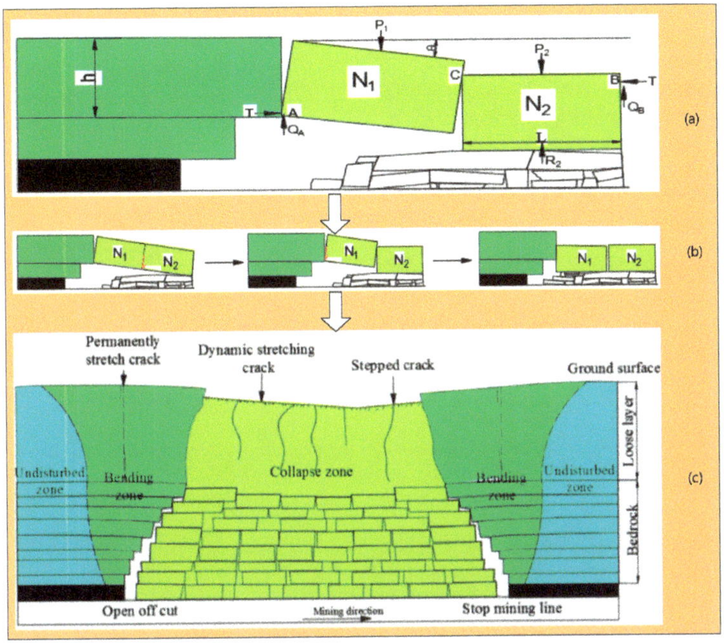

Figure 13. Formation mechanism of dynamic crack. (**a**) Force analysis; (**b**) Process of rock damage; (**c**) Schematic of final damage.

5. Conclusions

Understanding the formation mechanism of surface damage differences in deep and shallow mining is vital for disaster prevention and environmental protection. In this work, the formation mechanism was systematically analyzed by combining a similarity simulation experiment and geo-mining conditions of the relevant working face. The conclusions are summarized as follows:

(1) Under the condition of deep mining, the damage height of rock strata increases with the mining degree, showing a S-shaped development trend. However, the damage height of rock strata has an exponential development trend, with the mining degree under the condition of shallow mining.

(2) Under the mining of shallow coal seams, permanently stretched cracks above the mining boundary and dynamic cracks above the goaf on the surface are found. However, only permanently stretched cracks are observed on the surface above the mining boundary under the mining of deep coal seams.

(3) The rock strata inside the mining boundary develop into a wedge-shaped block due to the mining of underground coal seams. It inclines under the action of the horizontal thrust of nearby rock strata and self-gravity, resulting in a permanently stretched crack above the mining boundary.

(4) The overburden movement in deep coal seam mining is "three zones" mode. The overburden movement in shallow coal seam mining is in a "two zones" mode. The overburden strata slide and become unstable in the form of a bench rock beam, resulting in stepped cracks and dynamic stretching cracks on the surface above the goaf.

Author Contributions: Conceptualization, W.Y.; methodology, W.Y.; software, J.G.; validation, S.Y.; formal analysis, W.Y.; investigation, S.Y.; resources, S.Y.; data curation, W.Y.; writing—original draft preparation, W.Y.; writing—review and editing, S.Y.; visualization, S.Y.; supervision, J.G.; project administration, W.Y.; funding acquisition, S.Y. All authors have read and agreed to the published version of the manuscript.

Funding: This research was funded by the Open Fund of the State Key Laboratory of Water Resource Protection and Utilization in Coal Mining grant number (WPUKFJJ2019-17), the National Natural Science Foundation of China grant number (42207534, U21A20108), the Youth Backbone Teacher Support Program of Henan Polytechnic University (2019XQG-07), and the Doctoral Fund Program of Henan Polytechnic University (B2017-07, B2019-03).

Institutional Review Board Statement: Not applicable.

Informed Consent Statement: Not applicable.

Data Availability Statement: Not applicable.

Conflicts of Interest: The authors declare no conflict of interest.

References

1. Yan, W.; Chen, J.; Tan, Y.; He, R.; Yan, S. Surface dynamic damage prediction model of horizontal coal seam based on the idea of wave lossless propagation. *Int. J. Environ. Res. Public Health.* **2022**, *19*, 6862. [CrossRef]
2. Vervoort, A. The Time Duration of the Effects of Total Extraction Mining Methods on Surface Movement. *Energies* **2020**, *13*, 4107. [CrossRef]
3. Ju, J.; Xu, J. Surface stepped subsidence related to top-coal caving longwall mining of extremely thick coal seam under shallow cover. *Int. J. Rock Mech. Min. Sci.* **2015**, *78*, 27–35. [CrossRef]
4. Orwat, J. Causes analysis of occurrence of the terrain surface discontinuous deformations of a linear type. *J. Phys.* **2020**, *1426*, 012016. [CrossRef]
5. Strzałkowski, P.; Scigała, R.; Szafulera, K.; Tomiczek, K. *Forecasting of Discontinuous Deformations of Surface Type on the Mining and Post-Mining Areas*; Publishing House of Silesian University of Technology: Gliwice, Poland, 2020; p. 127.
6. Vervoort, A. Impact of the closure of a coal district on the environmental issue of long-term surface movements. *AIMS Geosci.* **2022**, *8*, 326–345. [CrossRef]
7. Peng, S. *Coal Mine Ground Control*, 3rd ed.; West Virginia University: Morgantown, WV, USA, 2008.
8. Strzałkowski, P.; Ścigała, R.; Szafulera, K.; Kołodziej, K. Surface Deformations Resulting from Abandoned Mining Excavations. *Energies* **2021**, *14*, 2495. [CrossRef]
9. Strzałkowski, P. Predicting mining areas deformations under the condition of high strength and depth of cover. *Energies* **2022**, *15*, 4627. [CrossRef]
10. Guo, J.; Li, Q.; Xie, H. Monitoring of vegetation disturbance and restoration at the dumping sites of the baorixile open-pit mine based on the LandTrendr algorithm. *Int. J. Environ. Res. Public Health.* **2022**, *19*, 9066. [CrossRef]
11. Radutu, A.; Vlad-Sandru, M.-I. Review on the Use of Satellite-Based Radar Interferometry for Monitoring Mining Subsidence in Urban Areas and Demographic Indicators Assessment. *Min. Rev./Rev. Minelor.* **2023**, *29*, 42–62. [CrossRef]
12. Rezaei, M.; Hossaini, M.F.; Majdi, A. A time-independent energy model to determine the height of destressed zone above the mined panel in longwall coal mining. *Tunn. Undergr. Space Technol.* **2015**, *47*, 81–92. [CrossRef]
13. Salmi, E.F.; Nazem, M.; Karakus, M. Numerical analysis of a large landslide induced by coal mining subsidence. *Eng. Geol.* **2017**, *217*, 141–152. [CrossRef]
14. Bai, E.; Guo, W.; Tan, Y. Negative externalities of high-intensity mining and disaster prevention technology in China. *Bull. Eng. Geol. Environ.* **2019**, *78*, 5219–5235. [CrossRef]
15. Tajdu´s, K.; Sroka, A.; Misa, R.; Hager, S.; Rusek, J.; Dudek, M.; Wollnik, F. Analysis of Mining-induced delayed surface subsidence. *Minerals* **2021**, *11*, 1187. [CrossRef]
16. Modeste, G.; Doubre, C.; Masson, F. Time evolution of mining-related residual subsidence monitored over a 24-year period using InSAR in southern Alsace, France. *Int. J. Appl. Earth Obs. Geoinf.* **2021**, *102*, 102392. [CrossRef]
17. Declercq, P.-Y.; Dusar, M.; Pirard, E.; Verbeurgt, J.; Choopani, A.; Devleeschouwer, X. Post Mining Ground Deformations Transition Related to Coal Mines Closure in the Campine Coal Basin, Belgium, Evidenced by Three Decades of MT-InSAR Data. *Remote Sens.* **2023**, *15*, 725. [CrossRef]
18. Li, P.; Zhu, X.; Ding, X.; Zhang, T. An Experimental Study of Industrial Site and Shaft Pillar Mining at Jinggezhuang Coal Mine. *Appl. Sci.* **2023**, *13*, 2340. [CrossRef]
19. Wang, Z.; Li, W.; Wang, Q. Monitoring the dynamic response of the overlying rock-soil composite structure to underground mining using BOTDR and FBG sensing technologies. *Rock Mech. Rock Eng.* **2021**, *54*, 5095–5116. [CrossRef]
20. Dong, L.; Zhu, H.; Yan, F.; Bi, S. Risk field of rock instability using microseismic monitoring data in deep mining. *Sensors* **2023**, *23*, 1300. [CrossRef] [PubMed]
21. Dudek, M.; Tajdu´s, K. FEM for prediction of surface deformations induced by flooding of steeply inclined mining seams. *Geomech. Energy Environ.* **2021**, *28*, 100254. [CrossRef]
22. Bo, H.; Li, Y.; Tan, X.; Dong, Z.; Zheng, G.; Wang, Q.; Yu, K. Estimation of Ground Subsidence Deformation Induced by Underground Coal Mining with GNSS-IR. *Remote Sens.* **2022**, *15*, 96. [CrossRef]
23. Fan, K.; Li, W.; Wang, Q. Formation mechanism and prediction method of water inrush from separated layers within coal seam mining: A case study in the Shilawusu mining area China. *Eng. Fail. Anal.* **2019**, *103*, 158–172. [CrossRef]

24. Yan, W.; Chen, J.; Yan, Y. A new model for predicting surface mining subsidence: The improved lognormal function model. *Geosci. J.* **2019**, *23*, 165–174. [CrossRef]
25. Yan, W.; Dai, H.; Chen, J. Surface crack and sand inrush disaster induced by high-strength mining: Example from the Shendong coal field, China. *Geosci. J.* **2018**, *22*, 347–357. [CrossRef]

sustainability

Review

Research Findings on the Application of the Arch Structure Model in Coal Mining, a Review

Quansheng Li [1], Junting Guo [1], Cun Zhang [2,*], Yuliang Yang [2], Jianqi Ma [2] and Zhaopeng Ren [2]

1 State Key Laboratory of Water Resource Protection and Utilization in Coal Mining, Beijing 102209, China
2 School of Energy & Mining Engineering, China University of Mining & Technology (Beijing), Beijing 100083, China
* Correspondence: cumt-zc@cumtb.edu.cn

Abstract: Studying the movement law and failure mechanism of overburden is important to underground safety production, aquifer protection, surface subsidence and ecological protection. A commonly used model, the arch structure model, for studying overburden movement is systematically reviewed in this paper. First, the arch structure in the mining field is divided into an unconsolidated layer arch, beam arch (hinged arch), and overburden arch according to arching medium. On this basis, the research progress and existing problems of these three arch structure models are discussed according to the research means, including theoretical modeling, numerical simulation, similar simulation, and field measurements. The application of the arch structure model focuses on the prediction and prevention of mining pressure and surface subsidence, and there are relatively few means to actively regulate overburden arch. Thus, three control methods of the arch structure in underground coal mining are proposed: the preset arch structure method before coal mining, process control arch methods during coal mining, and end reinforcement control arch methods after coal mining. Finally, the main research focus on the arch structure model in the future is discussed.

Keywords: underground coal mining; arch structure model; unconsolidated layer arch; beam arch; overburden arch; arch structure control

1. Introduction

The structure model of overlying strata in mining fields can be used to describe strata movement and predict mine disasters [1–3]; it is the foundation of safe mining in underground coal mines. The overburden structure theories and hypotheses proposed for mine pressure and rock strata control mainly include: key strata theory [4–6], transfer rock beam hypothesis [7], masonry beam theory (articulated rock block hypothesis) [8], hypothesis of pre-cracked beam [9], cantilever beam hypothesis [10], and the stress arch hypothesis [11]. These theories and hypotheses have been widely used and developed in coal mining. For example, the key strata theory has been applied in working face periodic pressure and support resistance prediction [4], the development of water flowing fracture zones [6], filling, mining, and surface subsidence [12,13], rockburst prevention and control [14], pressure relief mining and gas extraction [15], water inrush prediction from the floor [16], and grouting in separated layers [17]. The transfer rock beam hypothesis has led to important achievements in the innovation of mining methods and mining technologies, as well as the control practice of major accidents and disasters, such as the design of mining methods without coal pillars, the calculation of support resistance, the prediction of three-zone height, the prevention and control of rockburst, coal, and gas outburst, etc. [7,18]. Masonry beam theory, preformed fracture beam hypothesis and cantilever beam hypothesis all reveal the fracture and migration characteristics of overburden rock above the working face with the beam model, and provide the prediction model of mining pressure of the working face under specific geological conditions and mining parameters. The arch structure model, evolved on the basis of the beam model, can not only analyze the near-field mine pressure law, but also explain the characteristics of overburden and surface migration in the far field, which is widely used in the analysis of overburden fracture migration in coal mining. For example, in the process of coal seam mining, due to the existence of the stress arch, the adjacent coal seams will generate a stress increase zone and stress relief zone. If the adjacent coal seam is in the stress relief zone, the gas seepage capacity of the coal seam will be greatly improved, which is conducive to the gas drainage of the coal seam [19,20]. This paper classifies the overburden of the arch shell, discusses the related research findings and the application of the overburden arch shell structure, and, on this basis, puts forward the possible research focus in the future.

2. Formation of the Arch Structure Model Hypothesis

In the process of arch formation analysis and its application development, it can be divided into two aspects. On the one hand, it is applied to the stress distribution of the surrounding rock of coal mine roadways, and provides guidance for roadway support design. On the other hand, the overburden stress arch studied in this paper mainly focuses on the stress change and fracture migration structure of overburden in the mining field. This section mainly discusses the formation and development of the overburden stress arch model hypothesis.

The phenomenon of surrounding rock arch formation was first summarized by Ritter [21], demonstrating that the stress in the tunnel soil layer does not increase with the increase in buried depth. Subsequently, Terzaghi, Fayol, Engesser, Handy et al. [22,23] referred to such arched bearing structures in the soil (unconsolidated body) as the arch effect. Therefore, in the beginning, the arch effect is mainly used in granular materials (such as tunnel excavation in soil), and is rarely used to explain the bearing structure of rock strata. However, it provides a reference for the analysis of stable structures of rock strata in coal mining. Hack and Gillitzer [24] were the first to apply arch structure to the analysis of mining pressure in coal mines. Based on the "stress arch" hypothesis proposed by Apehc [25], they used the stress arch hypothesis to explain the evolution of abutment pressure during the advancing process of the coal face, as shown in Figure 1. This hypothesis can reliably explain the pressure relief area of surrounding rock, but the specific characteristics of the arch, migration, and fracture of overburden rock, and the support resistance of the mining field, are not further involved.

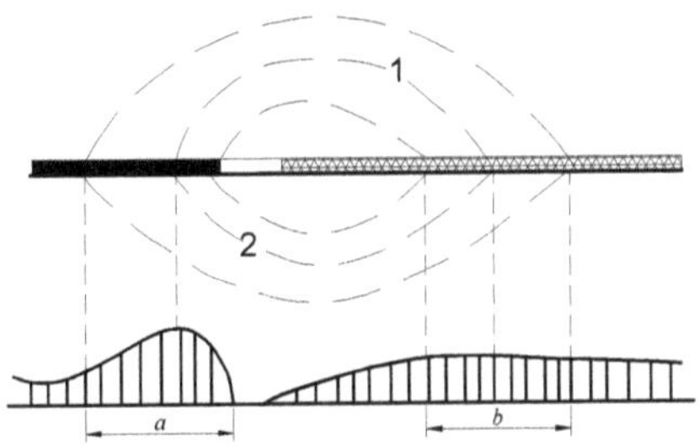

Figure 1. Stress arch in mining face. a–Front arch foot; b–rear arch foot; 1–roof arch structure; 2–floor arch structure.

Subsequently, Ime [26] proposed the hypothesis of stress arch and its influence on confining stress more systematically, and believed that the existence of the stress arch in the roof formed the pressure relief area in the arch, and the stress rise area in the arch rock strata. Dinsdale [27] pointed out that after the coal seam mining, the surrounding rock would form an oval pressure ring (similar to Figure 1), and presented the correlation between the pressure ring and the mining width of the working face and the surface subsidence. This was the first time researchers analyzed the mining subsidence problem using the arch effect. After that, scholars gradually began to pay attention to the location of the overburden arch, arch formation conditions, and other characteristics of the arch itself. Stemple [28] put forward that the condition of arch formation is that the buried depth must reach the critical depth. For example, the requirement of fracture arch formation at the present stage is that the buried depth of the working face should be greater than the height of the fracture zone. Denkhaus [29] provided the evolution model of the stress arch, believing that the stress arch moves forward with periodic roof caving, and the arch foot is the crushed rock compacted in goafs. Subsequently, with the development of computing ability, numerical analysis methods based on the finite element stress vector model and the discrete element caving model were gradually applied to the study of arch characteristics [30]. Chekan [31] and Luo et al. [32]. conducted in-depth research on the height to span ratio and shape characteristics of arches. However, with the withdrawal from coal mining in some developed countries, the development of an overlying arch hypothesis has gradually stagnated, especially after entering the 21st century; the development of the arch hypothesis and its related applications is, therefore, mainly important in China.

Zhao [33] introduced the natural balanced arch in analyzing the stress distribution of roadways. Miao [34] further analyzed the correlation between the natural balance arch and the stability of roadway surrounding rock, and presented the corresponding relationship between the pressure measurement coefficient and surrounding rock arch shape. It can be seen that the stress arch was mainly applied to the stress distribution of roadway surrounding rock at first, and the research on the stress arch of overlying rock in mining fields further promoted the application of the arch hypothesis in coal mining in China. Cao [35] put forward the composite stress arch model of the roof. Zou [11] presented a new interpretation of the stress arch according to the actual situation of coal mining in China. Kang [36] explained the process of roof breaking into the arch with the help of discrete element numerical simulation, and verified the role of arch structure in maintaining the stability of overburden. In the 21st century, with the rapid development of the Chinese coal mining industry, the theoretical model is continuously in-depth, numerical simulation technology is widely used, and the laboratory and field monitoring methods are constantly updated. There are more and more studies on the arch stress structure model. For example, in the characteristics of the arch structure model, it mainly involves the arch forming and breaking conditions, arch trace, arch stress distribution, arch thickness, arch-bearing capacity, two-dimensional arch, and three-dimensional shell, etc. In the aspect of coal mining parameters, mainly including mining height, working face length, advance speed, backfill mining, caving mining, strip mining, room and pillar mining, etc. In terms of the geological conditions of coal mines, it mainly involves the dip angle of the coal seam, the lithology (strength) of the roof and floor, the overburden medium (unconsolidated layer), and the depth of the coal seam. In terms of the application of the arch model, it involves the prediction of surface subsidence, the calculation of pressure coming from the working face, and the working resistance of the support, the design of hydraulic fracturing in the roof, the pressure relief gas extraction, and the prevention of water inrush from the floor, etc. Therefore, abundant achievements have been made in the research of the arch structure model, which provides a basis for the research and analysis of overburden migration, mining pressure behavior, gas extraction, and surface subsidence in coal mining.

3. Construction of Arch Structure Model

3.1. Model of Unconsolidated Layer Arch Model

The geological conditions of the thick unconsolidated layer are widely distributed in some mining areas in East China and North China, such as Huainan, North Anhui, Huaibei, Yanzhou, Xinwen, Kailuan, and other mining areas. The maximum thickness of the unconsolidated layer in 48 coal mines is 752 m, the minimum is 100 m, and the average is 319 m. It is found, from this measurement, that in the mining of the shallow coal seam, the load of the thick unconsolidated layer (mainly viscous unconsolidated layer) does not act on the main roof at once, and there is an arch-bearing structure, as shown in Figure 2.

Therefore, the load of the unconsolidated layer on bedrock cannot be simply equal to the weight of the unconsolidated layer, and it is necessary to study the evolution process of the unconsolidated layer arch model. At present, studies on the unconsolidated layer arch mainly focus on the geological conditions of the shallow buried thick unconsolidated layer, and analyze the influence of the unconsolidated layer arch on surface subsidence, fracture development, and the unconsolidated layer on bedrock load [37,38].

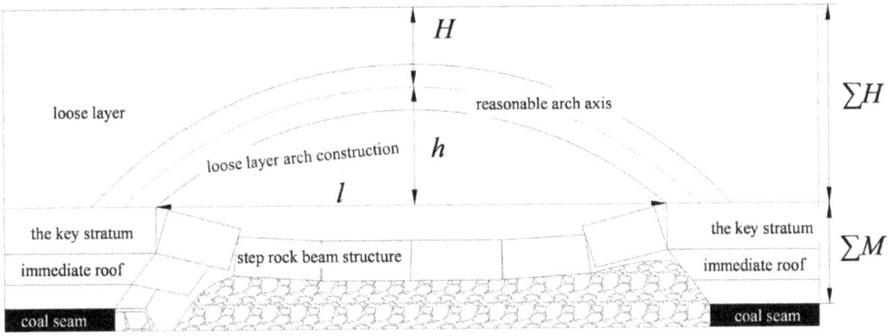

Figure 2. Unconsolidated layer arch structure in mining overburden. *l*–Span of arch structure; *h*–height of arch structure; *H*–burial depth of arch crown; $\sum H$–thickness of unconsolidated layer; $\sum M$–thickness of bedrock.

3.2. Arch Structure Model of Overlying Rock

(1) Linear arch (beam arch)

Beam theory is widely used in the analysis of overburden fracture and migration in coal mining. It is believed that once the flexural and tensile strength of the beam is exceeded, the yield failure will expand along the beam, and the effective thickness of the beam will be reduced due to transverse fracturing, which is an important conclusion of classical beam theory. However, the span of the arch structure formed during underground mining, surface subsidence, and transverse beam fracture is much longer than that predicted by classical beam theory. This behavior can be attributed to the formation of a linear arch in the beam, referred to in this paper as a beam arch (or hinged arch). On the basis of specific geometric structure, the principal stress traces in a thick rock layer show an arch distribution. As shown in Figure 3, the resistance acting on both ends of the beam leads to the formation of stress arches in the beam [39]. In a given range of the thickness to span ratio, the beam arch can bear a higher load than that predicted by traditional beam theory. At present, the beam arch model is mainly applied to analyze the strong mining pressure phenomenon caused by thick and hard roof structures.

(2) Overburden arch (stress arch, fracture arch)

At present, the overburden arch (three-dimensional "shell") model is mainly divided into the fracture arch and stress arch models [40–43]. The fracture arch is generally considered to be the top boundary of the overburden where the fracture occurs, so it usually has no thickness; the stress arch refers to the arch stress zone formed by the deflection of the overburden principal stress in the mining field, with a certain thickness [44], as shown in Figure 3. The double-arch model of overburden in mining fields is the mechanical manifestation of rock mass self-bearing characteristics and the compressive stress arch effect. Considering the traditional subzones of overburden in coal mining (caving zone, fracturing zone, and bending zone), the fracture arch can be considered as the range of the caving zone and fracturing zone, or as the boundary of the area where the caving zone and fracturing zone are located.

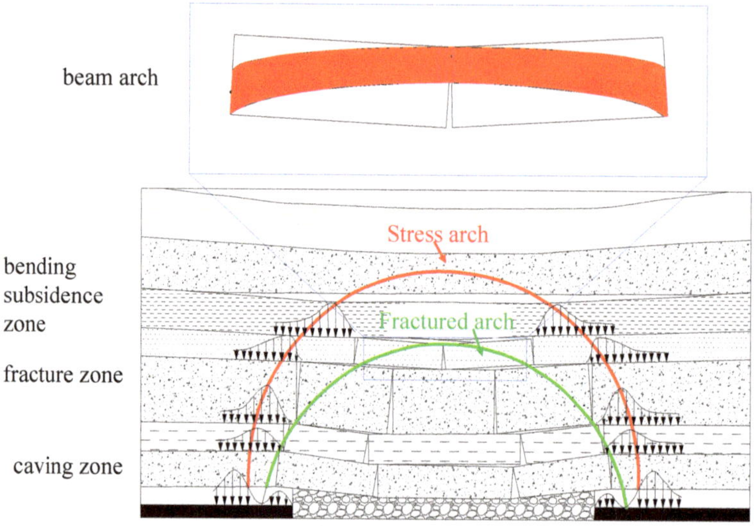

Figure 3. Overburden arch structure based on beam arch structure.

The stress arch is generally located in the upper part of the fracture arch, and is the bearing structure of the whole overburden. In fact, the stress arch is mainly caused by the stress relief of the fracture arch, which transfers the original bearing stress to the upper rock layer, and then forms the stress increasing area. Therefore, the stress arch has a certain thickness, which is the area enclosed by the stress in the overlying rock greater than the stress in the original rock. In addition, the stress arch can also be considered as the "far-field arch", which controls the migration of the whole overburden and surface [43]. As shown in Figure 4, compared with the inclined fracture arch and stress arch of the working face (Figure 3), the near-field arch belongs to the evolution structure of an arch in the direction of advance. The near-field arch is mainly formed by periodic fracture of the main roof, and moves forward with the advance of the working face, accompanied by arch formation and arch breaking. The far-field arch is generally composed of the key stratum. With the advance of the working face, the span of the far-field arch in the direction of advance is constantly expanded, but the arch height is basically unchanged, and gradually evolves into a flat-top arch.

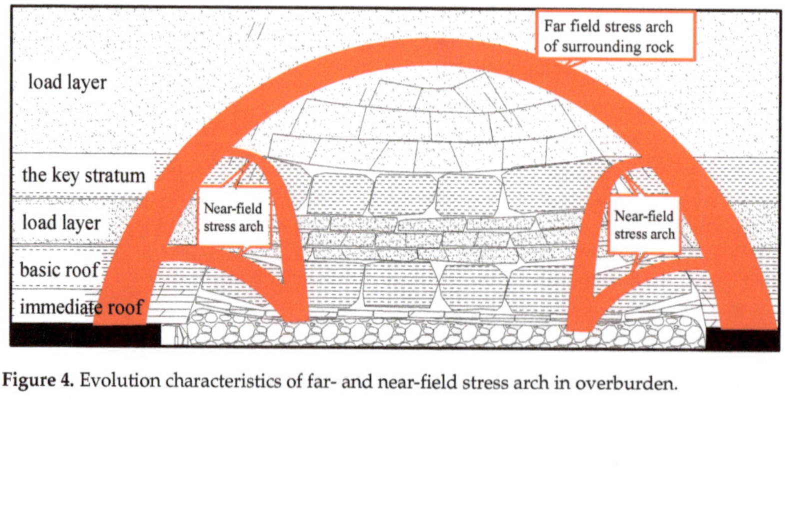

Figure 4. Evolution characteristics of far- and near-field stress arch in overburden.

4. Research Findings and Its Application of Unconsolidated Layer Arch Model

4.1. Research Findings of Unconsolidated Layer Arch

(1) Theoretical model

On the basis of arch theory, the unconsolidated layer arch structure model was established. The unconsolidated layer is regarded as a homogeneous medium in the model. The formation conditions of the unconsolidated layer arch structure were determined by calculating the equation of shape characteristic, the equation of sagittal ratio, and the equation of thickness, and the mechanism of arch formation and its influencing factors were revealed. The expression of the stress arch based on Figure 5 is shown in Equation (1) [37,38]:

$$\begin{cases} x^2 - L_{arch}x + \lambda y^2 + 1/(4H_{arch})({L_{arch}}^2 - 4\lambda {H_{arch}}^2)y = 0 \\ \frac{H_{arch}}{L_{arch}} \geq \frac{\sqrt{(C+\gamma h_0 \tan\varphi)^2 + \lambda\gamma^2 h_0^2} - \gamma h_0 \tan\varphi - C}{2\lambda\gamma h_0} \\ \delta_{arch} = \frac{\gamma h_0 L_{arch} \tan(45^\circ + \varphi/2)\cos\varphi}{\gamma(h_0 + iL_{arch})\tan(45^\circ+\varphi/2) + 2C} \end{cases} \quad (1)$$

where x and y are the arch trace coordinates of the unconsolidated layer; $I = H_{arch}/L_{arch}$; λ is the lateral pressure coefficient; φ (°) and C (MPa) are friction angle and cohesion in the unconsolidated layer, respectively; and γ is the bulk density of bedrock, kN/m. After determining the arch forming condition and calculating the arch baseline equation, the load of the unconsolidated layer arch acting on the bedrock can be further determined.

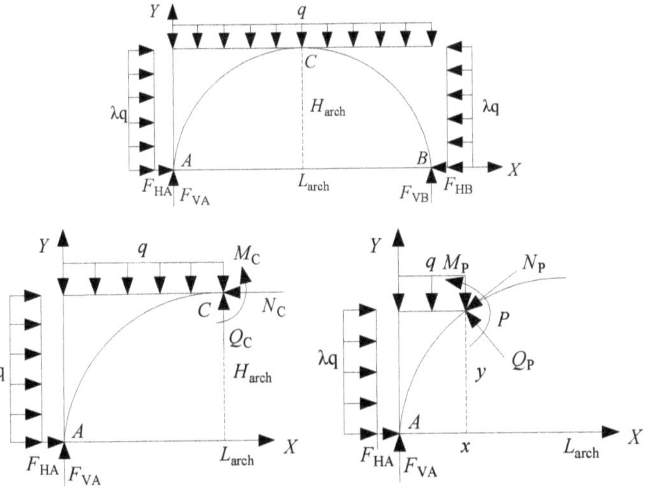

Figure 5. Mechanical structure diagram of unconsolidated layer arch. H_{arch}–Arch height; L_{arch}–arch span; λ–lateral pressure coefficient; q–load; F–support force of arch foot; M–bending moment; N–axial force; Q–shear force.

(2) Numerical and similarity simulation

Based on the construction of the theoretical model, the morphological characteristics of the unconsolidated layer arch structure and the dynamic evolution process of arch structure in the unconsolidated layer were described by PFC, UDEC discrete element numerical simulation software, and similar material simulation [37,45]. PFC is a particle-specific discrete element simulation software, and the particle morphology can effectively simulate the characteristics of the unconsolidated layer. Therefore, PFC^{2D} numerical simulation software can be used to analyze the spatial evolution law of the unconsolidated layer arch. In the process of working face mining, a stress chain in the form of an arch is assembled in the unconsolidated layer, and the shape of the unconsolidated layer arch is represented by the inner and outer envelope of the force chain zone, as shown in Figure 6a. The

UDEC numerical simulation generally uses Voronoi blocks to simulate the unconsolidated layer, and describes the arch structure of the unconsolidated layer through the envelope of the principal stress concentration area. As the width of the working face gradually increases, the envelope of principal stress concentration gradually rises, and the degree of principal stress concentration gradually increases, but the height to span ratio of the unconsolidated layer arch structure scarcely changes, as shown in Figure 6b. Through two numerical simulation results, it can be seen that the shape and stress distribution of the unconsolidated layer arch can be well presented by numerical simulation. Similar physical simulation is similar to numerical simulation, which can reveal the spatial evolution characteristics of the unconsolidated layer arch structure in the process of working face mining from the shape of arch and the layout of stress meter, and then reveal the load transfer effect based on unconsolidated layer arch structure.

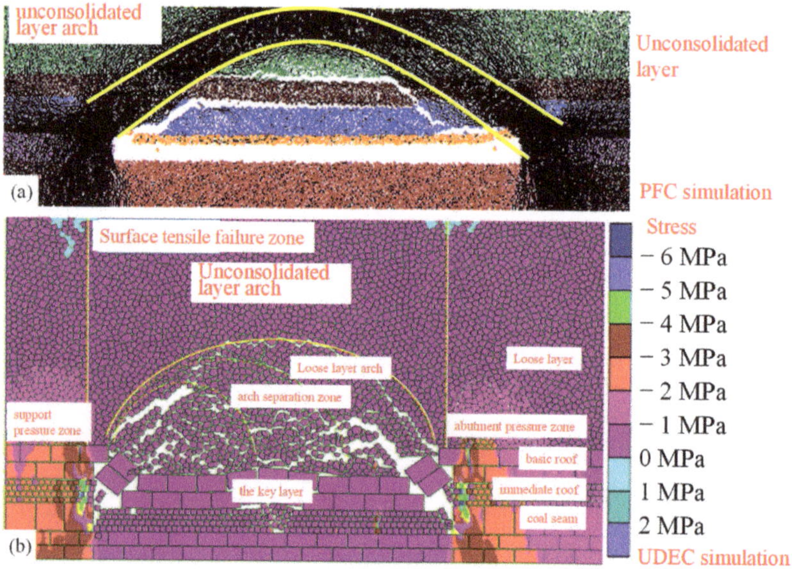

Figure 6. DEM evolution characteristics of arch structure in the unconsolidated layer.

(3) Field measurement verification

At present, the main purpose of studying the unconsolidated layer arch is to better obtain the influence mechanism of the unconsolidated layer arch on load transfer, rock movement, surface subsidence, overburden fracture, and instability, that is, the unconsolidated layer arch can be considered as a part of the overburden arch. Liu et al. [46] studied the movement and deformation law of overlying under the condition of the thick unconsolidated layer by physical simulation, which showed that the edge of the surface movement basin converged slowly, and the boundary angle was small when mining under the thick unconsolidated layer. Thus, the existence of the unconsolidated layer arch could be verified indirectly by field measurement results of surface subsidence. Through field measurement, Li [47] obtained that the surface of the thick unconsolidated layer is sensitive to the influence of mining, the surface movement, and deformation, the recession time is long, and the subsidence velocity and coefficient are large, which can also verify the existence of unconsolidated layer arch. Ju et al. [48] took the surface subsidence caused by coal mining under the thick unconsolidated layer of the Shuozhou mining area as the background, and revealed the surface subsidence mechanism of mining under the thick unconsolidated layer. In conclusion, the field measurement verification of the unconsolidated layer arch is essentially based on surface subsidence characteristics to indirectly verify the existence of unconsolidated layer arch.

4.2. Application of the Unconsolidated Layer Arch

The unconsolidated layer arch structure is mainly used in predicting the height of the water-conducting fracture zone in the thick unconsolidated layer, determining the working resistance of the stope support, and predicting the surface subsidence. Huang et al. [49–51] found the arch effect in the mining of a shallow, buried thick unconsolidated layer coal seam through similar simulation test, and determined the height of the unloaded arch during the initial pressure using the arch structure model, and provided the calculation formula of the stress component and load factor of arch structural failure. Furthermore, the criterion of sand arch fracture and the calculation method of the specific location of surface tensile fracture are provided. Zuo et al. [52,53] constructed an arch structure model of thick unconsolidated overburden and provided the calculation formula of surface subsidence range. Wang et al. [37,38,54,55] determined the calculation formula of working face support resistance based on the arch structure of the unconsolidated layer. Based on the influence of unconsolidated layer arch structure on overburden and surface movement, the characteristics of overburden movement and surface subsidence based on the coupling effect of the unconsolidated layer arch structure and key stratum structure are revealed, and the surface subsidence law of coal mining under the thick unconsolidated layer is obtained. On the basis of mastering the relationship between overburden and the unconsolidated layer, the strip filling mining technology based on controlling the arch structure of the unconsolidated layer and the stability of key blocks is developed, the critical mining width and critical filling rate are determined, and the surface subsidence control under thick unconsolidated layer is realized.

It can be seen from the above examples that the application of the unconsolidated layer arch model is mainly based on the geological conditions of the thick unconsolidated layer, and the model of the unconsolidated layer arch is used to predict the surface subsidence and the mining pressure of the working face.

5. Research Findings on the Application of the Overburden Arch Model

5.1. Research Findings of Overburden Arch Model

(1) Theoretical model

The theoretical model studies on beam and arch were mostly concentrated in mining. Cook et al. [56] applied the theory of beam and arch to the fracture analysis of overlying strata for the first time, trying to explain the rockburst phenomenon of deep gold mines in South Africa. Then, Sterling [57], Brady and Brown [58], Wold and Pala [59], Seedsman [60], Pells and Best [61], Sofianos and Kapenis [62], Diederichs and Kaiser [63], and Nomiko et al. [64] conducted a series of laboratory studies and field verification on this theory, and further developed the beam arch theory. The following assumptions are generally made when the beam arch model is constructed: (a) the beam is subjected to horizontal, symmetric, and uniformly distributed loads; (b) the supporting arch is rigid, and the hardness of the contact point is large enough; (c) Poisson's ratio of rock mass is 0; (d) before the beam is deflected, there is no lateral stress limiting the beam; (e) the loading environment is in a plane strain state. Brady and Brown [58] demonstrated the basic principles and characteristics of beam arch formation: (a) the roof strata forms hinged blocks due to natural penetrating fractures or mining-induced fractures, hence the continuous elastic beam or plate theory cannot be applied; (b) under the action of gravity load, the roof rock block forms the hinged beam arch by the lateral binding force; (c) the beam arch can be elastic in a certain range (that is, the transverse thrust–vertical deflection curve is linearly reversible), and the upper limit is close to the peak transverse load capacity; (d) for a beam arch with low span to thickness ratio, the most likely failure mode is shear failure at the arch support; (e) for a beam with high span to thickness ratio, the span stability is limited by the bendability degree of the beam, and there is no obvious splitting phenomenon in the middle and both ends; (f) for a beam arch with low rock strength or moderate span to thickness ratio, it may be unstable due to the crushing or peeling of the middle or both ends of the beam.

With the introduction of the key stratum theory, the key stratum, as a thick and hard rock, is more consistent with the beam arch theory. The thick and hard strata represented by the key strata are important load-bearing frameworks and force-transmitting structures, and their principal stresses are significantly higher than those of the adjacent strata, which play a dominant role in the macroscopic evolution of the mining stress field. Through mechanical analysis, Lou [39] showed that the force of thick and hard rock strata, represented by the key stratum, during bending deformation was significantly higher than that of neighboring rock strata with small thickness and low stiffness. Zhao et al. [65] quantitatively analyzed the variation characteristics of principal stress in each area of the beam arch in a single key stratum based on the geological conditions of the single key stratum in the Shendong mining area. Thus, the beam arch model is mainly combined with the articulated beam model, focusing on the evolution mechanism of the stress arch in thick and hard strata (key strata) of mining fields, and then studying the instability mechanism of the whole overburden structure.

The research on overburden arch is mainly established by applying the three-hinged arch model from the perspective of structural mechanics, and most of these are based on the compressive arch theory and Terzaghi soil arch effect. It is assumed that a semi-circular, semi-elliptical, and parabolic model can be used to deduce the morphological characteristic equation and identify the instability of the arch model [66]. In the model construction, it is basically similar to the steps of the unconsolidated layer arch, realizing the modeling of arch trace, and applying it to the prediction of mine pressure and mining subsidence. The theoretical model of the overburden stress arch can be summarized as follows:

① Before the construction of the stress arch model, the arch track shape is assumed, such as semi-circle, semi-ellipse, and parabola shape, and then the stress arch model under the assumed track shape is produced, and the stress load analysis under the relevant state is carried out.

② Although the overburden of a coal mine has obvious bedding characteristics, the strengths of the rock strata are also obviously different. The overburden is assumed to be homogeneous in the construction of the stress arch model, which is basically consistent with the modeling of the unconsolidated layer arch. However, the actual rock strata are inhomogeneous layered rock masses, including hard and soft rock strata. In addition to the large difference in the lithology of each layer, there are obvious weak planes between layers. To solve this problem, Lu et al. [67] proposed the recursive method of the fractured arch, that is, the equation of the fractured arch of the lower strata was solved first, and the equation of the upper layer was established on the basis of the solution of the next layer. This actually reverts to the calculation of the fracture arch of a single rock stratum, and extends the beam arch model to the stress arch model.

③ In the process of setting the boundary stress of the overburden arch, uniform load is generally applied above the arch, and the size is proportional to the buried depth, which is consistent with the boundary conditions of the unconsolidated layer arch (Figure 5). Considering the large depth span of the overburden stress arch, the loads on the left and right sides of the arch are developed from the initial uniform load to the trapezoidal load considering the buried depth. The magnitude is the vertical stress multiplied by the lateral pressure coefficient at its location. From the setting of stress boundary, it can be seen that the stress inside the arch is not taken into account in the process of constructing the arch structure model. However, there is rock mass support under both the stress arch and the fracture arch. Therefore, the current theoretical model of the overburden arch is more inclined to caving due to the stress boundary conditions.

④ The construction of the stress arch model is mainly to obtain the trace equation of the arch. The main parameters involved in the arch construction process include coal seam burial depth (H), arch height (h), arch span ($2a_0$), overburden firmness coefficient

f (or overburden cohesion, internal friction angle), lateral pressure coefficient λ, and bulk density γ; the trace equation of arch can be expressed as Equation (2).

$$q_0x^2 = 2a_0fq_2z + \left[\frac{(q_2-q_1)z^3}{h} + 2q_1z^2\right]\frac{2q_2+q_1}{3(q_2+q_2)} \quad (2)$$

where $q_0 = \gamma(H-h)$; $q_1 = \lambda\gamma(H-h)$; $q_2 = \lambda\gamma H$. The trace equation of "stress arch" can be obtained by replacing x and z in Equation (2) with span and arch height. After the trace equation of the arch is obtained, the bearing state of the arch can be further deduced, and the conditions for forming and breaking the arch can be judged.

⑤ There is no method to determine the arch thickness of overburden stress. The unconsolidated layer arch model proposed by Wang et al. [37] provides the calculation equation of arch thickness, which is mainly determined according to the range of abutment pressure in advance of the arch foot, and assumes that the thickness of the arch and the arch foot is the same. However, this is not applicable to the model of the heterogeneous overburden arch, and the thickness of arch obtained by numerical simulation is not consistent.

(2) Numerical simulation and similarity simulation

Aiming at many defects in the process of theoretical model construction, such as stress boundary setting, rock lithology, and arch trace pre-setting, numerical simulation can be used to study the evolution characteristics of overburden arch in the process of coal seam mining because it is intuitive and close to the field engineering situation. The numerical methods used for the simulation of the overburden arch mainly include a finite element and discrete element. The finite element is mainly FLAC^{3D}, while the discrete element is mainly UDEC and PFC software, consistent with the unconsolidated layer arch. The arch formation characteristics are judged by observing the deflection of principal stress in the process of coal seam mining (Figure 7) [43]. Because the numerical simulation can visually display the magnitude and distribution of stress, the existence of the stress arch can be verified, and the arch thickness can be further obtained. For example, the overburden stress arch forming index k ((maximum principal stress − maximum principal stress of original rock)/maximum principal stress of original rock) is used to partition the overburden, and then the main location of stress arch is determined (Figure 8) [68].

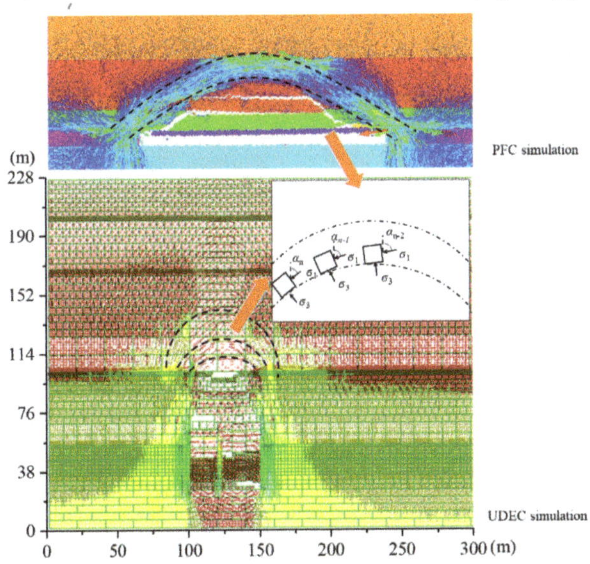

Figure 7. Principal stress deflection into arch mechanism.

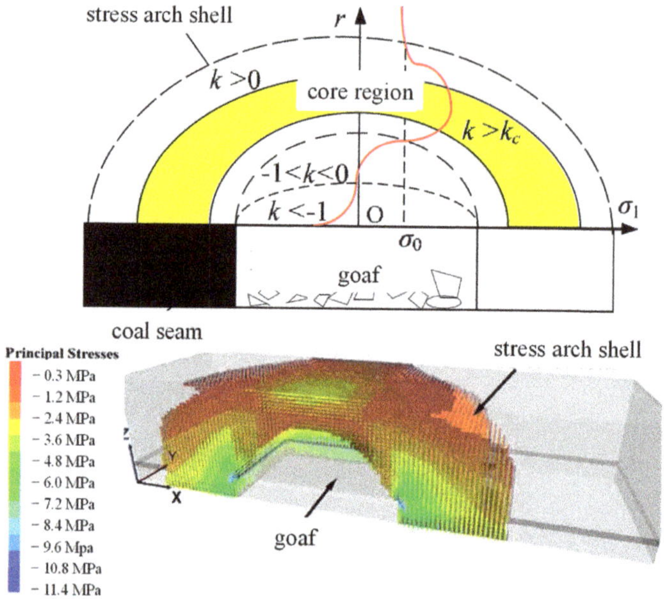

Figure 8. Stress arch zoning and FLAC^{3D} simulation results.

The distribution evolution characteristics of the stress arch and its correlation with mining technology and mining parameters are mainly obtained based on FLAC^{3D}. Xie et al. Researchers [40,41,69–72] have conducted a large number of FLAC^{3D} simulation studies, and some believed that there is a stress arch structure in the mining field, and thus have analyzed the influence of lithology, mining height, and working face length on the development law of the stress arch. Xie et al. [73] used FLAC^{3D} to analyze the distribution and evolution characteristics of the stress arch in the process of backfill mining. Luo et al. [74] used FLAC^{3D} simulation to classify the mining-induced stress concentration arch and pressure relief zone, which is helpful to further analyze the characteristics of gas migration. However, FLAC^{3D} model is generally a continuum model, which cannot simulate rock fracture and caving. In the simulation process, plastic zone is often used to simulate the development form of the arch model, as shown in Figure 9 [75,76]. It can be seen from the figure that the shape of the plastic zone is two-dimensional arch (or three-dimensional shell) in the stage of insufficient mining, and two-dimensional "saddle" (or three-dimensional "basin") when it reaches the state of full mining.

Based on discrete element 2D simulation software such as UDEC and PFC, the evolution of stress arch and fracture arch, and the influence of overburden fracture on arch evolution, are mainly obtained. Xie et al. [77] used PFC numerical simulation software to simulate and analyze the evolution characteristics of force chain in surrounding rock of mining field, and the simulation results are similar to Figure 6a. Through UDEC simulation, Du et al. [29] concluded that the existence of stress arches prevented the upward development of mining-induced fractures, and slowed down the mining pressure in the working face. Zhang et al. [78] used UDEC numerical simulation software to analyze the stress distribution characteristics of the "double-arch bridge" structure in the roof of the goaf, and obtained the stress transfer law. Two-dimensional software can only simulate the shape of the arch of the working face tendency or strike, but cannot realize the description of three-dimensional space. In addition, the particle size or block division of the two-dimensional model directly affects the fracture or caving characteristics of the overlying rock. Taking UDEC as an example, the thickness of the rock block and the division of the vertical fractures all affect the caving shape of the rock layer.

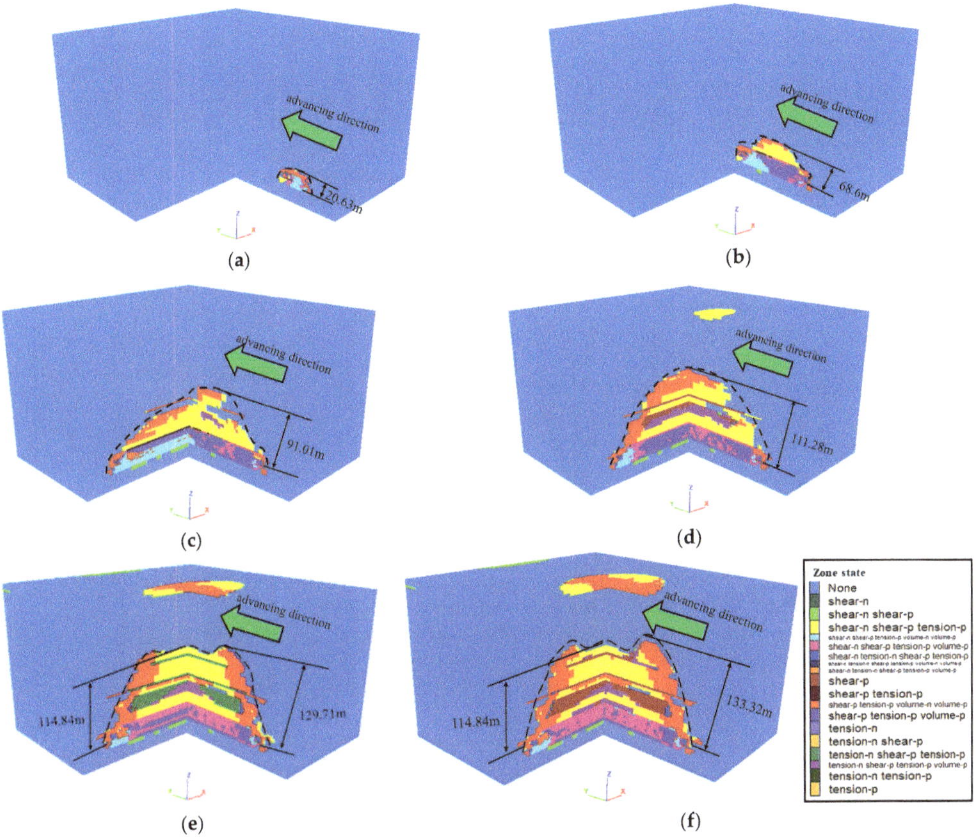

Figure 9. Development process of overburden damage zone. (**a**) advance 50 m; (**b**) advance 150 m; (**c**) advance 200 m; (**d**) advance 250 m; (**e**) advance 280 m; (**f**) advance 300 m.

Thus, numerical simulation can well provide the location, thickness, and evolution characteristics of the arch model. However, the stress arch model simulated by finite elements ignores the influence of rock fracture, while the stress or fracture arch model simulated by discrete elements depends on the division of blocks. In view of the above problems, the finite discrete element method can solve them well [79,80]. However, this method is rarely used to simulate arch structure at present. This method is more used for simulation of rockfalls, particle breakage, coal pillar instability, etc. [81–84]. In addition to numerical simulation, similarity simulation is also the main means to study overburden arch structure. At present, the similarity simulation is mainly two-dimensional, which is similar to the discrete element simulation. However, the similarity simulation mostly focuses on the description of fracture arch morphology and evolution process, which makes it difficult to realize the simulation analysis of stress arch [85–87]. Therefore, generally, similarity simulation and numerical simulation are used to verify each other in the process of arch structure model analysis [88].

(3) Field measurement verification

Although there are many studies on the theoretical analysis, numerical simulation, and similar simulation of overburden arch, and many research results have been achieved, there are few literatures on the field measurement of the existence of overburden arch. On the one hand, the field measurement is difficult, expensive, and has a certain impact on production; on the other hand, there is a lack of targeted monitoring means. Therefore, at present, the method of field measurement and verification of overburden arch is basically the measurement of the height of fracture zone. The default is that at the top boundary of fracture zone development is the location of fracture arch. The method of combining the measurement of flushing fluid loss and drilling television observation is the most widely used field test method [89,90]. This method can accurately measure the height of water-conducting fracture zone, but it needs a lot of time, manpower and money. Transient electromagnetic instrument, three-dimensional seismic and other methods can effectively detect the permeability of mining-induced overburden fractures, and comprehensively analyze the changes of apparent resistivity and aquifer water level to predict the evolution law of water-conducting fracture zones [91]. Sun et al. comprehensively used a variety of technical means to detect the height of the "two zones" of overburden, and believed that the accuracy of borehole color TV and borehole flushing fluid loss was significantly better than that of transient electromagnetic geophysical exploration [92].

5.2. Overburden Arch Structure Model Application

To obtain the position, bearing capacity, fracture, and instability conditions of the overburden arch by means of theoretical analysis, numerical simulation, and similar simulations is mainly to study the law of fracture and migration of overburden in coal mining. On this basis, mining problems such as mine pressure in the working face, height of water-conducting fracture zone, surface subsidence, and surface fracture development can be predicted. Fracture arch is mainly used to predict the height of the water-conducting or gas-conducting fracture zone. The stress arch is mainly used to determine the pressure relief range and predict the overburden fracture migration. Fu et al. [93] provided the calculation method of the working resistance of the support using the stress arch model constructed according to the geological characteristics of deep and weak bonding. According to the characteristics of shallow, buried extremely soft roof, Li et al. [94] explained the phenomena of primary pressure and periodic pressure using the stress arch model. Luo et al. [74] used the stress arch model to divide the stress concentration area and pressure relief area, and further analyzed the migration characteristics of pressure relief gas. Wang et al. [95] proposed the composite stress arch and explained the relationship between overburden migration and surface subsidence. Xu et al. [96] believed that there was no stress arch stage, single-stress arch stage, or double-stress arch stage in shallow, buried high-intensity mining, and predicted the periodic development law of surface fractures. The above research shows that the current research and application of the overburden arch are mainly focused on mastering the evolution characteristics of overburden arch and its influence on the pressure migration of overburden, so as to carry out related passive treatment.

Therefore, how to control the development and evolution characteristics of overburden arch actively, so as to control the overburden fracture migration and mining pressure evolution, is an important content to study the control of overburden arch structure on overburden fracture migration. By studying the control effect of overburden arch on separation, Su et al. [97,98] proposed the grouting technology of overburden separation to ensure the stability of overburden arch and control the surface subsidence. Liu et al. [99,100] proposed arch structure grouting and subsidence control technology for abandoned mined-out areas of coal mine, and proposed corresponding grouting parameters and locations. In addition to maintaining the stability of the arch, under certain geological conditions, the long-term unbroken overburden arch will also cause dynamic hazards. In view of the phenomenon of hard and thick roof being difficult to break and resulting in strong mining pressure, Xia et al. [101], based on the overburden arch model, proposed the technology of roof hydraulic fracturing for arch breaking.

6. Prospects for the Study of the Arch Structure Model

This paper systematically summarizes the research findings of unconsolidated and overburden arch models in coal mining, and divides the arch structure model into three categories according to the region and research object: unconsolidated layer arch, beam arch, and overburden arch. The unconsolidated layer arch is mainly aimed at the geological conditions of the thick unconsolidated layer. Since the geological conditions of the unconsolidated layer are similar to those of the Terzaghi soil arch effect, the soil arch theory can effectively explain the stress distribution characteristics in the unconsolidated layer and the load action of the unconsolidated layer on bedrock. Beam arch is put forward for single-layer thick and hard rock, and combined with the relevant characteristics of articulated beam, the theory of articulated arch is put forward. The formation and breakage of the unconsolidated layer arch are judged. The overburden arch is mainly proposed for the whole overburden structure, and can be divided into stress arch and fracture arch. The fracture arch is generally considered as the top boundary of the fracture zone in coal mining, while the stress arch is considered as the stress increasing area in the overburden. In general, the formation of the beam arch is the key to the formation of the overburden arch. Based on the main research results, the research findings of arch structure model and the future research focus are shown in Table 1.

Table 1. Main research findings of the arch structure model.

Main Aspects	Specific Parameter Characteristics		Research Findings
Arch structure characteristics	Arch forming condition		The arch forming condition of the unconsolidated layer arch is that the thickness of the unconsolidated layer is greater than the sum of the maximum rise height and thickness of the unconsolidated layer arch; The arching condition of the overburden fracture arch is that the overlying bedrock thickness of the working face is greater than the height of the fracture zone in the full mining stage; The arch forming condition of the overburden stress arch is sufficient horizontal stress; The arch forming condition of the beam arch is that a single, thick, hard rock layer can be broken to form an articulated structure
	Arch breaking condition		The instability modes of the arch: compression instability mode, tension instability mode, shear instability mode, and composite instability mode
	Trace shape of arch		Arch trace mainly includes elliptic, semi-circular, parabolic, catenoid, hyperbolic function shape, etc. (3D ellipsoid or spherical). With the continuous advancing of the working face, arch shape evolves dynamically
	Stress distribution of arch		The area where the arch axis is located is the stress increasing area; The outer part of the arch is the stress area of the original rock; The inner part of the arch is the pressure relief area
	Arch thickness		In numerical simulation, it is considered to be the range of stress increasing area, while in theoretical model, it is mainly the arch foot thickness
Mining process parameters	Room and pillar mining method		The evolution characteristics of stress arch in room-and-pillar mining are similar to those in roadway surrounding rock
	Strip mining method		The strip mining stress arch is similar to the roadway stress arch, but the width of the separated coal pillar affects the independence of the adjacent strip stress arch
	Longwall mining method	Fully mechanized top coal caving	The protective effect of stress arch formed by fully mechanized caving mining is greater than that of fully mechanized mining
		Filling mining	Compared with caving mining, the height of the arch structure is greatly reduced, and the shape of the arch is flat
		Mining height	With the increase in mining height, the arch height increases
		Length of working face	The stress concentration degree increases with the increase in working face length, the arch height increases gradually, and the flat rate increases
		Advancing speed	The advancing speed can improve the bearing capacity of the arch
		Width of coal pillar	The coal pillar width affects the superposition of the stress arch of two adjacent longwall working faces, and there is a critical coal pillar width. When the coal pillar width is larger than the critical coal pillar width, the stress arch of the two working faces are independent of each other

Table 1. *Cont.*

Main Aspects	Specific Parameter Characteristics	Research Findings
Geological conditions of rock stratum	Arching medium	The unconsolidated layer arch is formed under the condition of the thick unconsolidated layer. Single-layer thick hard rock formation beam arch; The overburden arch is mainly proposed for the whole overburden structure, and it can be divided into stress arch and fracture arch
	Lithology	The lithology is strengthened, the height of the stress arch is reduced and tilted, the bearing capacity of arch is enhanced, and the risk of dynamic disaster is enhanced
	Dip angle	In the coal seam with high dip angle and sharp dip angle, the stress arch is symmetrical arch in strike and asymmetrical arch in dip angle
	Burial depth	When the shallow coal seam is mined, the bearing capacity of the arch is lost after it extends to the surface. The arch structure will persist when the depth reaches a certain level
Application	Ground pressure prediction of working face	It can calculate the working resistance of hydraulic support and predict periodic weighting
	Roadway layout	Roadway layout inside the arch structure range
	Prediction and control of surface subsidence	The subsidence range and maximum subsidence value can be calculated according to the fracture arch; The surface subsidence rate and fracture development pattern can be predicted according to the periodic instability of the unconsolidated layer arch
	Roof aquifer control	Ensure that the maximum roof fracture arch height is below the aquifer
	dynamic disaster control in coal mine	The instability of far-field stress arch is the main cause of rockburst, and the roof pre-fracturing should be carried out in the peak area of stress arch
	Optimization of gas drainage technology	Gas extraction drilling holes should be arranged in the pressure relief area under the stress arch
Research focus	Theoretical research	Complicated stress boundary conditions, rock layer heterogeneity, irregularity of arch line, and the unequal thickness of the arch should be considered in the arch structure model
	Numerical simulation	Application of finite discrete elements or coupling of multiple simulation methods
	Similar simulation	Update of stress monitoring methods
	Field measurement	Overburden migration monitoring methods of space–sky–surface integration

Theoretical modeling, numerical simulation, similarity simulation, and field measurements have been used to study the overburden and unconsolidated layer arch, and rich results have been obtained, which have been applied to mine pressure prevention and mining subsidence control. According to the research methods, there are still some problems in the research process of the overburden and unconsolidated layer arch: (1) The theoretical model can provide the arch trace equation and then judge the bearing characteristics of the arch. However, there are some problems, such as the simplification of boundary stress conditions (mainly neglecting the abutment stress inside the arch), homogeneity assumption of strata, assumption of arch trace, and arch thickness calculation (the thickness of arch is not consistent). (2) Numerical simulation can readily provide the arch location, arch thickness, and evolution characteristics of the arch, but the stress arch simulated by a finite element ignores the influence of rock fracture, while the stress or fracture arch simulated by a discrete element depends on the division of blocks. The interaction between the unconsolidated layer and the overburden is rarely studied in the simulation process. The simulation method based on finite discrete element method (FDEM) can be used to simulate the arch of the overburden and unconsolidated layer. (3) Similar simulation can only qualitatively observe the development characteristics of the fracture arch, and it is difficult to determine the stress arch range. In future research, more effective stress monitoring methods with wider coverage is needed. (4) The field measurement is still focused on the borehole point monitoring at the height of the "three zones" of overburden rock, which only approximates the height of the fractured arch, and cannot detect the arch trace. Transient electromagnetic techniques and other methods, with which the accuracy is low, can only provide a rough range. In the future research, the migration characteristics of the arch can be comprehensively judged based on the space–sky–surface cooperative observation data, such as surface subsidence and movement, migration detection of deep base point of overburden, and mine pressure data of the working face [102].

The purpose of studying overburden and unconsolidated layer arches is to master the migration and fracture law of overburden induced by mining; on this basis, mining problems such as mine pressure in working face, height of the water-conducting fracture zone, pressure relief gas migration, and surface subsidence can be predicted. However, the application of overburden and unconsolidated layer arches focuses on the evolution law of arch structure, and there are relatively few active control methods for the overburden arch. Based on this, the overburden and unconsolidated layer arch control methods are divided into three categories: (1) Preset arch structure method before coal mining, including mining technology (filling mining, room and pillar mining, strip mining), mining parameter (mining height, working face length, advancing speed) design; The author has discussed this technology in detail in the literature [103,104]. (2) Process control arch methods during coal mining, e.g., grouting reinforcement arch protection (separated layer grouting, goaf grouting). Specifically, the arch trace equation under specific mining conditions can be calculated in advance to further provide the judgment conditions for arch fracture and stability. The grouting reinforcement area is proposed to ensure the stability of rock strata. In addition, some mining areas have thick and hard roof rock strata, and it is difficult to break the stable arch structure formed in the mining process, resulting in a dynamic risk. Under this condition, roof fracturing and arch breaking (hydraulic fracturing, deep hole blasting) are required to release the accumulated energy. (3) End reinforcement control arch methods after coal mining, in order to prevent and control the disturbance and instability in the reuse process of coal mine goaf, or the ground uplift caused by the rise of underground water level [105,106], arch grouting reinforcement is carried out for the goaf to prevent surface deformation and movement. As shown in Figure 10, the ground grouting holes surround the top boundary of the fracture zone in the goaf, and the boundary holes are arranged on the edge of the subsidence area to form the arch foot by grouting. The middle hole is located in the bending subsidence zone, and the grouting range is shown in Figure 10. According to the literature [99], the spacing of the grouting holes in goafs can be calculated with Equation (3).

$$a = 2R - \Delta b, \Delta b \geq F_s \frac{P_{\max}}{\sigma_c} \tag{3}$$

where: Δb is the spread width (m) of the grouting body; F_s is the safety factor, generally greater than 1.0; $P_{\max}$ is the maximum pressure of the overlying rock and soil layer (kN); σ_c is the compressive strength (kPa) of the grouting stone body.

There is relatively few reports and research on post-mining arches in coal mines. With the proposal of the strategy of carbon neutralization and carbon reaching peak, coal mining transformation is imperative in China, and the research focus on the abandonment mine ecological restoration and the space utilization of goaf will become a research hotspot [107–109]. Therefore, the future research on overlying rock arches in coal mine goafs mainly involves three aspects: (1) Determination of the arch structure location in coal mine goaf and its long-term (ten years and above) stability judgment. (2) After the mine is closed, due to no drainage, the water level of the goaf rises, long-term flooding and water level uplift are very important to the stability of the overlying arch. (3) The influence of engineering disturbances, such as pumped storage power stations, underground storage, engineering research center, surface industrial tourism, and ecological restoration on the stability of goaf arches in the process of mine reuse should be focused on.

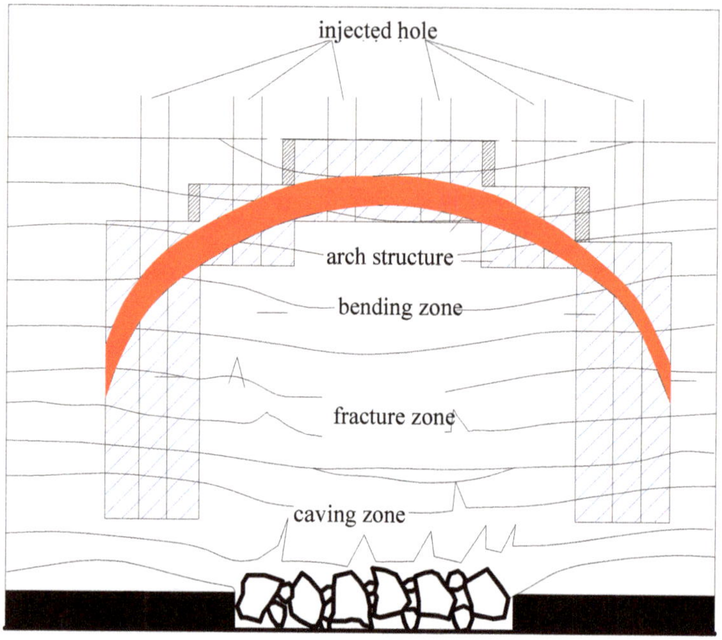

Figure 10. Schematic diagram of arch structure grouting in goaf.

7. Conclusions

(1) In order to fully understand the conduction mechanism of mining damage, the current research results of overburden movement are systematically reviewed with the arch structure model, and the research results are divided into the unconsolidated layer arch, beam arch, and overburden arch, according to the differences in characteristics of overburden combination and medium. The unconsolidated layer arch is proposed according to the condition of the thick unconsolidated layer. The beam arch is proposed for a single layer of thick hard rock. The overburden arch is proposed for the whole overburden structure, which is divided into stress arch and fracture arch.

(2) The existence of the overburden arch and unconsolidated layer arch was verified by theoretical model construction, numerical similarity, and simulation analysis. The morphological distribution and spatial evolution law of arch structure was provided, as well as the conditions of arch formation and arch breaking. The influence of mining parameters, lithology, dip angle, and mining method on the development of overburden arch structure was analyzed. On this basis, the working resistance calculation of working face support, the prediction of working face pressure, and the analysis of surface movement and subsidence were realized.

(3) The arch theoretical model of overburden and unconsolidated layers can provide the arch trace equation to judge the bearing characteristics of the arch. However, there are some limitations, such as simplification of boundary stress conditions, homogeneity assumption of strata, assumption of arch trace, and calculation of arch thickness. Numerical simulation can readily provide the location, thickness, and evolution characteristics of arch, but the stress arch simulated by finite element ignores the influence of rock fracture, while the stress or fracture arch simulated by discrete element depends on the division of blocks. The similarity simulation can only observe the development characteristics of fracture arch qualitatively, and it is difficult to determine the stress arch range. The field measurement mainly verifies the development characteristics of the fracture arch indirectly, through the development characteristics of the fracture zone.

(4) The application of overburden and unconsolidated layer arches focuses on the prediction and prevention of mine pressure and surface subsidence through the arch evolution model, and there are relatively few means to actively regulate the overburden arch. Thus, three control methods of arch structure in the underground coal mining were proposed: preset arch structure method before coal mining, process control arch methods during coal mining, and end reinforcement control arch methods after coal mining. Based on the current energy policy of China, the present paper puts forward the problems that may be encountered in the reuse process of abandoned mine: long-term stability, the impact of water immersion and water level rise, and engineering disturbance.

(5) Future research on arch structure modeling should focus on the following aspects: ① In the theoretical research, the support stress inside the arch should be considered in the stress boundary conditions. The heterogeneity of rock stratum and the non-uniformity of arch thickness should be considered. ② In the numerical simulation, the interaction between unconsolidated layer arch and overburden arch can be realized by finite discrete elements or coupling of multiple simulation methods. ③ In the similar simulation, the evolution method of stress arch can be obtained using the optical fiber method. ④ In field measurements, multiple methods designed for the space, sky, and surface can be used to monitor overburden migration characteristics.

Funding: Financial support for this work is provided by the National Natural Science Foundation of China (52104155), the Natural Science Foundation of Beijing (8212032), and the Open Fund for the State Key Laboratory of Coal Mining Water Resources Protection and Utilization (WPUKFJJ2019-5).

Institutional Review Board Statement: Not applicable.

Informed Consent Statement: Not applicable.

Data Availability Statement: Some or all data, models, or codes that support the findings of this study are available from the corresponding author upon reasonable request.

Conflicts of Interest: The author declares that there is no conflict of interest regarding the publication of this article.

References

1. Chen, C.; Hu, Z. Research advances in formation mechanism of ground crack due to coal mining subsidence in China. *J. China Coal Soc.* **2018**, *43*, 810–823.
2. Wang, J.; Yang, S.; Wei, W.; Zhang, J.; Song, Z. Drawing mechanisms for top coal in longwall top coal caving (LTCC): A review of two decades of literature. *Int. J. Coal. Sci. Technol.* **2021**, *8*, 1171–1196. [CrossRef]
3. Gao, R.; Kuang, T.; Zhang, Y.; Zhang, W.; Quan, C. Controlling mine pressure by subjecting high-level hard rock strata to ground fracturing. *Int. J. Coal. Sci. Technol.* **2021**, *8*, 1336–1350. [CrossRef]
4. Miao, X.; Qian, M. Advance in the key strata theory of mining rockmass. *J. China Univ. Min. Technol.* **2000**, *29*, 25–29.
5. Qian, M. Theory and Practice of Surrounding Rock Control in Stope. *Ground Press. Strat. Control* **1999**, *Z1*, 12–15.
6. Qian, M.; Miao, X.; Xu, J. Theoretical study of key stratum in ground control. *J. China Coal Soc.* **1996**, *3*, 2–7.
7. Song, Z.; Hao, J.; Shi, Y.; Tang, J.; Liu, J. An overview of connotation and development of practical ground pressure control theory. *J. Shandong Univ. Sci. Technol.* **2019**, *38*, 1–15.
8. Li, C.; Zhang, Y.; Zuo, J.; Tang, S.J.; Liu, S.F. Floor failure mechanical behavior and partition characteristics under the disturbance of voussoir beam instability in deep coal mining. *J. China Coal Soc.* **2019**, *44*, 1508–1520.
9. Xu, J. Strata control and scientific coal mining—a celebration of the academic thoughts and achievements of Academician Minggao Qian. *J. Min. Saf. Eng.* **2019**, *44*, 1508–1520.
10. Qian, M.; Xu, J. Behaviors of strata movement in coal mining. *J. China Coal Soc.* **2019**, *38*, 1–15.
11. Zou, X. A new interpretation of the pressure arch hypothesis. *Mine Press.* **1989**, *1*, 67–68.
12. Zhang, J.; Li, J.; An, T.; Huang, Y. Deformation characteristic of key stratum overburden by raw waste backfilling with fully-mechanized coal minning technology. *J. China Coal Soc.* **2010**, *35*, 357–362.
13. Yu, Q.; Zhang, H.; Deng, W.; Zou, Y. Analysis of influencing factors of surface skewed subsidence based on key strata theory. *J. China Coal Soc.* **2018**, *43*, 1322–1327.
14. Jiang, F.; Wen, J.; Bai, W.; Wang, G.; Li, M. Rock burst risk in surrounding abscission layer of overlying high key strata in deep strip mining mines. *J. China Univ. Min. Technol.* **2018**, *47*, 40–47.
15. Wang, H.; Fan, X.; Jia, J.; Hu, G.; Yuan, Z. Effects of key strata on protection when exploiting a steep-incline underlying a protecting stratum. *J. China Univ. Min. Technol.* **2011**, *40*, 23–28.

16. Lu, H.; Meng, X.; Zhang, Y.; Yao, D. Mechanical analysis of water barrier performance of floor layered structure key stratum on coal face. *J. China Univ. Min. Technol.* **2020**, *49*, 1057–1066.
17. Zhu, W.; Xu, J.; Lai, W.; Wang, Z. Research of isolated section-grouting technology for overburden bed separation space to reduce subsidence. *J. China Coal Soc.* **2007**, *32*, 458–462.
18. Song, Z. Development status of mining engineering discipline in China and discussion on issues of its further development. *Hazard Control. Tunn. Undergr. Eng.* **2019**, *1*, 7–12.
19. Yang, W.; Lin, B.Q.; Yan, Q.; Zhai, C. Stress redistribution of longwall mining stope and gas control of multi-layer coal seams. *Int. J. Rock Mech. Min. Sci.* **2014**, *72*, 8–15. [CrossRef]
20. Zhang, C.; Bai, Q.; Chen, Y. Using stress path-dependent permeability law to evaluate permeability enhancement and coalbed methane flow in protected coal seam: A case study. *Geomech. Geophys. Geo-Energy Geo-Resour.* **2020**, *6*, 1–25. [CrossRef]
21. Johann, W.R. *Die Statik der Tunnel Gewolbe*; Springer: Berlin/Heidelberg, Germany, 1879.
22. Terzaghi, K. *Theoretical Soil Mechanics*; John Wiley & Sons, Inc.: New York, NY, USA, 1943.
23. Handy, R.L. The arch in soil arching. *J. Geotech. Eng.* **1985**, *111*, 302–318. [CrossRef]
24. Liu, Z.; Wang, Y. *Present Situation of Strata Control and Subsidence Prevention Technology in Longwall Mining*; Shanxi Science Education Press: Taiyuan, China, 1986.
25. Apehc, B.A. *Rock and Ground Surface Movements*; Coal Industry Press: Beijing, China, 1989.
26. IME. Third progress report of investigation into cause of fall sand accidents due to falls in board and pillar whole workings roof fracture control in boards. *Trans. Inst. Min. Eng.* **1936**, *90*, 241–242.
27. Dinsdale, J.R. Ground Pressures and Pressure Profiles Around Mining Excavations. *Colliery Eng.* **1935**, *12*, 406–409.
28. Stemple, D.T. *A Study of Problems Encountered in Multiple-Seam Coal Mining in the Eastern United States*; Virginia Polytechnic Institute: Blacksburg, VA, USA, 1955.
29. Denkhaus, H.G. Critical review of strata movement theories and their application to practical problems. *J. S. Afr. Inst. Min. Metall.* **1964**, *64*, 310–332.
30. Sansone, E.C.; Da Silva, L.A.A. Numerical modeling of the pressure arch in underground mines. *Int. J. Rock Mech. Min. Sci.* **1998**, *4*, 436. [CrossRef]
31. Chekan, G.J.; Listak, J.M. *Design Practices for Multiple-Seam Longwall Mines*; US Department of the Interior, Bureau of Mines: Washington, DC, USA, 1993.
32. Luo, J.L. *Gateroad Design in Overlying Multi-Seam Mines*; Virginia Tech: Blacksburg, VA, USA, 1997.
33. Zhao, P. The shape of natural equilibrium arch on roadway roof considering the original horizontal stress. *J. China Inst. Min. Technol.* **1978**, 56–63. Available online: https://xueshu.baidu.com/usercenter/paper/show?paperid=aa424efb8fe7feb60b1a7ccd69c6303b&site=xueshu_se (accessed on 14 October 2022).
34. Miu, X. Natural Equilibrium Arch and Stability of Roadway Surrounding Rock. *Ground Press. Strat. Control* **1990**, 55–57.
35. Cao, S. Discussion on mechanical structure of compound arch in surrounding rock of stope. *J. Chongqing Univ. Nat. Sci. Ed.* **1989**, *12*, 72–78.
36. Kang, L.; Qi, Q.; Ning, Y. Study on the Discrete Element Method of Roof Fracture Mechanism. *Coal Min.* **1995**, *1*, 33–37.
37. Wang, F.; Chen, S.; Xu, J.; Ma, B.; Ding, X.; Liu, Z. Research on ground control based on arch structure in unconsolidated layers theory. *Coal Sci. Technol.* **2020**, *48*, 130–138.
38. Wang, F.; Xu, J.; Xie, J. Effect of arch structure in unconsolidated layers on failure of the overlying strata. *J. China Univ. Min. Technol.* **2019**, *5*, 975–983. [CrossRef]
39. Lou, J. Influence mechanism of beam-arch binary structure and strata characteristics on fracture and stress evolution of overlying strata in stope. *Int. J. Rock Mech. Min. Sci.* **2002**, *39*, 79–94.
40. Xie, G. Mechanical characteristics of fully mechanized top-coal caving face and surrounding rock stress shell. *J. China Coal Soc.* **2005**, *30*, 309–313.
41. Wang, L.; Xie, G.; Wang, J. Numerical investigation on the influence of surrounding rock stress shell on fractured field. *J. China Coal Soc.* **2015**, *9*, 2009–2014.
42. Li, Z.; Yang, K.; Hua, X.; Li, Y.; Liu, Q. Disaster-causing mechanism of instability and "macroscopic-big-small" structures of overlying strata in longwall mining. *J. China Coal Soc.* **2020**, *45*, 541–550.
43. Zhao, Y.; Yu, J.; Zhou, H.; Zhao, K.; Xiao, H. Characterization of pressure arching effect of arch shell surrounding rock considering deviation of principal stress axis. *Chin. J. Geotech. Eng.* **2021**, *43*, 1842–1850.
44. Feng, J.; Zhou, Y.; Li, H.; Liu, C. Three kinds of basic structure of working face in near horizontal coal seam. *J. China Coal Soc.* **2016**, *41*, 2576–2587.
45. Du, F.; Bai, H. Mechanism research of overlying strata activity with fully mechanized caving in thin bedrock with thick alluvium. *J. China Coal Soc.* **2012**, *37*, 1105–1110.
46. Liu, Y.X.; Dai, H.Y.; Jiang, Y.D. Model test for mining-induced movement law of rock and soil mass under thick unconsolidated layers. *Caikuang Yu Anquan Gongcheng Xuebao/J. Min. Saf. Eng.* **2012**, *29*, 700–706.
47. Li, D.; Xu, G.; Yu, H. Study on Features of Surface Dynamic Movement and Deformation Caused by Coal Mining Under Thick Alluvium. *Coal Sci. Technol.* **2014**, *7*, 103–106.
48. Ju, J.; Xu, J. Surface stepped subsidence related to top-coal caving longwall mining of extremely thick coal seam under shallow cover. *Int. J. Rock Mech. Min. Sci.* **2015**, *78*, 27–35. [CrossRef]

49. Huang, Q.; Zhang, P. Study on dynamic load distribution on key roof blocks of under thick sandy soil stratum. *Chin. J. Rock mech. Eng.* **2004**, *23*, 4179–4182.
50. Huang, Q. Study on Loading Distribution Law on Key Roof and Its Structure Upon Mining Face Under Thick Sandy Layer. *J. China Univ. Min. Technol.* **2005**, *34*, 289–293.
51. Huang, Q.; Zhang, A.; Dong, A. Mathematical model of "arch beam" of thick sandy soil layer movement in shallow seam. *Rock Soil Mech.* **2009**, *30*, 2722–2726.
52. Zuo, J.; Sun, Y.; Qian, M. Movement mechanism and analogous hyperbola model of overlying strata with thick alluvium. *J. China Coal Soc.* **2017**, *42*, 1372–1379.
53. Zuo, J.; Sun, Y.; Wang, J.; Shi, Y.; Wen, J. Mechanical and numerical analysis of "analogous hyperbola" movement of overlying strata after full mining extraction. *J. Min. Saf. Eng.* **2018**, *35*, 71–77.
54. Wang, F.; Xu, J.; Chen, S.; Ren, M. Arch structure in unconsolidated layers and its effect on the overlying strata movement. *J. Min. Saf. Eng.* **2019**, *36*, 497–504.
55. Wang, F.; Jiang, B.; Chen, S. Surface collapse control under thick unconsolidated layers by backfilling strip mining in coal mines. *Int. J. Rock Mech. Min. Sci.* **2019**, *113*, 268–277. [CrossRef]
56. Cook, N.G.W.; Klokow, J.W. *Practical Rock Mechanics for Gold Mining*; Chamber of Mines of South Africa: Johannesburg, South Africa, 1972.
57. Serling, R.L. The ultimate load behavior of laterally constrained rock beams. In Proceedings of the 21st US Symposium on Rock Mechanics (USRMS), Rolla, MI, USA, 27–30 May 1980.
58. Brady, B.H.G.; Brown, E.T. *Rock Mechanics: For Underground Mining*; Springer Science & Business Media: Berlin/Heidelberg, Germany, 1993.
59. Wold, M.; Pala, J. The response of powered supports and pillars to initial longwalling under a strong main roof. In Proceedings of the Conference on Ground Movement and Control Related to Coal Mining, Wollongong, Australia, August 1986.
60. Seedsman, R. Back-analysis of roof conditions in the Great Northern Seam, Newcastle Coal Measures, Australia, using voussoir beam theory. *Int. J. Min. Geol. Eng.* **1987**, *5*, 15–27. [CrossRef]
61. Pellsp, J.N.; Best, R.J. Aspects of primary support design for tunnels in the Sydney Basin. Transactions of the Institution of Engineers, Australia. *Civ. Eng.* **1991**, *33*, 57–66.
62. Sofianos, A.I.; Kapenis, A.P. Numerical evaluation of the response in bending of an underground hard rock voussoir beam roof. *Int. J. Rock Mech. Min. Sci.* **1998**, *35*, 1071–1086. [CrossRef]
63. Diederichs, M.S.; Kaiser, P.K. Tensile strength and abutment relaxation as failure control mechanisms in underground excavations. *Int. J. Rock Mech. Min. Sci.* **1999**, *36*, 69–96. [CrossRef]
64. Nomikos, P.P.; Sofianos, A.I.; Tsoutrelis, C.E. Structural response of vertically multi-jointed roof rock beams. *Int. J. Rock Mech. Min. Sci.* **2002**, *39*, 79–94. [CrossRef]
65. Zhao, Y.; Zhou, C.; Zahng, X.; Yuan, C. The pressure arching effect and distribution characteristics of fractured strata of single key layer under shallow buried condition. *J. China Coal Soc.* **2020**, *45*, 1–11.
66. Zhang, Y.; Wu, Y.; Luo, S.; Xie, P. Evolution process and characteristics of overlying strata macro support structure. *J. China Univ. Min. Technol.* **2020**, *49*, 280–288.
67. Lu, G.; Tang, J.; Huang, S. Derivation method of arch line equation of stope fracture zone. In Proceedings of the Progress in Research on Soft Rock Engineering and Deep Disaster Control in China—Proceedings of the Fourth Symposium on Deep Rock Mechanics and Engineering Disaster Control and the Centennial Anniversary Academic Conference of China University of Mining and Technology, Beijing, China, 26–28 September 2009.
68. Du, X.; Song, H.; Chen, J. Numerical simulation of the evolution of the pressure arch during coal mining. *J. China Univ. Min. Technol.* **2011**, *40*, 863–867.
69. Xie, G.; Yang, K. Study of macro stress shell evolving characteristics of rock surrounding face. *Chin. J. Rock Mech. Eng.* **2010**, *29*, 2676–2680.
70. Xie, G.; Wang, L. Effect of longwall length on mechanical characteristics of surrounding rock stress shell in mining face. *J. China Coal Soc.* **2008**, *33*, 1336–1340.
71. Xie, G.; Wang, L. Lithologic effect on the mechanical characteristics of mining-induced stress shell. *J. China Coal Soc.* **2013**, *1*, 44–49.
72. Xie, G. In fluence of mining thickness on mechanical characteristics of working face and surrounding rock stress shell. *J. China Coal Soc.* **2006**, *31*, 6–10.
73. Xie, S.; Hao, M.; Chen, D.; Yue, S.; Pan, H. Study on the distribution and evolution characteristics of spherical stress shell in backfilling mining stope. *J. Min. Saf. Eng.* **2019**, *36*, 16–23.
74. Luo, F.; Cao, S.; Li, G.; Li, Y. Evolution of mine-induced stress concentration shell and stress relief body and its gas migration. *J. Min. Saf. Eng.* **2018**, *35*, 155–162.
75. Zhao, J.; Konietzky, H.; Herbst, M.; Morgenstern, R. Numerical simulation of flooding induced uplift for abandoned coal mines: Simulation schemes and parameter sensitivity. *Int. J. Coal. Sci. Technol.* **2021**, *8*, 1238–1249. [CrossRef]
76. Zhang, C.; Ren, Z.; Han, P.; He, X.; Chen, J. Development characteristic of the water-conducting fracture zone in thick overburden working face with extra- large mining height in western mining area. *J. Min. Sci. Technol.* **2022**, *7*, 333–343.

77. Xie, G.; Fan, H.; Wang, L. Evolution law and engineering application of surrounding rock force chain in shallow coal seam working face. *J. China Coal Soc.* **2019**, *44*, 2945–2952.
78. Zhang, J.; Wang, B.; Bai, W.; Chen, C.; Zhang, J. Stability study of the double-arch structure formed by short-distance discontinuous mining in shallow strata. *J. China Univ. Min. Technol.* **2021**, *50*, 598–605.
79. Lisjak, A.; Garitte, B.; Grasselli, G.; Müller, H.R.; Vietor, T. The excavation of a circular tunnel in a bedded argillaceous rock (Opalinus Clay): Short-term rock mass response and FDEM numerical analysis. *Tunn. Undergr. Space Technol.* **2015**, *45*, 227–248. [CrossRef]
80. Ma, G.; Chen, Y.; Yao, F.; Zhou, W.; Wang, Q. Evolution of particle size and shape towards a steady state: Insights from FDEM simulations of crushable granular materials. *Comput. Geotech.* **2019**, *112*, 147–158. [CrossRef]
81. Liu, G.; Li, J.; Wang, Z. Experimental verifications and applications of 3D-DDA in movement characteristics and disaster processes of rockfalls. *Rock Mech. Rock Eng.* **2021**, *54*, 2491–2512. [CrossRef]
82. Zhang, C.; Zhao, Y.; Bai, Q. 3D DEM method for compaction and breakage characteristics simulation of broken rock mass in goaf. *Acta Geotech.* **2022**, *17*, 2765–2781. [CrossRef]
83. Ma, K.; Liu, G. Three-dimensional discontinuous deformation analysis of failure mechanisms and movement characteristics of slope rockfalls. *Rock Mech. Rock Eng.* **2022**, *55*, 275–296. [CrossRef]
84. Zhang, C.; Zhao, Y.; Han, P.; Bai, Q. Coal pillar failure analysis and instability evaluation methods: A short review and prospect. *Eng. Fail. Anal.* **2022**, *138*, 106344. [CrossRef]
85. Lai, X.; Dai, J.; Li, C. Analysis on hazard characteristics of overburden structure in steeply inclined coal seam. *J. China Coal Soc.* **2020**, *45*, 122–130.
86. Yang, D.; Guo, W.; Tan, Y.; Wang, Y.; Ma, X. Lithology and fissure characteristics of overburden in high-intensity mining. *J. China Coal Soc.* **2019**, *44*, 786–795.
87. Wang, Y.; Guo, W.; Bai, E.; Zhang, Z.; Kang, Y. Characteristics and mechanism of overlying strata movement due to high-intensity mining. *J. China Coal Soc.* **2018**, *43*, 28–35.
88. Xie, G.X.; Chang, J.C.; Yang, K. Investigations into stress shell characteristics of surrounding rock in fully mechanized top-coal caving face. *Int. J. Rock Mech. Min. Sci.* **2009**, *46*, 172–181. [CrossRef]
89. Wang, X.; Xu, J.; Han, H.; Ju, J.; Xing, Y. Stepped development characteristic of water flowing fracture heightwith variation of mining thickness. *J. China Coal Soc.* **2019**, *44*, 3740–3749.
90. Yang, D.; Guo, W.; Zhao, G.; Tan, Y.; Yang, W. Height of water-conducting zone in longwall top-coal caving mining underthick alluvium and soft overburden. *J. China Coal Soc.* **2019**, *44*, 3308–3316.
91. Zhang, H.; Zhu, Z.; Huo, L.; Chen, J.; Huo, B. Overburden failure height of superhigh seam by fully mechanized caving method. *J. China Coal Soc.* **2014**, *39*, 816–821.
92. Sun, Q.; Mou, Y.; Yang, X. Study on "two-zone" height of overlying of fully-mechanized technology with high mining height at Hongliu Coal Mine. *J. China Coal Soc.* **2013**, *38*, 283–286.
93. Fu, B.; Fang, E.; Tu, M.; Deng, R. Study on load-transfer mechanism and strata behavior characteristics in deep-buried weak-caking roof. *J. Min. Saf. Eng.* **2019**, *36*, 24–29.
94. Li, X.; Chen, P. Study on regularity of structural behaviours around coal face under the shallow-buried loose roof. *Chin. J. Rock Mech. Eng.* **2004**, *23*, 3305–3309.
95. Wang, J.; Zhao, Z.; Hou, Z. Study on the catastrophic collapse of surface land induced by mining under a shallow and hard strata. *J. China Coal Soc.* **2007**, *10*, 1051–1056.
96. Sun, K.; Fan, L.; Xia, Y.; Li, C.; Chen, J.; Gao, S.; Wu, B.; Peng, J.; Ji, Y. Impact of coal mining on groundwater of Luohe Formation in Binchang mining area. *Int. J. Coal. Sci. Technol.* **2021**, *8*, 88–102. [CrossRef]
97. Su, Z.; Yu, G.; Yang, L. Application of mechanical model to deformation of covered rock separation strata. *Chin. J. Geotech. Eng.* **2002**, *24*, 778–781.
98. Su, Z.; Liu, W. Study and Application of New Grouting Technology for Separated Strata to Slacken the Surface Subsidence. *China Saf. Sci. J.* **2001**, *11*, 21–24.
99. Liu, Y.; Liu, F.; Zhang, T.; Zhan, Y. Influence of gob treatment effect on mechanical parameters of grouting material. *Low Temp. Archit. Technol.* **2017**, *39*, 104–108.
100. Liu, F.; Liu, Y.; Zhang, T.; Zhan, Y.; Li, X. Study on the mechanism of subsidence control with arch structure grouting in goal mine gob area. *Saf. Environ. Eng.* **2015**, *22*, 22–27.
101. Xia, B.; Zhang, X.; Yu, B.; Jia, J. Weakening effects of hydraulic fracture in hard roof under the influence of stress arch. *Int. J. Min. Sci. Technol.* **2018**, *28*, 951–958. [CrossRef]
102. Zhang, C.; Zhao, Y.; He, X. Space-sky-surface integrated monitoring system for overburden migration regularity in shallow-buried high-intensity mining. *Bull. Eng. Geol. Environ.* **2021**, *80*, 1403–1417. [CrossRef]
103. Li, Q.; Zhang, C. Damage conduction model of high intensity mining in western mining area based on conservation of mining space and its application. *J. Min. Saf. Eng.* **2021**, *38*, 1–8.
104. Li, Q.; Guo, J.; Zhang, K. Damage conduction mechanism and key technology of source loss reduction of intensive coal mining in the west. *J. China Coal Soc.* **2022**, *46*, 3636–3644.
105. Vervoort, A. Uplift of the surface of the earth above abandoned coal mines. Part A: Analysis of satellite data related to the movement of the surface. *Int. J. Rock Mech. Min. Sci.* **2021**, *148*, 104896. [CrossRef]

106. Zhao, J.; Konietzky, H. An overview on flooding induced uplift for abandoned coal mines. *Int. J. Rock Mech. Min. Sci.* **2021**, *148*, 104955. [CrossRef]
107. Yuan, L. Comprehensive development and utilization of abandoned mine resources, help achieve the goal of "emission peak and carbon neutralization". *Sci. Technol. Rev.* **2021**, *39*, 1.
108. Chen, M.; Chen, X.; Hu, Z.; Fan, T.; Zhang, S.; Liu, Y. Contribution of root respiration to total soil respiration during non-growing season in mine reclaimed soil with different covering-soil thicknesses. *Int. J. Coal. Sci. Technol.* **2021**, *8*, 1130–1137. [CrossRef]
109. Yuan, L.; Yang, K. Further discussion on the scientific problems and countermeasures in the utilization of abandoned mines. *J. China Coal Soc.* **2021**, *46*, 16–24.

sustainability

MDPI

Article

Mining Subsidence Prediction Model and Parameters Inversion in Mountainous Areas

Bang Zhou [1], Yueguan Yan [1,*], Huayang Dai [1], Jianrong Kang [2], Xinyu Xie [1] and Zhimiao Pei [1]

[1] College of Geoscience and Surveying Engineering, China University of Mining & Technology, Beijing 100083, China; bqt1900204056@student.cumtb.edu.cn (B.Z.); dhy@cumtb.edu.cn (H.D.); 2010200327@student.cumtb.edu.cn (X.X.); 2010200305@student.cumtb.edu.cn (Z.P.)

[2] School of Geography, Geomatics and Planning, Jiangsu Normal University, Xuzhou 221116, China; kangjianrong@jsnu.edu.cn

* Correspondence: yanyueguan@cumtb.edu.cn

Abstract: Coal mining in mountainous areas is general in China, especially in Shanxi Province. Under the influence of topography in mountainous areas, surface collapses and landslides caused by underground mining happen at a certain frequency and threaten human lives and assets. Accurate prediction of the movement and deformation of mining subsidence in mountainous areas facilitates the prevention and control of geological disasters. The probability integral method is an official prediction method for mining subsidence prediction in China, while it is lacking in the prediction accuracy in mountainous areas due to the inherent topography. Therefore, a practical prediction model based on slopes slip combined parameters optimization was proposed in this study. The slip subsidence and slip horizontal movement were deduced based on the probability integral method considering the topography (slope angle $\alpha < 30°$) and geological conditions (loess covered) to build the prediction model. The dynamic step fruit fly optimization algorithm (DSFOA) was applied for parameters inversion about the probability integral method in the proposed prediction model, while the other parameters in the proposed model were determined by mechanics analysis based on the nature of losses. The determination of parameters is more efficient, objective and reasonable, so that the prediction accuracy can be improved. The measured data of the working panel 22,101 located in Taiyuan, Shanxi Province was verified by this practical model, and the result shows that the mean square error of subsidence and the horizontal movement was decreased to 71 mm and 276 mm, respectively, hence, the applicability of the proposed mining subsidence prediction model in mountainous areas is verified. This work will contribute to a comprehensive understanding on the law of surface movement and provide theoretical guidance for surface damage prevention and control in mountainous mining areas.

Keywords: probability integral method; slopes slip; mountainous areas; dynamic step fruit fly optimization algorithm; parameters inversion

1. Introduction

China is the second largest economy and the largest developing country in the world, and consumes a huge amount of energy. As a non-renewable resource, coal provides an important energy carrier for industrial production and social life in China. Even though the demand ratio of coal decreases due to the gradual use of the green energy, it still plays an important role in the structure of world energy [1]. According to the China Renewable Energy and Sustainable Energy Development Report, the proportion of coal will not fall below 50% by 2050. In China, the distribution of coal resources in the west such as Shanxi, Shaanxi, and Inner Mongolia is greater than that in the east. What's more, the coal in the east is being gradually depleted, making the amount of mining in the west increase.

Coal mining has effectiveness in gaining energy for economic construction, but it would threaten human safety and social stability by causing surface collapse, landslides, destruction of agricultural cultivated land, ecological damage [2,3], especially in the west of China due to the fragile ecological environment [4–6], surface damage under the influence of mining is mainly characterized by mining-induced fractures, as shown in Figure 1. For surface cracks caused by coal mining, Yang found that the ground cracks were formed ahead of the working face and developed rapidly during the period of the breakage of the immediate roof by field monitoring methods [7]. By means of field measurement and similar material simulation experiments, Zhou found that ground cracks are caused by strata deformation, but the formation of ground cracks can promote the deformation of strata. They also found that the horizontal tension deformation is the most stable factor for predicting ground cracks regardless of excavation speed [8]. By means of unmanned aerial vehicles, ground penetrating radar, and real-time kinematic and manual measurements, He and Zhao found that ultra large mining height longwall panel extraction causes a greater ground subsidence and horizontal deformation than those of the conventions, resulting in more severe damage in the permanent margin area [9]. Lian used field investigation, theoretical analysis and the particle flow code 2D numerical simulation method to study the movement of overburdened and discontinuous ground deformation of mining areas with a thick loess layer and a thin bedrock [10]. These adverse effects are closely related with the surface movements and discontinuous deformation caused by underground mining in mountainous areas. Therefore, it is important to study the law of surface movement in mountainous mining areas according to the mining subsidence theory [11], so that some necessary actions can be taken to protect the environment based on the precise movement laws.

(a) (b)

Figure 1. Mining subsidence surface damage: (**a**) Step crack; (**b**) Collapse pit.

Mining subsidence prediction is the core of mining subsidence theory, and includes theoretical methods, empirical methods, and experimental methods. The theoretical methods are elastoplastic theory, stochastic medium theory and the numerical calculation method, which have complex formulas and too many parameters related to the nature of rock strata which are not practical for engineering. The empirical methods need too much measured data and the method is useful for a typical mine; they do not have generality. The experimental methods mainly include similar material simulation experiments which are expensive and time-consuming. Compared with the methods above, the probability integral method is an official prediction method in China for its simpler mathematical expressions and clear-meaning parameters. It was first introduced to the strata movement research by the Polish scholar J. Litwiniszyn in the 1950s. Later, it was developed by Chinese scholars Liu Baochen and Liao Guohua. The probability integral method is based on the characteristics of horizontal and homogeneous media. It is accurate to predict coal mining whose coal seam dip angle is less than 5° in plain areas [12,13].

Due to the inherent landforms of mountainous areas, the surface movement and deformation caused by underground coal mining includes continuous deformation such as cracks and landslides, causing the situation to be very complicated [14,15]. The probability integral method is lacking in the prediction accuracy in mountainous areas. In this regard, further research on the prediction model of surface movement in mountainous mining areas is essential.

In the past decades, considerable research efforts have been devoted to the research on the prediction model of mining subsidence in mountainous areas. He and Kang analyzed the measured data of multiple ground movement observation lines in the mountainous areas represented by the Yangquan mining area in Shanxi, China. They compared data with the data in the plain areas and concluded that movement and deformation of the surface are related to the inclination direction of the surface and the nature of the topsoil [16,17]; then, the movement of the mountainous surface is considered to be the superposition of the mining influence component and the mining slip component from the perspective of the vector [18]. The "slide influence function method" and "stress-strain" model is established for predicting the surface movement in mountainous areas; mathematical analysis and mechanical derivation were then conducted to build the relationship between movement, deformation and topographic features caused by the mining slips. Finally, they proposed a mathematical model for predicting surface movement and deformation in mountainous areas that it is suitable for horizontal or near-horizontal coal seam mining with the average slope angle of surface less than 30° [19]. However, the existing mining subsidence prediction model in mountainous areas still has some shortcomings, and surface cracks caused by mining mean that the predicted results have a large discrepancy with the measured data, as the mechanism for the influence of topography on the laws of surface movement and deformation in mountainous areas have not been studied clearly.

At the same time, there are a lot of parameters in a prediction model, and the accuracy of prediction results also depends on the rationality of the prediction parameters. For the acquisition of parameters in a model, conventional methods are susceptible to problems such as parameter divergence and local optimization. The problem of obtaining the expected parameters is still to be solved. A typology of the parameters inversion methods about the probability integral method is commonly summarized as two sets of algorithms, the traditional one and the intelligent optimization method. The tradition methods include the linear least square and pattern search method, which have good accuracy for the linear model. However, when facing the nonlinear model like the probability integral method studied in this paper, there are problems such as high initial value requirements, divergence during the solution, and easy to be in the local optimal solution mainly because of the correlation between parameters. The intelligent optimization methods, including genetic algorithm, particle swarm optimization (PSO), and simulated annealing algorithm (SAA) provide effective solutions in the non-linear model since they partially overcome the weakness of traditional methods in divergence. Zha and Li demonstrated these advantages of the GA by comparing with the performance of the least square and pattern search method in parameters inversion in the probability integral method under complex conditions [20, 21]. These advantages were also verified with other intelligent methods, including PSO [22], SAA [23], quantum annealing [24–26], and the fireworks algorithm [27], The invasive weed optimization [28].

Considering that problems still exist in the classic intelligent optimization algorithms, such as slow convergence speed, a large amount of calculation, and easy to fall into local optimum, it is necessary to innovate a new method in parameter inversion. The fruit fly optimization algorithm (FOA), as one of the effective population-based metaheuristic approaches proposed by Pan [29–31], has been successfully applied in many regions such as stochastic multi-objective integrated disassembly-reprocessing reassembly scheduling [32], forecasting and optimizing agrobacterium-mediated genetic transformation [33], solving the multi-functional heterogeneous UAV cooperative mission planning problem [34], and short-term traffic flow prediction [35], while fewer scholars connect the FOA with the parameters inversion in mining subsidence.

To improve the performance of the FOA associated with the initial fruit fly swarm location, the population size, step length. In order to increase the universality of the FOA, scholars have proposed many improved methods to make the algorithm improve in terms of the computational complexity, stability and accuracy [30,36–40]. For the step length, if it is large, the global optimization performance is improved, the local optimization performance is declined, and vice versa. The constant step length makes the FOA unable to converge to an optimal solution rapidly. In this paper, a dynamic step is proposed based on the FOA to improve the performance of the algorithm. The search step in each iteration is related to the ratio of the best smell concentration value of the current and previous generations.

Accurate prediction of the movement and deformation of mining subsidence in mountainous areas facilitates the prevention and control of geological disasters, therefore, it is necessary to build a mining subsidence prediction model in mountainous areas based on the probability integral method, taking the mountain topography (slope angle $\alpha < 30°$) and geological conditions (loess covered) into account. Simultaneously, it is important to select an effective parameters inversion method of the probability integral method; the prediction accuracy of mining subsidence prediction model in mountainous areas can be enhanced with the two aspects above. In this study, a practical prediction model in mountainous areas is proposed, and the probability integral method and slopes slip analysis are combined in the proposed model. The intelligent optimization algorithm and mechanics analysis are applied for the determination of parameters in the proposed model, the acquisition of parameters is more efficient, and the values of parameters are more objective and reasonable. The contributions of this study can be summarized as follows: (1) the theoretical model considering slopes slip (slip subsidence ΔW, slip horizontal movement ΔU) in mountainous areas based on the probability integral method is established, the effect of the influence of topography is considered in this model combined with the slope angle and nature of topsoil; (2) The more accurate parameters of the probability integral method in the proposed model are obtained by the algorithm of parameters inversion which is the dynamic step fruit fly optimization algorithm (DSFOA); and (3) The other parameters about slopes slip are determined by mechanics analysis based on the nature of soil, which is more objective and reasonable.

2. Methodology

2.1. Probability Integral Method

The probability integral method is based on the stochastic medium theory, which was first introduced to strata movement research by the Polish scholar J. Litwiniszyn in the 1950s. Later, it was developed by Chinese scholars Liu Baochen and Liao Guohua et al. into a probability integral method [41].

The movement law of the stochastic medium theory can be abstracted into the theoretical model shown in Figure 2. The granular medium can be comprehended as a small and uniform ball, assuming that the first layer of ball C is taken away, then the probability of the second layer of two balls A and B falling into the space is 1/2 separately, and so on the probability of the movement of balls in each layer. The probability distribution fitting a curve can be seen to be close to the standard normal distribution probability density curve.

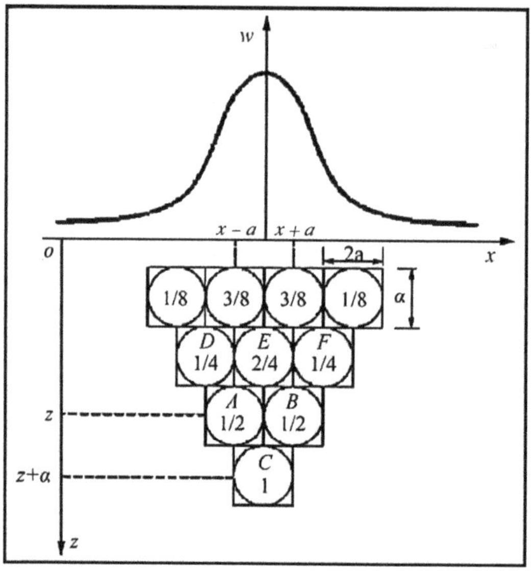

Figure 2. Principle of the probability integral model.

Referring to Figure 3a, considering the two-dimensional condition, surface subsidence caused by underground unit mining $W_e(x)$ is shown in Equation (1):

$$W_e(x) = \frac{1}{r} \cdot e^{-\pi \frac{x^2}{r^2}} \tag{1}$$

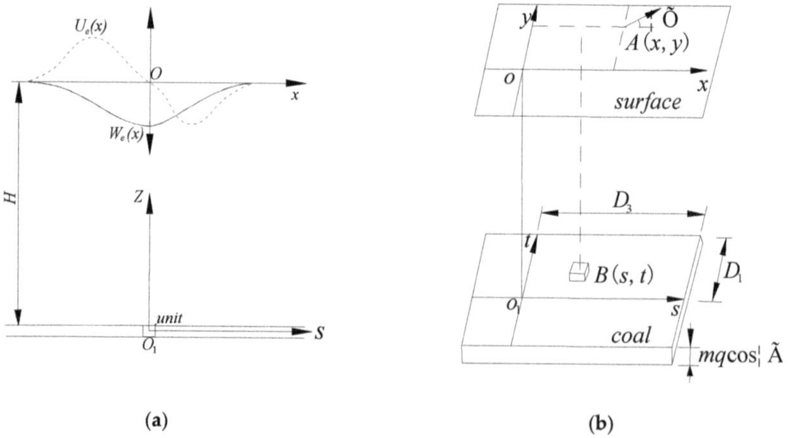

Figure 3. (**a**) Unit mining subsidence and unit horizontal movement; (**b**) Three-dimensional coordinate system.

Considering that the assumption of the volume is constant, shown in Equation (2), surface horizontal movement caused by underground unit mining $U_e(x)$ is shown is shown in Equation (3):

$$e = \varepsilon_x + \varepsilon_z = \frac{\partial U_e(x,z)}{\partial x} + (-\frac{\partial W_e(x,z)}{\partial z}) = 0 \tag{2}$$

$$U_e(x) = B \cdot \frac{dW_e(x)}{dx} = b \cdot r \cdot i(x) \tag{3}$$

where ε_x is the linear strain along the x direction, ε_z is the linear strain along the z direction. Minus "−" represents the w-axis is opposite to the Z-axis. $B = b{\cdot}r$, b is the horizontal movement coefficient, r is the major influence radius(m), $i(x)$ is the inclination.

As shown in Figure 3b, assuming that the coal seam is inclined, the calculation formula of subsidence value of surface point A(x,y) caused by mining unit B(s,t) is shown in Equation (4):

$$W_e(x,y) = \frac{1}{r^2} e^{-\pi \frac{(x-s)^2+(y-t+H\cdot\cot\theta_0)^2}{r^2}} \tag{4}$$

where r is the major influence radius (m); θ_0 is the mining influence propagation angle (°); H is the average mining depth (m).

According to the superposition principle, the subsidence of a point A caused by the whole mining area is:

$$W(x,y) = w_0 \iint_R W_e(x,y)dsdt \tag{5}$$

where $w_0 = mq\cos\gamma$ is the maximum surface subsidence value under the critical extraction, m is the thickness of coal seam (m), q is the subsidence coefficient, and γ is the average dip angle of coal seam (°), R represents the mining region.

For the rectangular working panel, the surface subsidence $W(x,y)$ is:

$$W(x,y) = w_0 \int_0^{D_3}\int_0^{D_1} W_e(x,y)dsdt = \frac{1}{w_0} \cdot w^0(x) \cdot w^0(y) \tag{6}$$

$$w^0(x) = \frac{w_0}{2} \cdot \left\{ erf(\frac{\sqrt{\pi}\tan\beta}{H}x) - erf[\frac{\sqrt{\pi}\tan\beta}{H}(x-l)] \right\} \tag{7}$$

$$w^0(y) = \frac{w_0}{2} \cdot \left\{ erf(\frac{\sqrt{\pi}\tan\beta}{H}y) - erf[\frac{\sqrt{\pi}\tan\beta}{H}(y-L)] \right\} \tag{8}$$

$$l = D_3 - s_3 - s_4 \tag{9}$$

$$L = (D_1 - s_1 - s_2)\frac{\sin(\theta_0+\alpha)}{\sin\theta_0} \tag{10}$$

$$r = \frac{H}{\tan\beta} \tag{11}$$

where D_1 and D_3 are the inclination and strike lengths of the working panel; s_1, s_2, s_3, and s_4 are the inflection point offsets of downhill, uphill, left, and right borders; tanβ is the main influence tangent.

The inclination $i(x,y,\varphi)$ of a point A(x,y) on the surface in φ direction (the angle that the positive x-axis counterclockwise to the specified direction, in Figure 3b is as follows:

$$i(x,y,\varphi) = \frac{\partial W(x,y)}{\partial x}\cos\varphi + \frac{\partial W(x,y)}{\partial y}\sin\varphi \tag{12}$$

For the horizontal movement $U(x,y,\varphi)$, the curvature $K(x,y,\varphi)$ and horizontal deformation $\varepsilon(x,y,\varphi)$ of a point A(x,y) in φ direction can be obtained by Equations (13)–(15).

$$U(x,y,\varphi) = bri(x,y,\varphi) \tag{13}$$

$$K(x,y,\varphi) = \frac{\partial i(x,y,\varphi)}{\partial x}\cos\varphi + \frac{\partial i(x,y,\varphi)}{\partial y}\sin\varphi \tag{14}$$

$$\varepsilon(x,y,\varphi) = brK(x,y,\varphi) \tag{15}$$

2.2. Fruit Fly Optimization Algorithm

The fruit fly optimization algorithm (FOA) is a new swarm intelligent optimization algorithm proposed by professor Pan Wen-Tsao based on the cooperative and competitive behavior of fruit flies in the food finding process.

The fruit fly itself is superior to other species in sensing and perception, especially in osphresis and vision. According to the process of fruit flies foraging through the senses, the basic principle of the fruit fly optimization algorithm [30] follows by two steps:

First is the rough positioning stage by osphresis: they judge the approximate position of the food based on scents floating in the air and fly towards the target.

Second is the precise positioning stage by vision: when fruit flies fly around the food and are able to visualize within the search range, they can accurately determine the specific location of the food, and they finally fly to the food.

The detail processes are considered as follows:

Step 1: Initialize parameters. First, the initial fruit fly swarm location (X_{-axis}, Y_{-axis}), the maximum iteration number (*maxgen*), the population size (*sizepop*), the random direction *rand*( ) and step length *k* should be considered.

$$\begin{cases} X_i = X_{-axis} + 2k \cdot rand() - k \\ Y_i = Y_{-axis} + 2k \cdot rand() - k \end{cases} \tag{16}$$

Step 2: Evolution starting. The generation = 0, the random flight path and the route for food finding of a single fruit fly are considered.

Step 3: Preliminary computations. The specific direction of the food is determined by calculating the distance $Dist_i$ of the individual fruit fly from the initial position. Subsequently, smell concentration judgment value S_i is determined. Then, the fitness function value (also called the smell concentration $Smell_i$) is calculated.

$$\begin{cases} Dist_i = \sqrt{(X_i - X_{-axis})^2 + (Y_i - Y_{-axis})^2} \\ S_i = 1/Dist_i \end{cases} \tag{17}$$

$$Smell_i = Function(S_i) \tag{18}$$

Step 4: Record the optimal individual and the location. Compare the smell concentration values of fruit flies in the current population and mark the fruit fly individual with the highest (or lowest) smell concentration as the optimal fruit fly individual. Record and keep the best smell concentration value *bestSmell* and the coordinates position (*X*, *Y*) of the optimal fruit fly individual. The fruit fly swarm will use vision to fly towards that position:

$$[bestSmell, bestIndex] = \min/\max(Smell_i) \tag{19}$$

$$\begin{aligned} Smellbest &= bestSmell \\ X_{-axis} &= X(bestIndex) \\ Y_{-axis} &= Y(bestIndex) \end{aligned} \tag{20}$$

Step 5: Iteration stops. Consider generation = generation + 1, the fitness function value is determined again. Judge if the smell concentration is superior to the previous iterative smell concentration. When the generation attains the maximum iteration number, and the optimized parameter value of the specific model can be reached. Otherwise, the optimization process should go back to Step 2.

2.3. Dynamic Step Fruit Fly Optimization Algorithm

The performance of the FOA is associated with the initial fruit fly swarm location, the population size, and step length. In order to increase the universality of the FOA, scholars have proposed many improved methods to make the algorithm improve in terms of computational complexity, stability and accuracy [31,37–41]. For the step length, if it is large, the global optimization performance is improved, the local optimization performance is declined, and vice versa. The constant step length makes the FOA unable to converge to an optimal solution rapidly. In this paper, the dynamic step fruit fly optimization algorithm (DSFOA) was applied for parameters inversion in the proposed prediction model to improve the performance of the algorithm. The search step in each iteration is related to the ratio of the best smell concentration value of the current and previous generations. The improved search step of the FOA is as follows (take the minimum smell concentration as an example).

Let g be the number of iterations, k_g be the search step length of the gth iteration, S'_g be the optimal smell concentration value of the gth iteration, $\Delta = |S'_g—S'_{g-1}|$;

When $g = 1$, k_1, S'_1 is calculated;

When $g \geq 2$, if $S'_g < S'_{g-1}$, it indicates that the current optimal smell concentration value is better than the previous generation optimal smell concentration value, the step should be shorted to improve the optimization accuracy, then

$$k_{g+1} = k_g(1 - \frac{S'_g}{S'_{g-1}}) \tag{21}$$

if $S'_g \geq S'_{g-1}$, it indicates that the optimal smell concentration value of the previous generation is better than or equal to the current optimal smell concentration value. The step should be enlarged to increase the search range and improve the global search ability.

$$k_{g+1} = k_g(1 + \frac{S'_{g-1}}{S'_g}) \tag{22}$$

The horizontal movement U is mainly affected by topography in mountainous areas, the distribution law of the horizontal movement is relatively complicated, the law of subsidence is relatively simple and clear due to its unvaried vertical downward direction. The prediction parameters q, $\tan\beta$, θ_0, s_i (i = 1, 2, 3, 4) related to the subsidence values were inversed first, and the horizontal movement coefficient b related to the horizontal movement U was obtained separately.

Set $p = [q\ tan\beta\ \theta_0\ s_1\ s_2\ s_3\ s_4]$, P is the search space of vector p (the value range of the prediction parameters), and the optimization principle which is used as the fitness function(f_p) is the minimum mean square error (MSE) between fitting values of the probability integral method whose parameters are obtained by DSFOA and the measured data of the corresponding n points. f_p is shown in Equation (23):

$$\min f_p = \sqrt{\frac{\sum\limits_{i=1}^{n} (W_i - w_i)^2}{n}} \tag{23}$$

where W_i is the measured subsidence values, w_i is the fitting subsidence values, and n is the number of observation points.

Set the search range and step length of the parameter vector p. The procedure of prediction parameters inversion of the probability integral method based on the subsidence values is as follows:

(1) Set the parameters such as the initial populations number and the maximum iteration number, and generate the initial population according to the range of the probability integral method parameter vector p. The position of initialization population is the initial solution of the N-dimensional vector, and N is the number of parameters optimized (N = 7).
(2) Calculate the fitness function value, combine the information of the working panel to the mining subsidence of the probability integral method and update the step.
(3) In each iteration process, judge and select the population with the optimal fitness value, and loop until the optimal solution is found.

Then, the horizontal movement coefficient b is inversed, and the fitness function(f_p) as in Equation (24):

$$\min f_p = \sqrt{\frac{\sum\limits_{i=1}^{n} (U_i - u_i)^2}{n}} \tag{24}$$

where U_i is the measured horizontal movement values, u_i is the fitting horizontal movement values, and n is the number of observation points.

2.4. *Prediction Model of Mining Subsidence in Mountainous Areas Considering Slopes Slip*

The surface of the mountainous areas affected by mining will produce discontinuous deformation, especially ground fissures. During the development of the surface movement, slopes slip can occur easily, which has an obvious impact on the law of surface movement and deformation in mountainous areas.

The formation mechanism of tensile mining fissures and slopes slip are combined to analyze a prediction model of mining subsidence in mountainous areas considering that slopes slip is established in this section.

2.4.1. Expression of Movement Vector and Slopes Slip Vector

The reason why the law of surface movement and deformation in mountainous areas is different from that of plain areas is mainly due to the effect of the additional slopes slip caused by topography. In order to explore the law of mining subsidence in mountainous areas, the movement vector is introduced, as shown in the right of Figure 4.

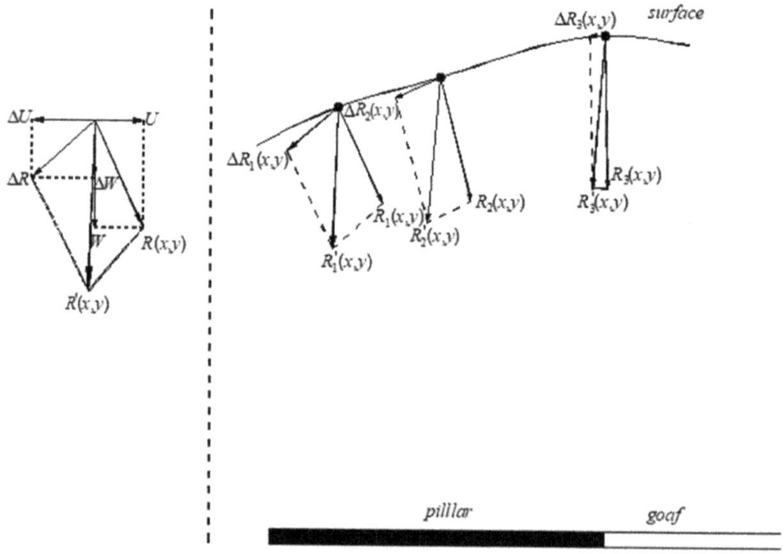

Figure 4. Movement vector and slopes slip vector.

The movement vector $R'_i(x,y)$ at a point P_i on the mountainous surfaces under the influence of coal mining can be regarded as the sum of the movement vector $R_i(x,y)$ produced by mining in the plain areas and slopes slip vector $\Delta R_i(x,y)$ on the mountainous surface. It can be expressed as the superposition principle shown in Equation (25):

$$\overrightarrow{R'(x,y)} = \overrightarrow{R(x,y)} + \overrightarrow{\Delta R(x,y)} \tag{25}$$

where:

$$\overrightarrow{R'(x,y)} = \overrightarrow{W'(x,y)} + \overrightarrow{U'(x,y)} \tag{26}$$

$$\overrightarrow{R(x,y)} = \overrightarrow{W(x,y)} + \overrightarrow{U(x,y)} \tag{27}$$

$$\overrightarrow{\Delta R(x,y)} = \overrightarrow{\Delta W(x,y)} + \overrightarrow{\Delta U(x,y)} \tag{28}$$

In the left part of Figure 4, $W'(x,y)$ and $U'(x,y)$ represent the measured subsidences and horizontal movements, $W(x,y)$ and $U(x,y)$ represent the subsidence and horizontal movements of plain areas under similar geological and mining conditions which can be evaluated with the probability integral method, $\Delta W(x,y)$ and $\Delta U(x,y)$ represent the slip subsidence and slip horizontal movement affected by mountain topography, which are deduced in Section 2.4.2.

2.4.2. Establishment of the Mining Subsidence Prediction Model

In this section, the slip subsidence $\Delta W(x,y)$ and the slip horizontal movement $\Delta U(x,y)$ are deduced based on the subsidence $W(x,y)$, the horizontal movement $U(x,y)$,the inclination $i(x,y,\varphi)$ and horizontal deformation $\varepsilon(x,y,\varphi)$ in the probability integral method, then the mining subsidence prediction model in mountainous areas is established based on the $W(x,y)$, $U(x,y)$, $\Delta W(x,y)$ and $\Delta U(x,y)$. According to Equation (29):

$$R(x,y) = \sqrt{W^2(x,y) + U^2(x,y)} \tag{29}$$

Since the slip vector $\Delta R(x,y)$ is the additional slip caused by the mining vector $R(x,y)$, the component of the mining vector $R_\alpha(x,y)$ along the inclination direction of the surface slope determines the magnitude of the slip vector $\Delta R(x,y)$ (Figure 5a).

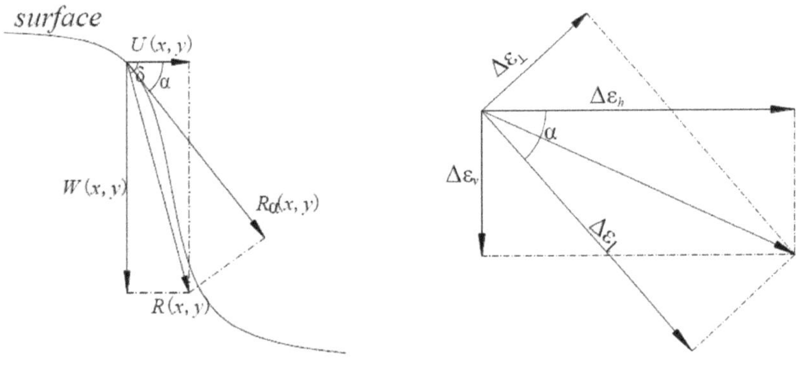

(a) (b)

Figure 5. (**a**) Slopes slip vector analysis; (**b**) Deformation of the slip vector along the l direction.

From Figure 5a, we can deduce:

$$\Delta R(x,y) = K_S \cdot R_\alpha(x,y) = K_S \cdot R(x,y)\cos(\delta - \alpha) \tag{30}$$

where Ks is the surface characteristic coefficient, which is related to the topography and surface soil properties, δ is the angle between the mining vector $R(x,y)$and the horizontal movement component $U(x,y)$, and its value is

$\delta = \tan^{-1}(W(x,y)/U(x,y))$, α is the slope angle of the surface.

Square both sides of the equation and make the partial derivatives of l on both sides:

$$\Delta R(x,y) \cdot \frac{\partial \Delta R(x,y)}{\partial l} = K_S{}^2 \cdot \cos^2(\delta - \alpha) \cdot R(x,y) \cdot \frac{\partial R(x,y)}{\partial l} \tag{31}$$

where l is the length along the inclination direction of the slope.

As $\partial l = \partial x/\cos\alpha$, from Equations (30) and (31), we have:

$$\frac{\partial \Delta R(x,y)}{\partial l} = K_S \cdot \cos(\delta - \alpha) \cdot \cos\alpha \cdot \frac{\partial R(x,y)}{\partial x} \tag{32}$$

Square both sides of the Equation (29) and make the partial derivative of x:

$$\frac{\partial R(x,y)}{\partial x} = \frac{W(x,y)}{R(x,y)} \cdot \frac{\partial W}{\partial x} + \frac{U(x,y)_\varphi}{R(x,y)} \cdot \frac{\partial U}{\partial x} \tag{33}$$

According to the probability integral method, $\partial W/\partial x = i(x,y)_x$, $\partial U/\partial x = \varepsilon(x,y)_x$, $\varphi = 0°$.

$$\frac{\partial R}{\partial x} = \frac{W(x,y)}{R(x,y)} \cdot i(x,y)_x + \frac{U(x,y)_\varphi}{R(x,y)} \cdot \varepsilon(x,y)_x \tag{34}$$

For $W(x,y)/R(x,y) = \sin\delta$, $U(x,y)/R(x,y) = \cos\delta$, we have:

$$\frac{\partial \Delta R}{\partial l} = K_S \cdot \cos(\delta - \alpha) \cdot \cos\alpha[\sin\delta \cdot i(x,y)_x + \cos\delta \cdot \varepsilon(x,y)_x] \tag{35}$$

We suppose $\Delta\varepsilon_l = \partial\Delta R/\partial l$, it is the deformation of the slip vector along the l direction.

According to the concept of the Poisson's ratio, the deformation along the direction perpendicular to the slope direction is:

$$\Delta\varepsilon_\perp = \mu \cdot \Delta\varepsilon_l \tag{36}$$

where μ is Poisson's ratio for the topsoil.

In Figure 5b, we have Equations (37) and (38):

$$\Delta\varepsilon_h = \Delta\varepsilon_l \cos\alpha + \Delta\varepsilon_\perp \sin\alpha = \Delta\varepsilon_l(\cos\alpha + \mu \cdot \sin\alpha) \tag{37}$$

$$\Delta\varepsilon_v = \Delta\varepsilon_l \sin\alpha - \Delta\varepsilon_\perp \cos\alpha = \Delta\varepsilon_l(\sin\alpha - \mu \cdot \cos\alpha) \tag{38}$$

Considering Equations (37) and (38), slip horizontal movement ΔU and slip subsidence ΔW are shown below:

$$\Delta U = \int \Delta\varepsilon_h dx = K_S \cdot \cos(\delta - \alpha) \cdot \cos\alpha[\sin\delta \cdot W(x,y) + \cos\delta \cdot U(x,y)_\varphi](\cos\alpha + \mu \cdot \sin\alpha) \tag{39}$$

$$\Delta W = \int \Delta\varepsilon_v dx = K_S \cdot \cos(\delta - \alpha) \cdot \cos\alpha[\sin\delta \cdot W(x,y) + \cos\delta \cdot U(x,y)_\varphi](\sin\alpha - \mu \cdot \cos\alpha) \tag{40}$$

2.4.3. Calculation of the Values of the Parameters in the Mining Subsidence Model

In Equations (39) and (40), parameters δ and K_s should be determined, as they are strongly associated with the geological and mining conditions and the nature of the topsoil. Parameters δ and K_s are important in the proposed mining subsidence model in mountainous areas, and they are discussed in this section.

Values about δ:

When the slope is shown in Figure 6a,c, $\delta = \tan^{-1}(W(x,y)/U(x,y))$; when the slope is shown in Figure 6b,d, $\delta = 180° + \tan^{-1}(W(x,y)/U(x,y))$.

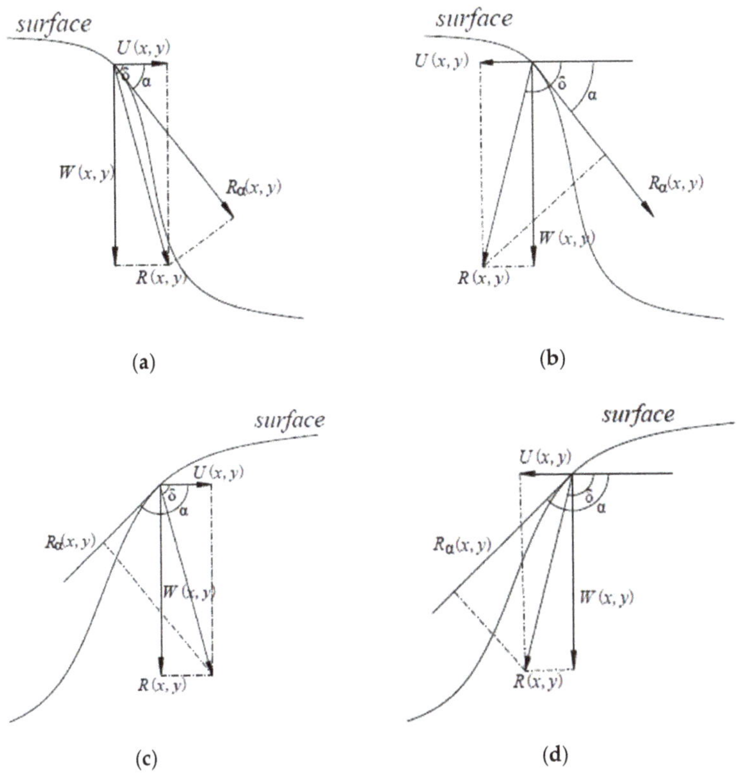

Figure 6. The computational diagram of different Slope directions.

If the direction of $W(x,y)$ is downward, $W(x,y) > 0$; if the direction of $U(x,y)$ is directed to the goaf, $U(x,y) > 0$, vice versa.

Values about surface characteristic coefficient K_s:

The surface soil cannot withstand tensile stress, therefore, when a small tensile stress is generated on the mountain surface under the influence of mining, then the cracked surface slope block will slip.

As shown in Figure 7a, the slip amount of the slope is analyzed.

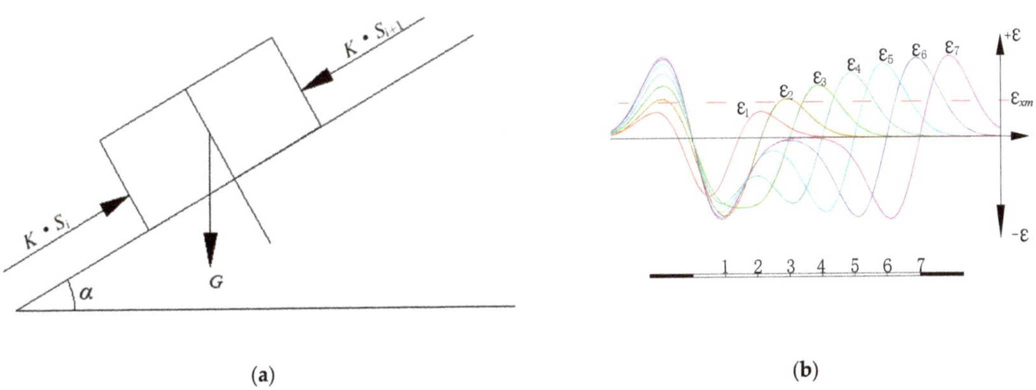

Figure 7. (**a**) Analysis of a unit block slips; (**b**) Variation of horizontal deformation.

The surface characteristic coefficient K_S reveals the influence degree of the topography on the amount of slip, which is mainly related to the topography and the nature of the soil.

Select a unit block i, the length along the slope direction is Δl, the height along the slope normal direction is Δh, the slip amount of the ith block is S_i, Suppose that ith block undergoes resistance from the $(i-1)$th block below itself when it slips and the resistance is proportional to the amount of ith block slips, which refers KS_i, and it also bears the force generated by the slip of the upper $(i+1)$th block, namely Ks_{i+1}. The block tends to balance after it slips:

$$G \cdot \sin\alpha + K \cdot S_{i+1} \cdot \Delta h - G\cos\alpha\tan\varphi - K \cdot S_i \cdot \Delta h = 0 \tag{41}$$

In Equation (41), K is the compressive modulus of the topsoil, φ is the internal friction angle of the soil, G is the weight of the unit block, $G = \rho g \cdot \Delta l \cdot \Delta h$, ρ is the natural density of the topsoil, g is the gravitational acceleration.

Let $\Delta S = S_{i+1} - S_i$:

$$\Delta\varepsilon_l = \frac{\Delta S}{\Delta l} = \frac{\rho g(\cos\alpha \cdot \tan\varphi - \sin\alpha)}{K} \tag{42}$$

Consider Equation (42):

$$K_S = \frac{\rho g(\cos\alpha \cdot \tan\varphi - \sin\alpha)}{K \cdot \cos(\delta-\alpha) \cdot \cos\alpha[\sin\delta \cdot i(x,y)_x + \cos\delta \cdot \varepsilon(x,y)_x]} \tag{43}$$

The value of $\varepsilon(x,y)_x$ in Equation (43) is the critical horizontal deformation value when the slope starts slipping, shown in Equation (44) [42].

$$\varepsilon_{xm} = \frac{2C}{E} \cdot \tan(45^\circ - \frac{\varphi}{2}) \tag{44}$$

During the process of the coal mining, if the tensile deformation of a point does not exceed ε_{xm} or the horizontal deformation is always compression deformation, the maximum of the horizontal deformation is selected, as shown in Figure 7b.

According to the probability integral method, from Figure 7b, when the coal unit 1 is mined, the curve of horizontal deformation is shown in ε_1, when the coal unit 1 and the coal unit 2 are mined, the curve of horizontal deformation is shown in ε_2, and the other coal units are similar. When the horizontal deformation of a point on the surface exceeds ε_{xm} for the first time during the process of coal mining, the slip subsidence $\Delta W(x, y)$ and the slip horizontal movement $\Delta U(x,y)$ are calculated. When $\varepsilon(x,y)_x$ is determined, $i(x,y)_x$ is also determined.

3. Case Study

3.1. Study Area

The study coal mine working panel 22,101 is located in Taiyuan, Shanxi Province, China, the location of which is shown in Figure 8.

The surface above the working panel is high in the west and low in the east, with the elevation from +1290 to +1305 m. The terrain is relatively open, with a southwest dip angle of about 4°. The working panel adopted comprehensive mechanized mining and an all collapse method to manage the roofs. Its geological mining conditions were designed by a length along the strike D_3 = 440 m, and a length along the inclination D_1 = 140 m, the mining height m = 3.18 m, the dip angle γ = 1~3°, and the average mining depth H = 220 m. Two observation lines were laid along the strike of the main section and the inclination of the main section, respectively. Forty-five monitoring points were laid along the two observation lines shown in Figure 9, the image of a point is shown in Figure 10. Each monitoring point was filled with concrete and buried with prefabricated piles in the soil pit. The kind of data at these 45 points includes the plane position of each point (x,y) and the elevation of each point (h), traverse measurement was carried out according to the accuracy requirement of 5, "and triangulation elevation traverse was measured at the same time.

The Leica T402 total station instrument was used for the measurement of the 45 points. The nominal accuracy of angle measurement is 2″, and the distance measurement accuracy is 3 mm + 2 ppm.

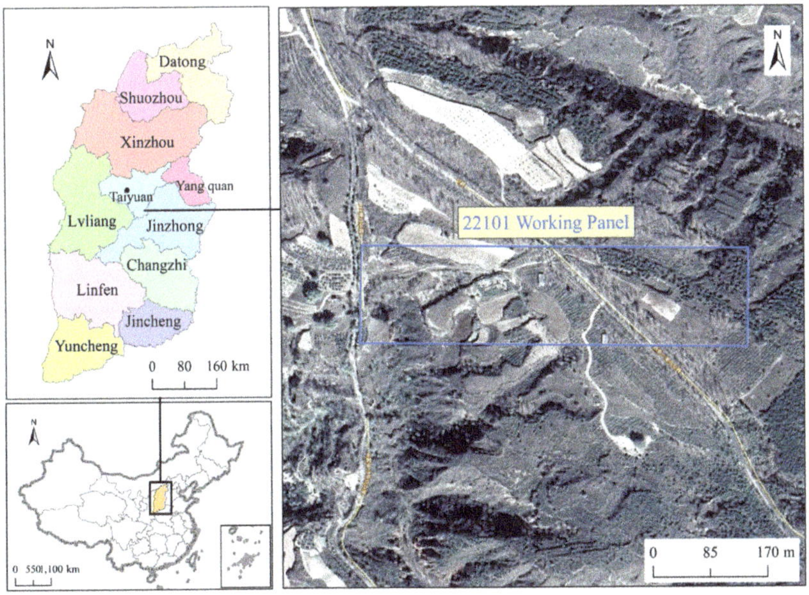

Figure 8. Location of the working panel 22,101.

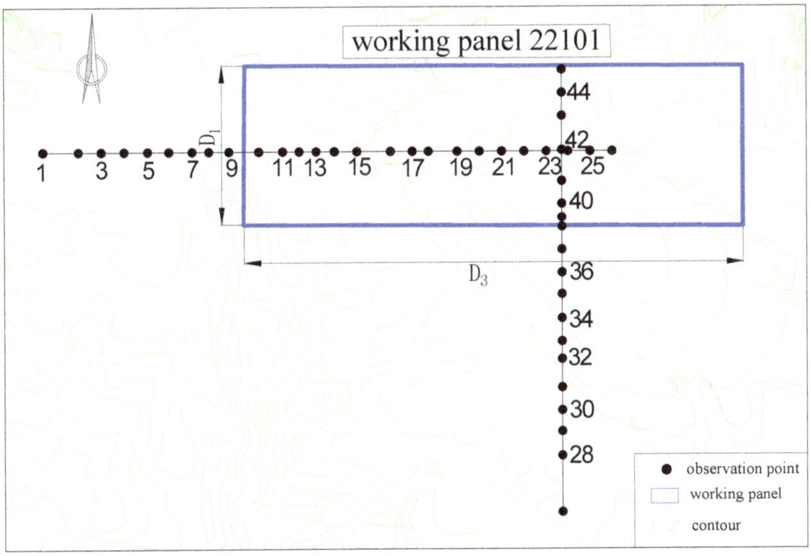

Figure 9. The Layout of working panel 22,101 and surface monitoring points.

Figure 10. A monitoring point of the study area.

In order to study the law of surface movement and deformation in mountainous areas according to the contours of the study area, the surface elevations along two observation lines are extracted based on the AutoCAD software. The ground surface section planes of two observation lines are shown in Figure 11.

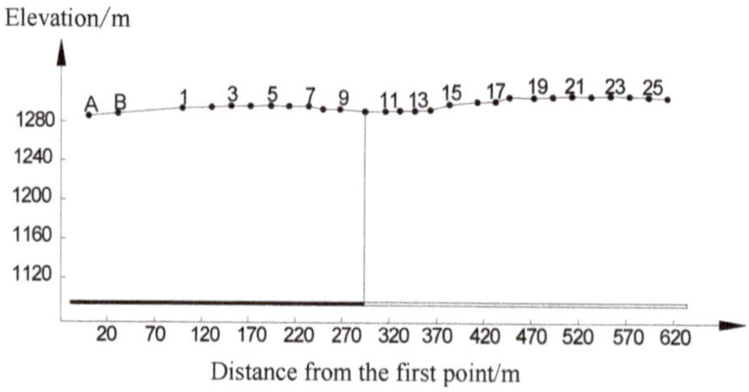

(a)

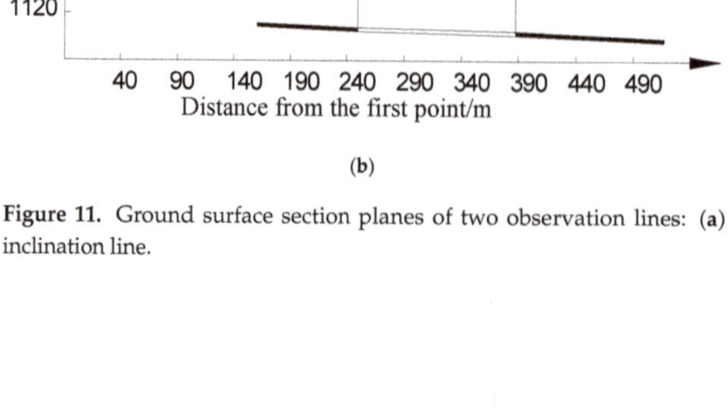

(b)

Figure 11. Ground surface section planes of two observation lines: (**a**) The strike line, (**b**) The inclination line.

3.2. Information of Topsoil

The upper layer of the working panel 22,101 is the Quaternary Loess of Malan. It is gray yellow sub clay, with large sand content, small tensile strength, with good water permeability. The parameters of geotechnical physical and mechanical experiments are shown in Table 1. The critical horizontal deformation value ε_{xm} *is* 3.2 mm/m.

Table 1. Parameters of geotechnical physical mechanics of topsoil.

Name	Natural Density(Kg/m^3)	Cohesion (KPa)	Poisson's Ratio	Elasticity Modulus (MPa)	Compression Modulus (MPa)	Internal Friction Angle (°)
Loess of Malan	1420	43	0.4	17.5	7	24

3.3. Analysis of the Results

3.3.1. Inversion Results of Prediction Parameters Based on DSFOA

The parameter ranges of the probability integral method about the study area are set as follows: the subsidence coefficient q is 0.3∼1, the main influence tangent tanβ is 1∼3, the mining influence propagation angle θ_0 is 80∼90°, the inflection point offsets s1, s2, s3, s4 are −30~30 m.

The initialization parameters of DSFOA are set as the population size *sizepop* 5, 10, and 20, the number of cycles is 50, 100, and 500. The step length of the subsidence coefficient q is 0.1, 0.2, the step length of the main influence angle tangent tanβ is 0.2, 0.3, the step length of the mining influence propagation angle θ_0 is 5, 8, and the inflection point offset s is 10, 20, 30. The value range of the horizontal movement coefficient b is given from 0.1 to 0.5, the population size *sizepop* is 5, 10, and 20, and the number of iterations is 50, 100, and 500. The step length is 0.1 and 0.2.

The prediction parameters of 22,101 working panel based on DSFOA are shown in Table 2. The prediction parameters inverted by traditional methods in Table 2 are obtained based on the empirical formulas offered by the mine company. Subsidence coefficient q = 0.79 ± 0.1, main influence angle tangent tanβ = 2.4, mining influence propagation angle θ_0 = 90°-$k_0\gamma$ (k_0 is the constant), the inflection point offsets $s_1 = s_2$ = 0.05 H, $s_3 = s_4$ = 0.1 H, b = 0.35.

Table 2. Comparison of the parameters inversion result.

Algorithm	q	tanβ	θ_0	s_1	s_2	s_3	s_4	MSE/mm
traditional method	0.79	2.4	88	−12	−12	−22	−22	248.40
DSFOA	0.59	2.19	83.19	−2.43	−2.43	7.57	7.57	122.27

The comparison between the fitting subsidence values and the measured subsidence value is shown in Figure 12.

After the parameters inversion of DSFOA, b is 0.42. The comparison between the fitting horizontal movement values and the measured horizontal movement values is shown in Figure 13.

The fitting mean square error of the horizontal movement for the traditional method is 397 mm, and for the fitting mean square error of DSFOA it is 386 mm.

When the mean square error of subsidence $MSE_1 \leq$ (0.05~0.1) W_{max} is satisfied, the mean square error of horizontal movement $MSE_2 \leq$ (0.1~0.3) U_{max} is satisfied, and it means that the distribution law of the movement and deformation revealed by the method is reliable and feasible [41]. According to the measured data, the maximum subsidence value is 1898 mm at No. 18 point, and the maximum horizontal movement value is 1324 mm at No. 40 point, (0.05~0.1). W_{max} ranges from 94.9 mm to 189.8 mm. (0.1~0.3) U_{max} ranges from 132.4 mm to 397.2 mm, and it indicates that the prediction parameters of the probability integral method inversed by DSFOA are reliable and effective.

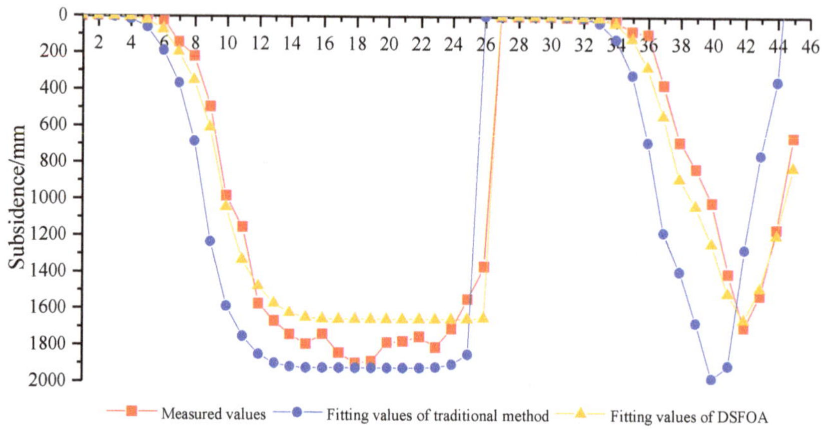

Figure 12. Comparison of the subsidence fitting values and measured values.

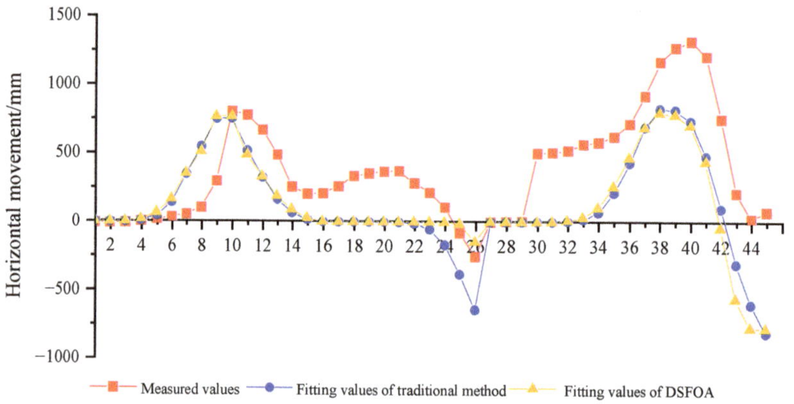

Figure 13. Comparison of the horizontal movement fitting values and measured values.

3.3.2. Prediction Results of the Proposed Model

According to the prediction model of mining subsidence in mountainous areas considering slopes slip based on the MATLAB program, first, the predictive subsidence W and the predictive horizontal movements U of 45 points are calculated, then the slip subsidence ΔW and slip horizontal movements ΔU of 45 points are evaluated ($W + \Delta W$) and ($U + \Delta U$) compared with the measured data W' and U,′ respectively, as shown in Figures 14 and 15.

The mean square error of the subsidence decreased from 122 mm to 71 mm, and the mean square error of the horizontal movement decreased from 386 mm to 276 mm. The applicability of the prediction model of mining subsidence in mountainous areas considering slopes slip is verified.

The effect of the subsidence prediction is perfect; for the horizontal movement prediction, the effect is improved, the effect of the horizontal movements on the strike line is superior to the horizontal movements on the tendency line, and we think the reason is the difference in the mining sufficiency degree of the two lines. For points from No. 7 point to No. 9 point, points from No. 30 point to No. 35 point, especially points from No. 43 point to No. 45 point, the effect of these points are not satisfied more or less, we consider for points from No. 7 point to No. 9 point and points from No. 43 point to No. 45 point, they are above the boundary of the gob, while for points from No. 30 point to No. 35 point, they are at the vicinity of surface movement boundary. The predicted values of horizontal movement are not accurate enough at the boundary due to the boundary effect of goaf.

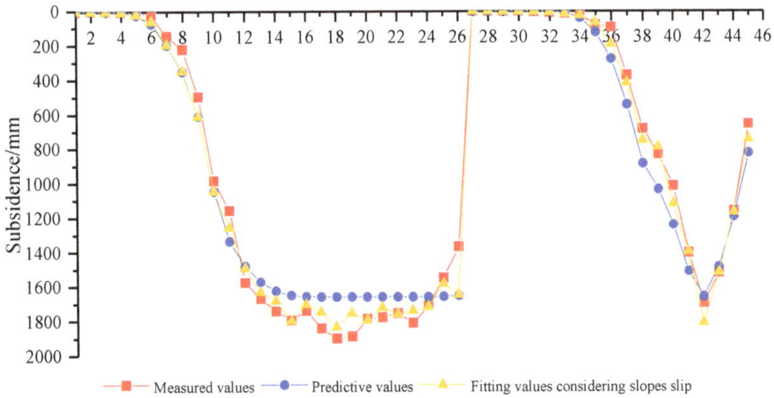

Figure 14. Comparison of the subsidence fitting values and measured values considering slopes slip.

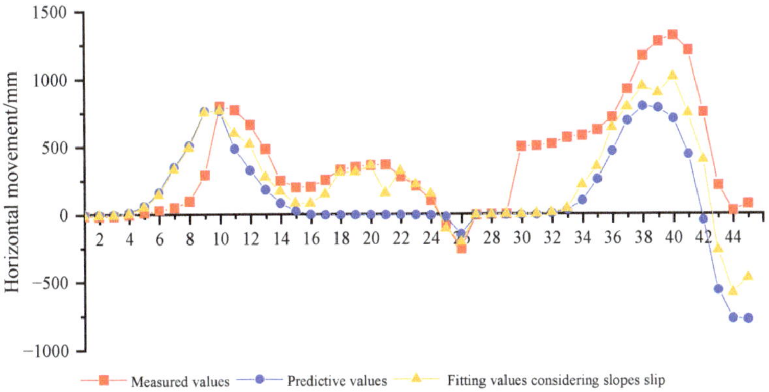

Figure 15. Comparison of the horizontal movements fitting values and measured values considering slopes slip.

4. Discussion

4.1. Prediction Results of the Existing Model in Mountainous Areas

According to Section 2.4, the reason that the law of the surface movements in mountainous mining areas is different from that in the flat areas is the superposed influences of the underground mining and the topography. Accurately predicting the surface subsidence can decrease the harmful effects of mining. In China, the existing prediction model widely used in mountainous regions was provided by He and Kang [14], as expressed in Equations (45) and (46).

$$\begin{aligned} W_s(x,y) = \; & W(x,y) + D_{x,y}\{P[x]\cos^2\varphi + P[y]\sin^2\phi \\ & + P[x]P[y]\sin^2\varphi\cos^2\varphi\tan^2\alpha_{x,y}\}W(x,y)\tan^2\alpha_{x,y} \end{aligned} \tag{45}$$

$$U_s(x,y)_\psi = U(x,y) + |D_{x,y}|W(x,y)\{P[x]\cos\psi\cos\varphi + P[y]\sin\psi\sin\varphi\}tg\alpha_{x,y} \tag{46}$$

where $W(x,y)$ and $U(x,y)$ are the subsidence and horizontal movement of plain areas under similar geological and mining conditions which can be evaluated with the probability integral method. $D_{x,y}$ is a coefficient reflecting the surface feature, the value of which can be obtained from Table 3. φ is the prediction direction for surface displacement, it is the angle that the positive x-axis counterclockwise to the specified direction shown in Figure 1, $\alpha_{x,y}$ is the slope angle, ψ is the angle between the slope direction of the ground surface and the main section, $P[x]$ and $P[y]$ are the mining landslide influence functions, as expressed in Equations (47) and (48).

$$P[x] = 1 + A \cdot [e^{-\frac{1}{2}(\frac{x}{r}+P)^2} + e^{-\frac{1}{2}(\frac{l-x}{r}+P)^2}] + W_m \cdot [e^{-t(\frac{x}{r}+P)^2} + e^{-t(\frac{l-x}{r}+P)^2}] \quad (47)$$

$$P[y] = 1 + A \cdot [e^{-\frac{1}{2}(\frac{y}{r}+P)^2} + e^{-\frac{1}{2}(\frac{L-y}{r}+P)^2}] + W_m \cdot [e^{-t(\frac{y}{r}+P)^2} + e^{-t(\frac{L-y}{r}+P)^2}] \quad (48)$$

where W_m is the max value of surface subsidence. A, P and t are the coefficients of landslide effect, with reference values of 2π, 2 and π, respectively. The other parameters are the same as the parameters in the probability integral method.

Table 3. Surface feature coefficient $D_{x,y}$ in the mountain areas.

Features of the Topsoil and Ground Vegetation	Feature Coefficient $D_{x,y}$	
	Concave Landforms	Convex Landforms
Sandy clay slope of <2 m thickness, with dense bushes or trees	−0.1~−0.2	+0.2~+0.3
Sandy clay slope of 2–5 m thickness, with bushes and open forest	−0.2~−0.3	+0.3~+0.6
Loess slope with >5 m thickness, with the plough and orchard	−0.3~−0.4	+0.6~+1.0
Collapsible loess slope with vertical joints and >5 m thickness, and growing of the plough	−0.4~−0.5	+1.0~+1.5

4.2. Comparison of the Proposed Model and the Existing Model

The results of the prediction model of mining subsidence in mountainous areas considering slopes slip proposed in this paper and the existing model are shown in Figures 16 and 17. The mean square error of the subsidence for the existing model is 110 mm, and the mean square error of the horizontal movement for the existing model is 343 mm. While for the proposed model, the mean square error of the predictive subsidence is 71 mm, the mean square error of the predictive horizontal movement is 276 mm, and the applicability of the proposed model is verified again.

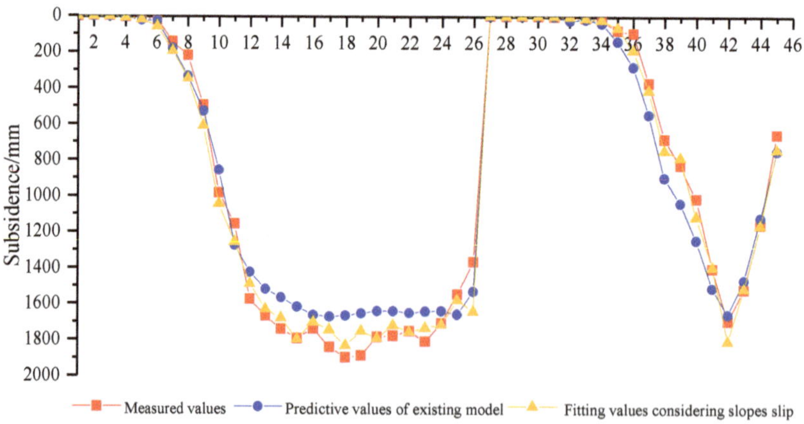

Figure 16. Comparison of the horizontal movements fitting values and measured values considering slopes slip.

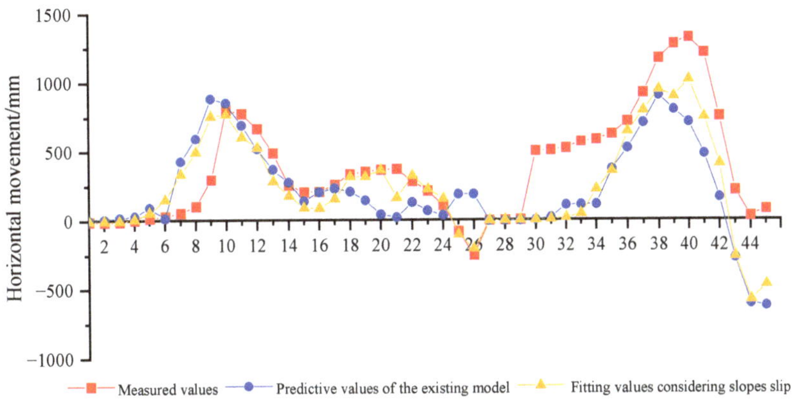

Figure 17. Comparison of the horizontal movements fitting values and measured values considering slopes slip.

Considering the relative position between the monitoring points and the working panel, we consider the points from No. 1 to No. 10, points from No. 11 to No. 26, points from No. 27 to No. 38, and points from No. 39 to No. 45.

For the points above the goaf, the predictive effect of the proposed model is superior to that of the existing model. For the values of the mining landslide influence functions, $P[x]$ and $P[y]$ are relatively small or even equal to 0 in the mining area center, so the predictive results of the existing model are similar to the results of the probability integral method. However, for the proposed model, the predictive effect is more superior, the main reason being that the points above the goaf are completely affected by the coal mining, when the horizontal deformation of these points can reach to the critical horizontal deformation ε_{xm}, then the slip subsidence ΔW and the slip horizontal movement ΔU can be calculated accurately. At the border of the pillar, the predictive effect of subsidence is good for both models, while the predictive effect of horizontal movement is relatively poor for both models; this is because the prediction accuracy of the horizontal movement for the probability integral method is imperfect at the boundary of the coal pillar due to the deficiency in principle. We can make a further improvement for the probability integral method.

In summary, the proposed model and the existing model were evaluated based on the measured data, and the proposed model achieved better performance than the existing model, which verified its effectiveness and superiority. Overall, the existing model is lacking in prediction accuracy due to the deficiency of its the mining landslide influence functions, and the determination of parameters in the mining landslide influence functions has relatively strong subjectivity. Furthermore, the existing model is lacking in mechanical analysis. However, the proposed model in this paper considers the slope slip caused by tensile deformation of soil under the influence of mining, the mechanical analysis is combined in the model, the parameters of probability integral method in the proposed model are optimized based on DSFOA, and the determination of the other parameters in the proposed model are more objective and the meaning of the parameters are clear. The proposed model can help to make a comprehensive understanding of the surface movement in mountainous mining areas and make a better prediction of the mining subsidence in mountainous areas so that it can help to evaluate geological hazards in order to prevent and control geological disasters.

5. Conclusions

Understanding the law of surface movement in mountainous mining areas is important for sustainable development and geological hazard assessment. In this study, the relevant issues of mining subsidence in mountainous areas were systematically analyzed. A new practical prediction model of mining subsidence in mountainous areas considering slopes slip is proposed. The prediction formula of slip subsidence ΔW and slip horizontal movement ΔU is deduced considering movement and deformation values (W, U, i, ε) in the probability integral method. Aiming at the contradiction between the step length and global optimal solution of the FOA, the dynamic step fruit fly optimization algorithm (DSFOA) is proposed to conduct the parameter inversion of the practical model so that the performance of parameter optimization for the FOA is improved. For the determination of parameters in the model, the acquisition of parameters of the probability integral method is more efficient and accurate; the determination of the other parameters about slopes slip in the proposed model are more objective and the meaning of the parameters are clear, while for the existing model it shows relatively strong subjectivity. It would make the large discrepancy between the predictive values and measured values.

Taking the measured data of the 22,101 working panel as a validation, the mean square error of subsidence value fitting is reduced from 248 mm to 122 mm, and the mean square error of horizontal movement is reduced from 397 mm to 386 mm. The slip subsidence ΔW and slip horizontal movement ΔU are calculated by using the predictive parameters obtained from the parameters inversion, and the slip fitting values are compared with the predicted value of probability integral method and the measured value. The mean square error of subsidence further decreased from 122 mm to 71 mm, and the mean square error of horizontal movement decreased from 386 mm to 276 mm. Compared with the existing model, the mean square error of subsidence further decreased from 110 mm to 71 mm, and the mean square error of horizontal movement decreased from 343 mm to 276 mm. It shows again the practicability of the mining subsidence prediction model considering slope slip in mountainous areas. The proposed model can help with the comprehensive understanding of the surface movement in mountainous mining areas and evaluate geological hazards so that prevention and control of geological disasters can be taken further.

Author Contributions: Conceptualization, J.K.; methodology, J.K.; software, Y.Y.; validation, B.Z.; formal analysis, H.D.; investigation, X.X.; resources, Z.P.; data curation, B.Z.; writing—original draft preparation, B.Z. and J.K.; writing—review and editing, Y.Y.; visualization, X.X.; supervision, H.D.; project administration, Y.Y.; funding acquisition, Y.Y. All authors have read and agreed to the published version of the manuscript.

Funding: This research was funded by the National Natural Science Foundation of China grant number [51574242], the Fundamental Research Funds for the Central Universities grant number [2022YJSDC19, 2022YJSDC20]. The APC was funded by [2022YJSDC19].

Institutional Review Board Statement: Not applicable.

Informed Consent Statement: Not applicable.

Data Availability Statement: Not applicable.

Conflicts of Interest: The authors declare that they have no conflict of interest.

References

1. Yuan, L. Scientific conception of precision coal mining. *J. China Coal Soc.* **2017**, *42*, 1–7. [CrossRef]
2. Zhang, C.; Zhao, Y.; Han, P.; Bai, Q. Coal pillar failure analysis and instability evaluation methods: A short review and prospect. *Eng. Fail. Anal.* **2022**, *138*, 106344. [CrossRef]
3. Zhang, J.; Zhang, Q.; Sun, Q.; Gao, R.; Germain, D.; Abro, S. Surface subsidence control theory and application to backfill coal mining technology. *Environ. Earth Sci.* **2015**, *74*, 1439–1448. [CrossRef]
4. Yang, Y.; Erskine, P.D.; Zhang, S.; Wang, Y.; Bian, Z.; Lei, S. Effects of underground mining on vegetation and environmental patterns in a semi-arid watershed with implications for resilience management. *Environ. Earth Sci.* **2018**, *77*, 605. [CrossRef]
5. Yang, Z.; Li, W.; Pei, Y.; Qiao, W.; Wu, Y. Classification of the type of eco-geological environment of a coal mine district: A case study of an ecologically fragile region in Western China. *J. Clean Prod.* **2018**, *174*, 1513–1526. [CrossRef]
6. Bai, E.; Guo, W.; Tan, Y. Negative externalities of high-intensity mining and disaster prevention technology in China. *Bull. Eng. Geol. Environ.* **2019**, *78*, 5219–5235. [CrossRef]

7. Yang, X.; Wen, G.; Dai, L. Ground Subsidence and Surface Cracks Evolution from Shallow-Buried Close-Distance Multi-seam Mining: A Case Study in Bulianta Coal Mine. *Rock Mech. Rock Eng.* **2019**, *52*, 2835–2852. [CrossRef]
8. Zhou, D.; Wu, K.; Bai, Z. Formation and development mechanism of ground crack caused by coal mining: Effects of overlying key strata. *Bull. Eng. Geol. Environ.* **2019**, *78*, 1025–1044. [CrossRef]
9. He, X.; Zhao, Y.; Yang, K. Development and formation of ground fissures induced by an ultra large mining height longwall panel in Shendong mining area. *Bull. Eng. Geol. Environ.* **2021**, *80*, 7879–7898. [CrossRef]
10. Lian, X.; Zhang, Y.; Yuan, H. Law of Movement of Discontinuous Deformation of Strata and Ground with a Thick Loess Layer and Thin Bedrock in Long Wall Mining. *Appl. Sci.* **2020**, *10*, 2874. [CrossRef]
11. Yan, W.; Chen, J.; Yang, W. On-Site Measurement on Surface Disturbance Law of Repeated Mining with High Relief Terrain. *Sustainability* **2022**, *14*, 3166. [CrossRef]
12. Liu, B.; Dai, H. Research Development and Origin of Probability integral method. *Coal Min. Technol.* **2016**, *21*, 1–3. [CrossRef]
13. Cui, X.; Deng, K. Research review of predicting theory and method for coal mining subsidence. *Coal Sci. Tech.* **2017**, *45*, 160–169. [CrossRef]
14. Jianjun, Z.; Xun, W.; Yanbing, S.; Jiangbo, W.; Lee, M.L.; Xue, Y. Deformation Behavior of Mining beneath Flat and Sloping Terrains in Mountainous Areas. *Geofluids* **2021**, *2021*, 6689966. [CrossRef]
15. Zhang, C.; Mitra, R.; Hebblewhite, B. Evaluation of valley closure subsidence effects under irregular topographic conditions. *Min. Technol.* **2013**, *122*, 172–183. [CrossRef]
16. He, W. Mountains surface movement under the influence of the mining. *Coal Sci. Tech.* **1981**, *9*, 23–29+62. [CrossRef]
17. He, W. Mountain surface movement and deformation prediction caused by mining. *Coal Sci. Tech.* **1983**, *11*, 46–52+60. [CrossRef]
18. He, W.; Kong, Z.; Kang, J. Mechanism and vector analysis of surface mining slip in mountainous areas. *Min. Surv.* **1991**, 21–25.
19. Wanlong, H.; Jianrong, K. Research on the laws of mountain surface movement and deformation. *J. China Coal Soc.* **1992**, *9*, 79–89.
20. Zha, J.; Feng, W.; Zhu, X. Research on Parameters Inversion in Probability integral method by Genetic Algorithm. *J. Min. Saf. Eng.* **2011**, *28*, 655–659. [CrossRef]
21. Li, P.; Peng, D.; Tan, Z.; Deng, K. Study of probability integration method parameter inversion by the genetic algorithm. *Int. J. Min. Sci. Technol.* **2017**, *27*, 1073–1079. [CrossRef]
22. Xu, M.; Zha, J.; Li, H. Parameters Inversion in Probability integral method by Particle Swarm Optimization. *Coal Eng.* **2015**, *47*, 117–119, 123. [CrossRef]
23. Wang, L.; Li, N.; Zhang, X.; Wei, T.; Chen, Y.; Zha, J. Full parameters inversion model for mining subsidence prediction using simulated annealing based on single line of sight D-InSAR. *Environ. Earth Sci.* **2018**, *77*, 161. [CrossRef]
24. Wei, T.; Guo, G.; Li, H. Fusing Minimal Unit Probability Integration Method and Optimized Quantum Annealing for Spatial Location of Coal Goafs. *KSCE J. Civ. Eng.* **2022**, *26*, 2381–2391. [CrossRef]
25. Wei, T.; Wang, L.; Jiang, K.; Wang, Y.; Yang, X. One spatial geometric characteristic identification method of a coal mine working face based on ground movement and deformation monitoring data. *Energy Sources Part A Recovery Util. Environ. Eff.* **2021**, *26*, 1–15. [CrossRef]
26. Wang, L.; Jiang, K.; Wei, T.; Jiang, C.; Zha, J.; Chi, S. Estimation of parameters of probability integral method model based on improved fireworks algorithm. *Surv. Rev.* **2020**, *53*, 366–382. [CrossRef]
27. Wang, L.; Jiang, K.; Wei, T. Development of a new inversion method for detecting spatiotemporal characteristics of coal mines based on earth observation technology. *Int. J. Appl. Earth Obs. Geoinf.* **2021**, *26*, 102346. [CrossRef]
28. Yang, J.; Liu, C.; Chen, T.; Zhang, Y. The invasive weed optimization–based inversion of parameters in probability integral model. *Arab. J. Geosci.* **2019**, *12*, 424. [CrossRef]
29. Pan, W.-T. A new Fruit Fly Optimization Algorithm: Taking the financial distress model as an example. *Knowl. Based Syst.* **2012**, *26*, 69–74. [CrossRef]
30. Pan, W.-T.; Zhu, W.-Z.; Ma, F.-X.; Zhong, Z.-C.; Yuan, X.-F. Modified fruit fly optimization algorithm of logistics storage selection. *Int. J. Adv. Manuf. Technol.* **2017**, *93*, 547–558. [CrossRef]
31. Zhang, Y.; Cui, G.; Wu, J.; Pan, W.-T.; He, Q. A novel multi-scale cooperative mutation Fruit Fly Optimization Algorithm. *Knowl. Based Syst.* **2016**, *114*, 24–35. [CrossRef]
32. Fu, Y.; Zhou, M.; Guo, X.; Qi, L. Stochastic multi-objective integrated disassembly-reprocessing-reassembly scheduling via fruit fly optimization algorithm. *J. Clean Prod.* **2021**, *278*, 123364. [CrossRef]
33. Hesami, M.; Alizadeh, M.; Naderi, R.; Tohidfar, M. Forecasting and optimizing Agrobacterium-mediated genetic transformation via ensemble model- fruit fly optimization algorithm: A data mining approach using chrysanthemum databases. *PLoS ONE* **2020**, *15*, e0239901. [CrossRef]
34. Luo, R.; Zheng, H.; Guo, J. Solving the Multi-Functional Heterogeneous UAV Cooperative Mission Planning Problem Using Multi-Swarm Fruit Fly Optimization Algorithm. *Sensors* **2020**, *20*, 5026. [CrossRef] [PubMed]
35. Yan, H.; Zhang, T.; Qi, Y.; Yu, D.-J. Short-term traffic flow prediction based on a hybrid optimization algorithm. *Appl. Math. Modell.* **2022**, *102*, 385–404. [CrossRef]
36. Li, Y.; Xu, F. Acoustic emission sources localization of laser cladding metallic panels using improved fruit fly optimization algorithm-based independent variational mode decomposition. *Mech. Syst. Signal Process.* **2022**, *166*, 108514. [CrossRef]
37. Qisong, Q.; Gening, X.; Xiaoning, F.; Jun, W. A new type bionic global optimization: Construction and application of modified fruit fly optimization algorithm. *Proc. Inst. Mech. Eng. Part B J. Eng. Manuf.* **2014**, *229*, 1614–1621. [CrossRef]

38. Sang, H.-Y.; Pan, Q.-K.; Duan, P.-Y. Self-adaptive fruit fly optimizer for global optimization. *Nat. Comput.* **2017**, *18*, 785–813. [CrossRef]
39. Pan, Q.-K.; Sang, H.-Y.; Duan, J.-H.; Gao, L. An improved fruit fly optimization algorithm for continuous function optimization problems. *Knowl. Based Syst.* **2014**, *62*, 69–83. [CrossRef]
40. Hou, Y.; Li, J.; Yu, H.; Li, Z. BIFFOA: A Novel Binary Improved Fruit Fly Algorithm for Feature Selection. *IEEE Access* **2019**, *7*, 81177–81194. [CrossRef]
41. He, G.Q.; Yang, L. *Mining Subsidence*; China University of Mining and Technology Press: Xuzhou, China, 1991; pp. 188–195.
42. Gao, C.; Xu, N. Research on surface crack depth and crack width caused by coal mining. *Coal Eng.* **2016**, *48*, 81–83+87. [CrossRef]

sustainability

MDPI

Article

Determination of Fractured Water-Conducting Zone Height Based on Microseismic Monitoring: A Case Study in Weiqiang Coalmine, Shaanxi, China

Wei Gao [1,2], Yingchun Li [3,*] and Qingyuan He [4]

1 Faculty of Electronic and Information Engineering, Xi'an Jiaotong University, Xi'an 710049, China; gw8828@sina.com
2 Shaanxi Yanchang Petroleum Sale Electric Co., Ltd., Xi'an 710065, China
3 State Key Laboratory of Coastal and Offshore Engineering, Dalian University of Technology, Dalian 116024, China
4 State Key Laboratory of Coal Resources and Safe Mining, China University of Mining and Technology, Xuzhou 221116, China; 5783@cumt.edu.cn
* Correspondence: yingchun_li@dlut.edu.cn

Abstract: The NO. 3 coal seam of the Weiqiang coalmine is faced with a water inrush threat caused by two large reservoirs and the upper confined aquifer. However, existing empirical formulas cannot accurately predict the height of the fractured water-conducting zone; thus, it is necessary to investigate the extent of the fractured water-conducting zone according to the geological conditions of the No. 3 coal seam. The microseismic monitoring system (MMS) was used to monitor the surrounding area of the 1311 working face in real time to obtain the microseismic event information during the fracture development. The detailed situation of the fractured water-conducting zone in three-dimensional space was determined by computer analysis. The results show that the height of the fractured water-conducting zone is 110 m, which is consistent with the range (105.4~120.4 m) measured by various field drilling results and the mechanical mechanism analysis based on the key strata theory (112 m). The achievement of microseismic monitoring provides a guarantee for safety production and prevention and control of mine water disaster in the 1311 working face of the No. 3 coal seam, and can also be applied and promoted in coalmines with similar geological conditions, with significant social and economic benefits.

Keywords: fractured water-conducting zone; microseismic monitoring; key strata; field drilling observation

1. Introduction

Coal is an important energy resource in China, and the annual output is gradually approaching 4 billion tons. By 2030, coal may still account for about 50% of primary energy consumption. Such a huge amount of mining and consumption has put forward higher requirements for the safe mining of coal resources [1–3]. Mine water inrush accidents represent one of the five major disasters in mining production. Since the 1950s, they have caused huge economic losses of more than 1 billion yuan [4]. In mining engineering, the overlying strata will move and deform in a large range due to the mining effect. At the same time, internal defects and fissures of the rock mass continue to develop, forming an important channel for water from the aquifer to the working face, or the water-conducting zone. Water inrush sources and water inrush channels in mines are two major factors that cause water disasters. The water-conducting zone is the main cause of the water inrush channel [5], which makes it easy for mine water inrush accidents to occur. Therefore, exploring the development of the water-conducting zone is the basic strategy for water disaster prevention and control.

Many scholars have carried out research to confront this problem. On the one hand, the development of the water-conducting zone is closely related to the movement mechanism of the overlying strata above the goaf, and theoretical research in this area is abundant. The key stratum theory proposed by Qian [6,7] emphasizes the controlling effect of key strata on rock formations, which is still widely used and studied. Jiang et al. [8], Xu et al. [9,10], and Qin et al. [11] further discussed the movement law of overlying strata and the method of discriminating key strata. Huang et al. [12] and Hu et al. [13] explored various factors affecting the development height of water-conducting zones. Xu et al. [14] gave a method for predicting the development height by the position of the key strata. Cao et al. [15] and Wang et al. [16] explored the development characteristics of water-conducting fractured zones. Additionally, Zhang et al. [17] and Chen et al. [18] studied the methodologies (theoretical models, numerical simulations, and field measurements) for estimating pillar system loads and strengths. The above research results have laid a solid theoretical foundation for the prediction of the water-conducting zone. On the other hand, there are various methods for predicting the development of water-conducting zones. At present, empirical formulas, numerical simulations, similar physical experiments, microseismic monitoring, and field drilling detection methods are widely used, and most projects use several methods to jointly determine the development height, which can give more accurate and objective results. In engineering, the empirical formulas provided by "The regulations for coal pillar retention and pressed coal mining of buildings, water bodies, railways and main roadways" (hereafter referred to as three lowers) cannot be applied to coal seam mining under some special conditions, and often incur larger errors [13]. Miao et al. [19], Wang et al. [5], Hu et al. [13], and Li et al. [20] pointed out this phenomenon and put forward theoretical formulas suitable for different mining conditions. Du et al. [21], Chen et al. [22], and Lv [23] used RFPA, UDEC, and other software to carry out numerical simulations on the development of the water-conducting zone heights, which is consistent with the actual monitoring results. Therefore, numerical simulations and similar physical experiments are often used to predict the laws of overburden caving zones and to predict development heights. In addition, in view of the water inrush problem faced by coal seams, the risk analysis of aquifers was carried out according to the on-site geological conditions, and it was pointed out that aquifers have an important impact on coal seam mining [24–27]. Recently, due to the development of new technologies such as computer technology and communication technology, microseismic monitoring has been gradually applied in mining engineering. A large number of researchers have combined microseismic monitoring technology with deep learning, numerical simulation, and other technologies, and applied them in coal mining projects in many places around the world [28–37]. D'Angiò et al. [28] proposed the Damping Ratio Anomalies Monitoring (DaRtAM) to detect the rock mass damaging. Ma et al. [29] combined microseismic monitoring and numerical simulation (RFPA) to analyze the water inrush problem of the Dongjiahe coal seam. Srinivasan et al. [30], Simser et al. [31], Occhiena et al. [32], Hassani et al. [33], and Mngadi et al. [34] have applied microseismic monitoring technology to mining sites around the world and guided the safe mining of coal resources. Many scholars have applied microseismic monitoring technology in the fracture monitoring of mine strata, the distribution of mining stress fields, and the movement law of overlying strata in the stope [38–44], which has deepened the cognition of the mining law of the overlying strata. Wang et al. [45], Jiang et al. [46], Liu et al. [47], and Han et al. [48] used microseismic monitoring technology to conduct field measurements and research on the development height of water-conducting zones and the height of fracture damage, and monitoring results show that microseismic monitoring has the advantages of rapid diagnosis, real-time monitoring, and early warning. On-site drilling detection is often limited by funds, manpower, time, and scale, and thus cannot show the characteristics of dynamic development of water-conducting fractured zones. However, as monitoring results are the most credible, on-site drilling detection is often used for the verification of other research methods.

In summary, although the current detection methods are diverse, they each have certain limitations. There are still many limitations in empirical and theoretical formulas, which are difficult to popularize and apply. Numerical simulations and physical experiments are generally only used as auxiliary research methods for conducting height prediction because the parameter settings are too ideal, the artificial parameter adjustment lacks objectivity, and it is difficult to consider the complex construction and geological conditions of the field. As a new technology, microseismic monitoring still needs to be further promoted and applied. On-site drilling detection cannot be carried out on a large scale due to the high cost. Therefore, in practical engineering, the strategy of "complementary advantages and multiple measures" is often adopted to improve the accuracy of the predicted value.

There are two large reservoirs near the No. 3 coal seam in the Weiqiang coalmine, which is facing the threat of water inrush from the roof confined aquifer. Once the water-conducting zone breaks through the aquifer during the mining process, it will cause serious economic losses. In addition, the empirical formula cannot accurately give the height of the water-conducting zone, and the height of the water-conducting fracture zone must be accurately determined according to the existing mining conditions of the No. 3 coal seam. In response to the above problems, microseismic monitoring equipment was installed in the 1311 working face of the No. 3 coal seam. After capturing the microseismic signal, the inversion analysis was obtained, and the monitoring results were verified by on-site drilling and mechanical mechanism analysis. The research results of this article will guide the rock formation control and water hazard prevention in the subsequent mining of the working face, ensure the safe and efficient mining of the mine, and can also be promoted in the surrounding mines, with significant economic and social benefits.

2. Weiqiang Coalmine Overview

The Weiqiang mining field is located in the northwest of the Yuheng mining area of the Jurassic coalfield in northern Shaanxi (Figure 1). The No. 3 coal seam is the mineable coal seam in the whole area of the mine field, among which the coal seam of the 1311 working face is stable and simple in structure. The thickness of the coal seam is between 2.85 and 3.48 m, the average coal thickness is 3.15 m, the actual mining height is 3.32 m, and the dip angle of the coal seam is less than 1°. The longwall fully mechanized mining method is adopted, and the roof is managed by all caving methods. The overlying strata of the No. 3 coal seam are mainly medium-thick-layered siltstone and thick-layered medium-grained feldspar sandstone, with a thickness of 10.2–26.93 m, including the sandstone aquifer at the bottom of the fourth member of the Yan'an Formation. The floor of the coal seam is mainly light gray medium-thick-layered siltstone, 3.82–10.86 m thick, with well-developed horizontal wave bedding.

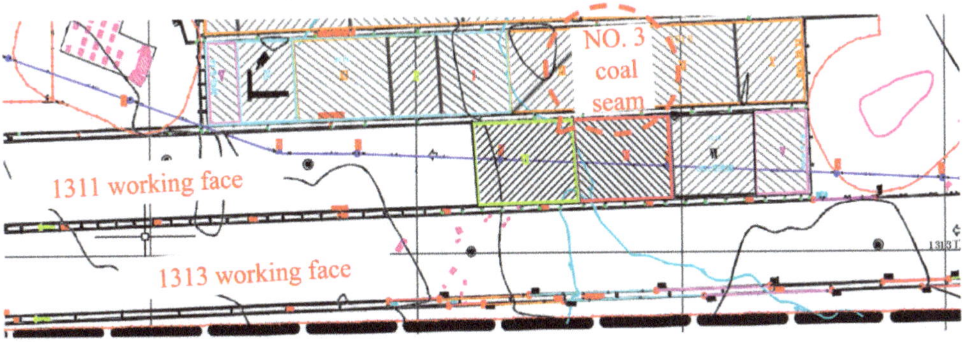

Figure 1. Schematic diagram of the working faces.

In the panel area where the mining area is located, there are many ground structures such as water systems, reservoirs, cultural relics, etc. In particular, there are two reservoirs in Dachuangou in the east of the coalmine, namely Hejia Village and Luotuobozigou Reservoir (Figure 2), and the total storage capacity is about 2.37 million m^3. According to the previous geological data, when the No. 3 coal seam is mined, the water disaster will mainly affect the confined aquifer of the sandstone fissures on the roof of the coal seam. At this time, the mining of large-area near-horizontal coal seams will inevitably cause large-scale movement of overlying rocks. Once the water-conducting zone develops to the water-resisting layer, it will have a high probability of water inrush disaster. Therefore, the determination of the development height of the water-conducting zone in the overlying rock on the 1311 working face plays an important role in evaluating the safe mining of the No. 3 coal seam.

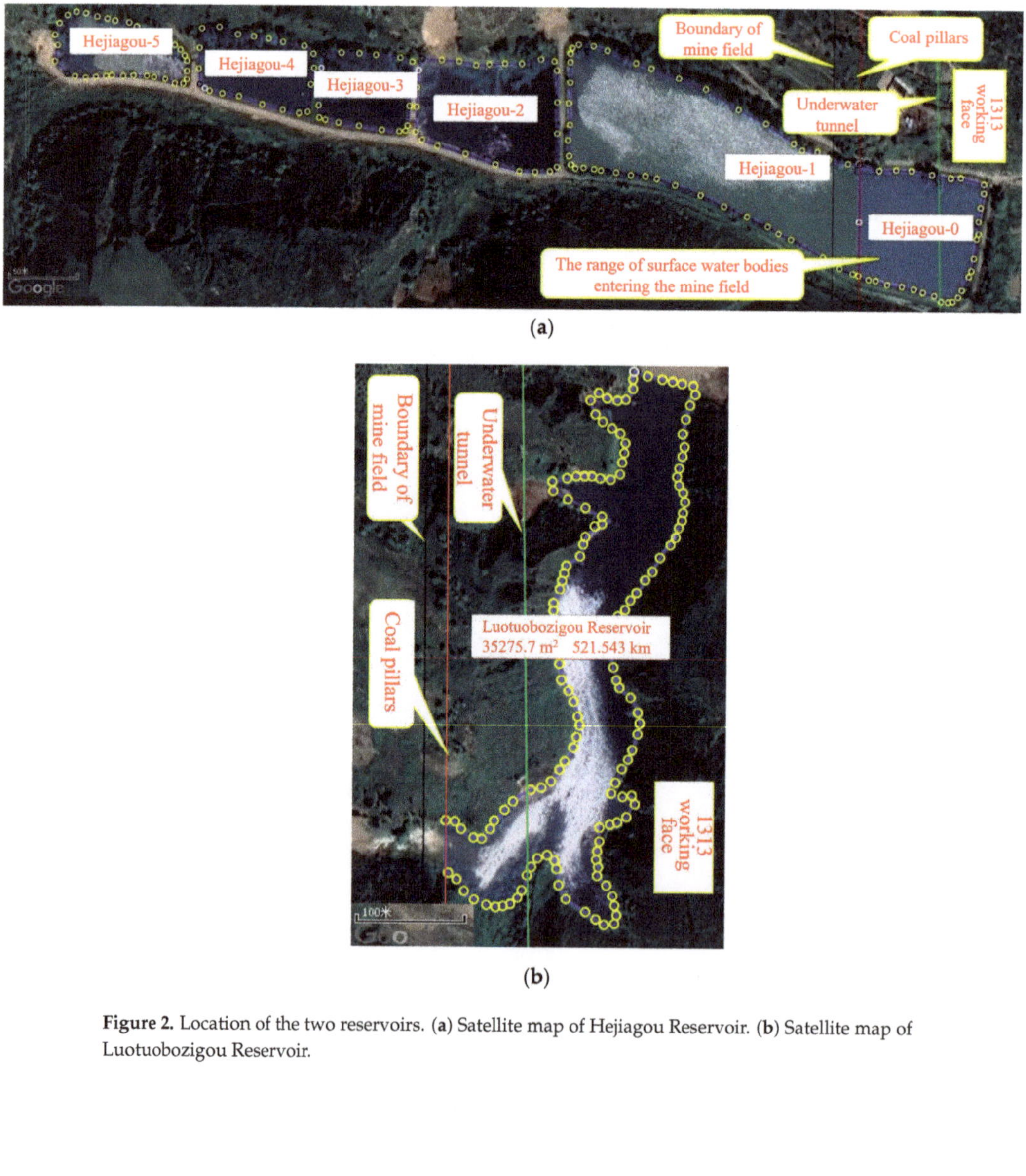

Figure 2. Location of the two reservoirs. (**a**) Satellite map of Hejiagou Reservoir. (**b**) Satellite map of Luotuobozigou Reservoir.

3. Microseismic Monitoring

3.1. Principle and Composition of MMS (Mining Microseismic Monitoring)

Under the action of mine pressure, the rock formations in the mining area show different degrees of deformation, cracking, and breaking. This macroscopic damage is actually the result of the gradual evolution of many internal defects and micro-fractures. During the mining process, the local strain energy in the surrounding rock is concentrated, resulting in the compaction of the old micro-cracks in the coal-rock body and the initiation of new micro-cracks. The microseismic event is that the above-mentioned surrounding rock continuously radiates energy outward during the process of microcrack generation and dislocation. Microseismic monitoring technology can collect real-time information of micro-earthquakes in the rock formation by reasonably arranging sensor arrays in the mining area, determining the location of micro-crack ruptures, and using related software to visualize micro-seismic events.

The MMS used on-site is an ESG MMS equipped with three-dimensional visualization software (MMS-View). The system is completed by the Canadian ESG Company, and is used to monitor the microseismic events of weak seismic waves in the process of deformation and rupture of rock mass. The ESG system consists of two parts: software and hardware. The software system consists of Paladin standard version monitoring system, HNAS software (real-time signal acquisition and recording), SeisVis software (3D visualization of events), and WaveVis software (waveform processing and event relocation). The hardware system consists of a 42-channel accelerometer, Paladin sensor interface box, Paladin seismic recorder, Paladin master control time server, software operation monitoring card WatchDog, and other facilities.

Based on computer technology, communication technology, GPS real-time precise positioning technology, and visualization software, MMS can determine the time, location, and magnitude of rock mass microseismic events in real time in the three-dimensional space of the stope rock mass, and then determine the range of rock mass deformation and give a safety assessment. At present, MMS has been effectively used in the stability analysis and monitoring and early warning of slopes, tunnels, mines, dams, and other projects.

3.2. Location Algorithms and Moment Magnitude

The damage of rock mass can be analyzed through microseismic monitoring. First of all, it is necessary to invert and locate the position of the microfracture accurately, and then obtain the parameters such as the moment magnitude of the microseismic event, so that the source parameters reflecting the stability of the rock mass can be calculated and the mechanism of rock formation can be explored. The positioning accuracy of the microseismic monitoring system is 10 m, which can meet the needs of the project. It should be noted that during the microseismic monitoring process, the system will collect various vibration signals, including environmental noise (blasting, car whistle, drilling, impact). The basic principle of signal noise reduction is to determine the type of noise according to the signal characteristics (frequency–amplitude curve) of different noises, so as to identify and determine the microseismic signal of rock mass rupture. For example, the signal characteristic of blasting is large amplitude, small attenuation, and long duration; the signal characteristic of whistle is small amplitude, long duration, and multi-peak; the signal characteristic of drilling is periodic variation and small amplitude; and the signal of impact is characterized by fast decay and short duration. Based on this principle, waveform analysis of different signals can be performed to identify effective microseismic signals and improve monitoring accuracy.

The MMS uses the widely used Geiger positioning method to locate microseismic events through an iterative process. Given the initial space–time coordinates of the test point, $\theta(x, y, z, t)$, the correction amount $\Delta\theta(x, y, z, t)$ is generated based on the least squares method, and the two are added to obtain new coordinates. If the new coordinates satisfy Formula (1), it is the source parameter; otherwise, it needs to iterate until it is satisfied.

$$\left[(x_i - x)^2 + (y_i - y)^2 + (z_i - z)^2\right]^{\frac{1}{2}} = v_p \cdot (t_i - t) \tag{1}$$

where (x, y, z) is the spatial coordinate of the test point, t is the time when the microseismic event occurs, (x_i, y_i, z_i) is the spatial coordinate of the ith sensor, t_i is the time when the sensor i captures the elastic wave P wave, and v_p is the P wave velocity. The initial values of the above-mentioned variables such as coordinates, time, and wave speed are all manually specified. The first-order Taylor expansion is performed near the time calculated from the spatial coordinates of the test point, and the arrival time of the P wave captured by sensor i is expressed as:

$$t_{o,i} = t_{c,i} + \frac{\partial t_i}{\partial x}\Delta x + \frac{\partial t_i}{\partial y}\Delta y + \frac{\partial t_i}{\partial z}\Delta z + \frac{\partial t_i}{\partial t}\Delta t \tag{2}$$

where $t_{c,i}$ is the time that the P wave obtained from the spatial coordinates of the test point is transmitted to the sensor i, and the other parameters are expressed as:

$$\frac{\partial t_i}{\partial x} = \frac{(x_i - x)}{v_p R};\ \frac{\partial t_i}{\partial y} = \frac{(y_i - y)}{v_p R};\ \frac{\partial t_i}{\partial z} = \frac{(z_i - z)}{v_p R};\ \frac{\partial t_i}{\partial t} = 1$$
$$R = \left[(x_i - x)^2 + (y_i - y)^2 + (z_i - z)^2\right]^{\frac{1}{2}}$$

For a microseismic system with N sensors, N equations can be established, and the matrix expression form is:

$$A\Delta\theta = B \tag{3}$$

where

$$A = \begin{bmatrix} \frac{\partial t_1}{\partial x} & \frac{\partial t_1}{\partial y} & \frac{\partial t_1}{\partial z} & 1 \\ \frac{\partial t_2}{\partial x} & \frac{\partial t_2}{\partial y} & \frac{\partial t_2}{\partial z} & 1 \\ & & & \\ \frac{\partial t_n}{\partial x} & \frac{\partial t_n}{\partial y} & \frac{\partial t_n}{\partial z} & 1 \end{bmatrix}, \Delta\boldsymbol{\theta} = \begin{bmatrix} \Delta x \\ \Delta y \\ \Delta z \\ \Delta t \end{bmatrix}, \boldsymbol{B} = \begin{bmatrix} t_{o,1} - t_{c,1} \\ t_{o,2} - t_{c,2} \\ \\ t_{o,n} - t_{c,n} \end{bmatrix}$$

Solve Equation (3) with the Gaussian elimination method to obtain the corresponding correction:

$$\boldsymbol{A}^T\boldsymbol{A}\Delta\boldsymbol{\theta} = \boldsymbol{A}^T\boldsymbol{B} \tag{4}$$

$$\Delta\boldsymbol{\theta} = \left(\boldsymbol{A}^T\boldsymbol{A}\right)^{-1}\boldsymbol{A}^T\boldsymbol{B} \tag{5}$$

After the correction amount is obtained by Formulas (4) and (5), $(\boldsymbol{\theta} + \Delta\boldsymbol{\theta})$ is used as a new test point to continue iterating until the given error requirement is met.

After the microseismic event is located, the source parameter calculation is carried out. The moment magnitude M_W can be calculated from the seismic moment M_0, which is used to measure the magnitude level. The seismic moment M_0 is determined by the low-frequency limit Ω_0 of the far-field displacement spectrum of the seismic wave:

$$M_0 = \frac{4\pi\rho_0 c_0^3 R\Omega_0}{F_c R_c S_C} \tag{6}$$

where ρ_0 is the density of the source medium, c_0 is the seismic wave velocity, R is the distance between the source position and the measuring point, F_c is the seismic wave radiation coefficient, R_c is the seismic wave free surface amplification coefficient, and S_c is the seismic wave field correction coefficient. In this case, the source cannot be accurately determined, so it is assumed that the radiation coefficients of P-wave and S-wave are both 0.52, and considering that the sensors are often arranged in the borehole, R_c and S_c are taken as 1, then:

$$M_0 = \frac{4\pi\rho_0 c_0^3 R\Omega}{F_c} \tag{7}$$

The relationship between the moment magnitude M_W and the seismic moment M_0 is:

$$M_W = \frac{2}{3}\log M_0 - 6.033 \quad (8)$$

According to the source parameters, the activity law of the overlying strata can be analyzed, and the development of the fractured water-conducting zone (fractured zone and caving zone) can be judged by combining the distribution characteristics of the "three-zones" (bending zone, fractured zone, and caving zone).

3.3. On-Site Monitoring System Layout

The monitoring position is selected near the 1311 working face of the No. 3 coal seam in the Weiqiang coalfield, and the data processing equipment and monitoring equipment of the MMS system are arranged up-hole and down-hole, respectively (Figure 3). Real-time monitoring, positioning, and analysis against the coal and rock mass micro-fractures are carried out in the mining-affected area of the working face, and the development height of the water-conducting zone is analyzed in combination with the deformation and stress characteristics of the "three zones".

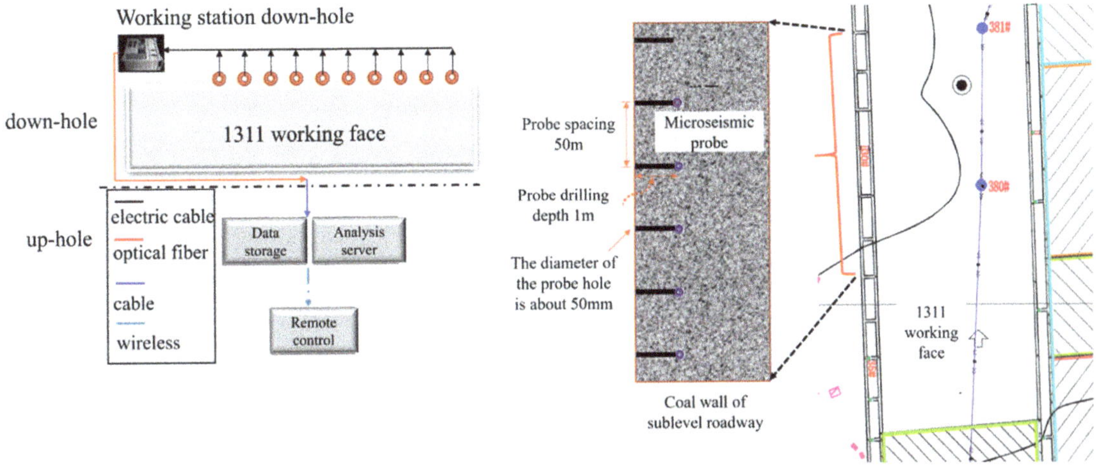

(**a**) (**b**)

Figure 3. Operation and monitoring plan of MMS in the 1311 working face. (**a**) Operation schematic diagram of MMS. (**b**) Field monitoring plan of MMS.

As shown in Figure 3, in order to achieve advanced monitoring, a microseismic monitoring system with a range of 400 m was arranged at the transport roadway (at a distance of 560 m from the 1311 working face), and the monitoring duration was 4 months. The site is equipped with six high-precision sensors, and the hole is located at the reserved coal pillar on the left side of the 1311 auxiliary transportation roadway, about 1.5 m away from the bottom plate. In order to improve the quality of on-site monitoring signals and reduce external construction noise interference, each sensor is placed at a depth of 1 m in the hole, and the distance between each sensor is about 50 m. During the sensor installation process, first, the sensor is sent to the bottom of the borehole and firmly attached to the rock wall at the bottom; then, an appropriate amount of cement mortar is slowly injected into the borehole to make the mortar cover the sensor; finally, when the mud begins to solidify, the installation tool is slowly pulled out of the hole, and the hole is filled with cement to fix the sensor at the bottom of the hole. The latitude and longitude coordinates and elevation of each probe position are shown in Table 1.

Table 1. Coordinates and elevation of sensors.

Sensor Number	N	E	Elevation (M)
1	037:59:55.915306	109:15:12.732641	788.218
2	037:59:57.532875	109:15:12.694932	787.749
3	037:59:59.157386	109:15:12.657013	788.916
4	038:00:00.510342	109:15:12.622709	786.871
5	038:00:02.140542	109:15:12.589681	786.828
6	038:00:03.759413	109:15:12.560642	785.881

On-site debugging of the monitoring system is carried out after the hardware equipment is installed: (1) for the initial positioning test, simply tap the anchored sensor and determine whether the difference between the event coordinates calculated by the microseismic event positioning and the actual sensor coordinates is within the error range; (2) set the P wave speed and frequency range to 5500–6500 and 2500–3000, respectively; (3) adjust the threshold to obtain high-quality microseismic events and reduce interference events, which can be dynamically adjusted by the proportion of staged microseismic events to the total events.

3.4. Analysis of Microseismic Monitoring Results

The microseismic activity characteristics from 24 October 2019 to 15 February 2020 were projected along the strike of the coal seam according to the periodic breaking time of the key strata. The distribution characteristics of microseismic activities in the roof and overlying strata during the fracture process of rock formations are analyzed, with emphasis on the aggregation characteristics of microseismic events along the strike.

Figure 4a,b shows the characteristics of microseismic activity before the low-level key strata are broken. Microseismic events are scattered in the immediate roof and low-level rock formations. After the initiation and gradual expansion of micro-cracks in the coal rock mass and low-level key strata, there is a tendency to form a microseismic event accumulation area in front of the coal wall.

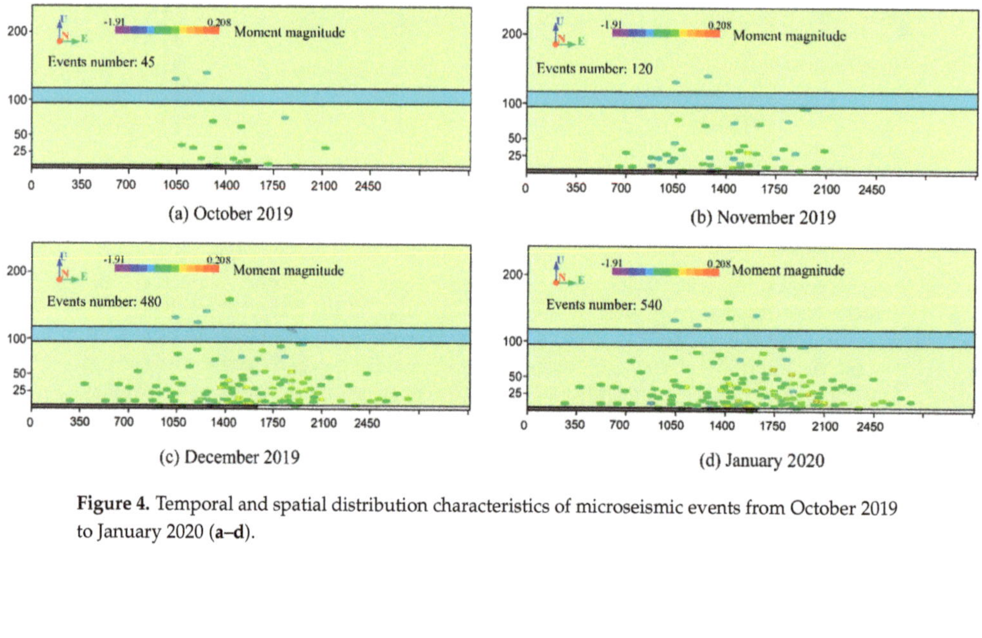

Figure 4. Temporal and spatial distribution characteristics of microseismic events from October 2019 to January 2020 (**a**–**d**).

As shown in Figure 4c, as the working face continues to advance, the microseismic events in front of the coal wall increase sharply and form a gathering area, where macroscopic fractures occur in the low-level key strata, and the working face suffers periodic pressure. At the same time, due to the slewing and sinking of the broken rock blocks above the working face, another microseismic event gathering area is formed in the rear goaf. The distribution density of microseismic events between the two gathering areas is relatively small, and the rock blocks above the working face still have good integrity after breaking. In addition, observed from the vertical direction, the accumulation area of the two microseismic events is linearly distributed, which is the fracture line of the stratum formed after the periodic compression of the low-level key strata, and most of the microseismic events initiated by the micro-seismic events of the mid-level key strata are also distributed in this range.

Figure 4d shows that the roof strata experienced a large area of caving and gyration subsidence after the low-level key strata were completely broken, and most of the newly added microseismic events in the strike direction were distributed between the broken key rock blocks, that is, between two clusters of microseismic events. At the same time, the microseismic events were further initiated and expanded in the mid-level key strata, and there was a tendency to form agglomeration areas on the extension line of the fracture line formed after the low-level key strata were broken.

It can be found that the microseismic events occur in pairs along the strike accumulation area during the fracture process of the median and low-level strata, and are located on the stratum breaking line and its extension line formed after the incoming pressure, that is, the two ends of the key stratum breaking. The broken key block and the weak rock formation above it will initiate and expand cracks during the process of gyration and subsidence, which induces microseismic events.

In summary, with the gradual progress of the 1311 working face, the micro-fractures in the overlying stratum continue to develop. From the analysis of the evolution results of microseismic activities and the development characteristics of the water-conducting zone in the "three zones", it is known that the distribution of microseismic events within 25 m of the overlying stratum is relatively dense, and the fissures are relatively developed, which is regarded as the upper boundary of the caving zone. In addition, the density of microseismic events in the range above 110 m is significantly weakened, which is regarded as the upper boundary of the water-conducting zone of the overlying stratum.

4. Field Drilling Verification

Drilling detection above the 1311 working face of the No. 3 coal seam (Figure 5) and the development range of the water-conducting zone were comprehensively determined by using four methods: simple hydrological observation method, engineering geological catalogue, geophysical logging, and downhole television. Figure 6 shows the results of the four kinds of detections implemented at YZK01.

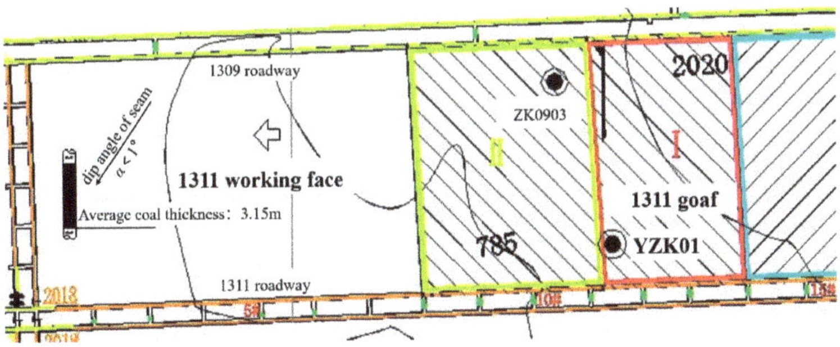

Figure 5. Schematic diagram of YZK01 plane position.

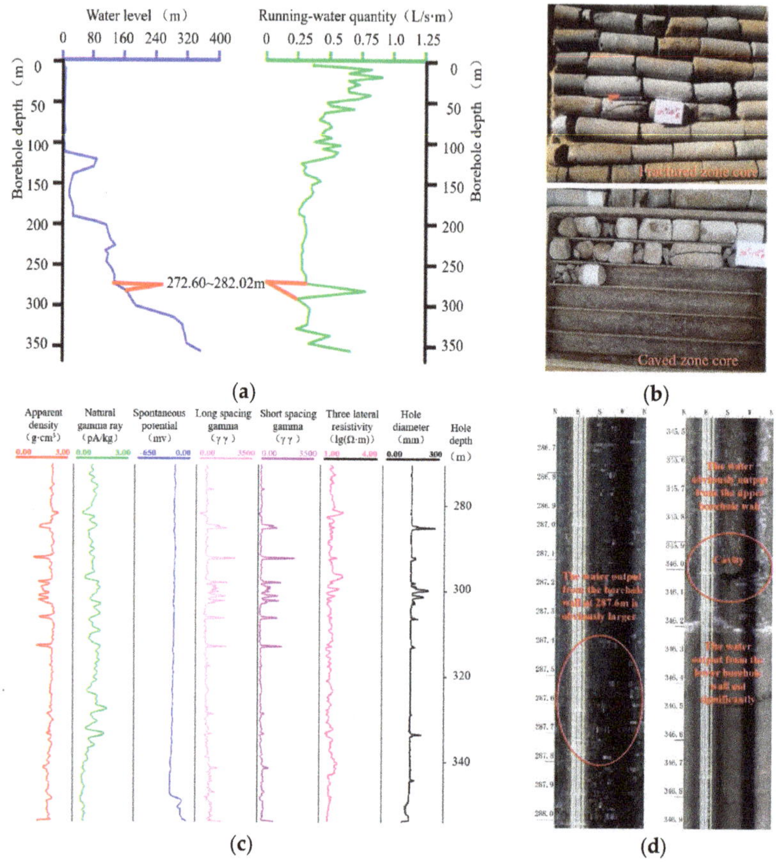

Figure 6. The height of "three-zones" measured by field drilling detection. (**a**) Simple hydrological observation method. (**b**) Engineering geological catalogue. (**c**) Geophysical logging. (**d**) Downhole television.

The simple hydrological observation method entails observing the consumption of flushing fluid, mud water level, and drilling footage during the drilling process. The engineering geology catalogue involves counting the RQD value, which is the quality index, by studying the integrity of the core and the development of fissures. Geophysical logging mainly analyzes rock formations from the perspective of apparent density, three lateral resistivity, spontaneous potential, long and short source distance gamma, natural gamma, acoustic wave, and other parameters that change with burial depth. The downhole television method uses intelligent panoramic imaging logging in the borehole to form a 360° image of the rock wall and identify the development of fractures. The results obtained by the above methods are shown in Table 2.

In summary, the development height of the water-conducting zone obtained by field drilling detection ranges from 105.4 m to 120.4 m.

Table 2. Field drilling detection results.

Method	Height of Water-Conducting Zone (m)	The Ratio of the Height of the Fractured Zone to the Mining Height
Simple hydrological observation method	120.4	36.27
Engineering geological catalogue	111.7	33.64
Geophysical logging	108	32.53
Downhole television	105.4	31.75

5. Mechanical Mechanism Analysis

During mining, the overlying rock layer is deformed and ruptured in a large range due to mining. At this time, the rock layer that controls all or part of the rock mass movement is called the key strata, and is generally also a hard rock layer. If the critical layer breaks, the deformation of all or part of the rock layer above it will remain the same; that is, the overall movement will occur. The one that controls all the upper rock layers is called the main key strata, and the breaking law of the main key strata directly affects the law of surface subsidence and movement, and the one that only plays a leading role in part of the rock layer is called the sub-key strata [6,7]. In actual engineering, there is only one main key strata, but the sub-key strata may have multiple layers due to the existence of multiple hard rock layers. At present, the key strata theory is widely used to analyze problems such as overlying rock motion control, pressure relief gas drainage, etc. It is an important means to understand the problems of fracture development and stope pressure.

The development height of the water-conducting zone in the overlying stratum of the 1311 working face is directly affected by the distribution and fracture of key strata. In response to this problem, based on the physical and mechanical parameters of each rock formation in the stope, the KSPB (Key Strata Judging Program) developed by China University of Mining and Technology was used to determine the position of the main/sub-key strata in the overlying strata. The calculation results are shown in Table 3.

It can be seen from Table 3 that there are six hard rock layers above the coal seam, one main key strata, and one sub-key strata. Among them, the main key strata is a siltstone layer with thickness of 15.5 m and depth of 281 m; the sub-key strata is a feldspar sandstone layer with thickness of 22.63 m and depth of 393 m. Xu et al. [14] pointed out that the position of the key strata of the overlying rock will affect the development height of the water-conducting zone. Only when the position of the key strata is less than a certain critical height from the mined coal seam, the fracture of the key strata will penetrate into the water-conducting zone, and at the same time, the ruptured fractures of the overlying strata will also penetrate into the water-conducting zone. However, the height of the main key strata of the No. 3 coal seam in the Weiqiang mine field is 112 m, which exceeds the critical height by 33.2 m (ten times the mining height). According to the calculation method proposed by Xu et al. [14], the height of the water-conducting zone can be determined as the range from above the coal seam top (393 m depth) to the bottom of the main key strata (281 m depth). Therefore, the height of the water-conducting zone is 112 m, and the ratio of the height of the fractured zone to the mining height is 33.73 m.

Table 3. Discrimination of key strata location.

Strata	No.	Depth (m)	Thickness (m)	Name	Key Strata	Hard Rock Formation	Three Zones
Q_21	1	68.00	68.00	Loess			bending zone
K_1zh	2	69.20	1.20	medium-grained sandstone			
	3	120.65	51.45	coarse-grained sandstone		NO. 5	
J_2a	4	127.90	7.25	sandy mudstone			
	5	132.40	4.50	fine-grained sandstone			
	6	136.80	4.40	sandy mudstone			
	7	140.00	3.20	coarse-grained sandstone			
	8	143.90	3.90	sandy mudstone			
	9	149.80	5.90	siltstone			
	10	153.50	3.70	sandy mudstone			
	11	157.10	3.60	siltstone			
	12	162.00	4.90	fine-grained sandstone			
	13	166.30	4.30	medium-grained sandstone			
	14	170.00	3.70	siltstone			
	15	173.50	3.50	sandy mudstone			
	16	175.00	1.50	medium-grained sandstone			
	17	180.40	5.40	sandy mudstone			
	18	182.00	1.60	medium-grained sandstone			
J_2z	19	186.60	4.60	siltstone			
	20	191.30	4.70	medium-grained sandstone			
	21	196.50	5.20	siltstone			
	22	207.50	11.00	coarse-grained sandstone			
	23	220.00	12.50	siltstone			
	24	221.80	1.80	medium-grained sandstone			
	25	243.30	21.50	siltstone		NO. 4	
	26	245.90	2.60	fine-grained sandstone			
	27	248.30	2.40	siltstone			
	28	252.00	3.70	medium-grained sandstone			
	29	259.40	7.40	fine-grained sandstone			
	30	264.10	4.70	siltstone			
	31	265.50	1.40	coarse-grained sandstone			
	32	281.00	15.50	siltstone	main key strata	NO. 3	
	33	282.80	1.80	medium-grained sandstone			fractured zone
	34	295.80	13.00	siltstone			
	35	298.10	2.30	medium-grained sandstone			
	36	305.80	7.70	sandy mudstone			
	37	310.00	4.20	siltstone			
	38	313.10	3.10	medium-grained sandstone			
	39	319.60	6.50	siltstone			
	40	321.30	1.70	sandy mudstone			

Table 3. *Cont.*

Strata	No.	Depth (m)	Thickness (m)	Name	Key Strata	Hard Rock Formation	Three Zones
	41	324.90	3.60	medium-grained sandstone			
	42	329.60	4.70	siltstone			
	43	336.00	6.40	medium-grained sandstone			
	44	341.00	5.00	fine-grained sandstone			
	45	358.28	17.28	coarse-grained sandstone		NO. 2	
J_2y	46	360.13	1.85	siltstone			
	47	361.90	1.77	mudstone			
	48	365.79	3.89	siltstone			
	49	366.67	0.88	mudstone			
	50	370.37	3.70	siltstone			
	51	393.00	22.63	feldspar sandstone	sub-key strata	NO. 1	caving zone
J_2y^3	52	396.32	3.32	coal			

6. Conclusions

Based on the microseismic monitoring technique applied to monitor the strata movement and breakage of the 1311 working face of the No. 3 coal seam in the Weiqiang coalmine, the following conclusions can be drawn:

(1) The MMS is used to monitor the microseismic events of the overlying rock in the mining area of the No. 3 coal seam in real time. The results show that the water-conducting zone gradually develops upward with the advancement of the 1311 working face, and the microseismic events at 110 m from the top of the coal seam are significantly weakening. Combined with the characteristics of the "three zones", the height of the water-conducting zone was determined to be 110 m.

(2) By comprehensively using four methods—the simple hydrological observation method, engineering geological catalogue, geophysical detection, and downhole television—the range of the water-conducting zone is about 105.4–120.4 m.

(3) Based on the key strata theory, the mechanical mechanism of the development of the water-conducting zone was analyzed. The KSPB was used to determine the position of the key strata, and the height of the water-conducting zone was determined to be developed to the bottom of the 15.5 m thick siltstone layer, and the height was 112 m.

(4) The microseismic monitoring is consistent with the theoretical results of on-site drilling detection and key strata theory, and can accurately determine the time, location, and nature of rock mass micro-ruptures. Compared with traditional prediction methods, it has the advantages of safety, high efficiency, accurate prediction, and easy studying of the law of overlying rock movement. Therefore, the promotion of microseismic monitoring technology is conducive to the safe production of mines, and has good social and economic benefits.

Author Contributions: W.G.: data curation, methodology, formal analysis, original draft; Y.L.: supervision, methodology, resources, review and editing; Q.H.: data curation, formal analysis, review and editing. All authors have read and agreed to the published version of the manuscript.

Funding: This research received no external funding.

Institutional Review Board Statement: Not applicable.

Informed Consent Statement: Not applicable.

Data Availability Statement: The data that support the findings of this study are available from the corresponding author (Y.L.), upon reasonable request.

Acknowledgments: The technical staff of the Weiqiang coalmine are appreciated for their field work.

Conflicts of Interest: The authors declare no conflict of interest.

Abbreviations

x_i, y_i, z_i	The spatial coordinate of the *i*th sensor
x, y, z	The spatial coordinate of the test point
t_i	The time when the sensor *i* captures the elastic wave P wave
t	The time when the microseismic event occurs
v_p	Velocity of P wave
$t_{c,i}$	The time that the P wave obtained from the spatial coordinates of the test point is transmitted to the sensor *i*
ρ_0	The density of the source medium
c_0	The seismic wave velocity
R	Distance between the source position and the measuring point
F_c	The seismic wave radiation coefficient
R_c	The seismic wave free surface amplification coefficient
S_c	The seismic wave field correction coefficient
M_W	The moment magnitude
M_0	The seismic moment

References

1. Qian, M.; Xu, J.; Wang, J. Further on the sustainable mining of coal. *J. China Coal Soc.* **2018**, *43*, 1–13.
2. Qian, M.; Xu, J. Behaviors of strata movement in coal mining. *J. China Coal Soc.* **2019**, *44*, 973–984.
3. Qian, M.; Mou, X.; Xu, J. Green mining of coal resources harmonizing with environment. *J. China Coal Soc.* **2007**, *32*, 1–7.
4. Yang, T.; Tang, C.; Tan, Z.; Zhu, W.; Feng, Q. State of the art of inrush models in rock mass failure and developing trend for prediction and forecast of groundwater inrush. *Chin. J. Rock Mech. Eng.* **2007**, *26*, 268–277.
5. Wang, X.; Xu, J.; Zhu, W. Influence of primary key stratum structure stability on evolution of water flowing fracture. *J. China Coal Soc.* **2012**, *37*, 606–612.
6. Qian, M.; Mou, X.; Xu, J. Theoretical study of key stratum in ground control. *J. China Coal Soc.* **1996**, *21*, 2–7.
7. Xu, J.; Qian, M. Study on the influence of key strata movement on subsidence. *J. China Coal Soc.* **2000**, *25*, 122–126.
8. Jiang, F. View point of spatial structures of overlying strata and its application in coalmine. *J. Min. Saf. Eng.* **2006**, *23*, 30–33.
9. Xu, J.; Lian, G.; Zhu, W.; Qian, M.-G. Influence of the key strata in deep mining to mining subsidence. *J. China Coal Soc.* **2007**, *32*, 686–690.
10. Xu, J.; Zhu, W.; Wang, X.; Ti, M.-S. Classification of key strata structure of overlying strata in shallow coal seam. *J. China Coal Soc.* **2009**, *34*, 865–870.
11. Qin, W.; Xu, J. Discussion on Key stratum distinguishing method of overlying strata based on the thin slab theory. *J. China Coal Soc.* **2010**, *35*, 194–197.
12. Huang, B.; Liu, Z.; Xu, J. Effect of little fault in working face on water conducted fissure height. *J. China Coal Soc.* **2009**, *34*, 1316–1321.
13. Hu, X.; Li, W.; Cao, D.-T.; Liu, M.-C. Index of multiple factors and expected height of fully mechanized water flowing fractured zone. *J. China Coal Soc.* **2012**, *37*, 613–620.
14. Xu, J.; Zhu, W.; Wang, X. New method to predict the height of fractured water-conducting zone by location of key strata. *J. China Coal Soc.* **2012**, *37*, 762–769.
15. Cao, Z.; Ju, J.; Xu, J. Distribution model of water-conducted fracture main channel and its flow characteristics. *J. China Coal Soc.* **2019**, *44*, 3719–3728.
16. Wang, X.; Xu, J.; Han, H.K.; Ju, J.F.; Xing, Y.T. Stepped development characteristic of water flowing fracture height with variation of mining thickness. *J. China Coal Soc.* **2019**, *44*, 3740–3749.
17. Zhang, C.; Zhao, Y.; Han, P.; Bai, Q. Coal pillar failure analysis and instability evaluation methods: A short review and prospect. *Eng. Fail. Anal.* **2022**, *138*, 106344. [CrossRef]
18. Chen, J.; Liu, P.; Liu, L.; Zheng, B.; Zhao, H.; Zhang, C.; Zhang, J.; Li, D. Anchorage performance of a modified cable anchor subjected to different joint opening conditions. *Constr. Build. Mater.* **2022**, *336*, 127558. [CrossRef]
19. Miao, X.; Cui, X.; Wang, J.; Xu, J. The height of fractured water-conducting zone in undermined rock strata. *Eng. Geol.* **2011**, *120*, 32–39. [CrossRef]

20. Li, P.; Wang, X.; Cao, W.; Zhang, D.; Qin, D.; Wang, H. Influence of Spatial Relationships between Key Strata on the Height of Mining-Induced Fracture Zone: A Case Study of Thick Coal Seam Mining. *Energies* **2018**, *11*, 102. [CrossRef]
21. Gui, H.; Zhou, Q.; Liao, D.; Kang, Q.; Fan, Z.; Zou, H. Prediction of maximum height of the fractured zone by stressing method for sub-level caving mining. *J. China Coal Soc.* **1997**, *22*, 41–45.
22. Chen, R.; Bai, H.; Feng, M. Determination of the height of water flowing fractured zone in overburden strata above fully-mechanized top-coal caving face. *J. Min. Saf. Eng.* **2006**, *23*, 220–223.
23. Lu, W. Measure and simulation for development height of water conducted crack zone in overburden roof. *J. Xi'an Univ. Sci. Technol.* **2014**, *34*, 309–313.
24. Yi, W.; Weifeng, Y.; Ming, L.; Liu, X. Risk assessment of floor water inrush in coal mines based on secondary fuzzy comprehensive evaluation. *Int. J. Rock Mech. Min. Sci.* **2012**, *52*, 50–55.
25. Zhang, J. Investigations of water inrushes from aquifers under coal seams. *Int. J. Rock Mech. Min. Sci.* **2005**, *42*, 350–360. [CrossRef]
26. Li, T.; Mei, T.; Sun, X.; Lv, Y.; Sheng, J.; Cai, M. A study on a water-inrush incident at Laohutai coalmine. *Int. J. Rock Mech. Min. Sci.* **2013**, *59*, 151–159. [CrossRef]
27. Huang, Z.; Jiang, Z.; Zhu, S.; Qian, Z.; Cao, D. Characterizing the hydraulic conductivity of rock formations between deep coal and aquifers using injection tests. *Int. J. Rock Mech. Min. Sci.* **2014**, *71*, 12–18. [CrossRef]
28. D'Angiò, D.; Lenti, L.; Martino, S. Microseismic monitoring to assess rock mass damaging through a novel damping ratio-based approach. *Int. J. Rock Mech. Min. Sci.* **2021**, *146*, 104883. [CrossRef]
29. Ma, K.; Sun, X.Y.; Tang, C.A.; Yuan, F.Z.; Wang, S.J.; Chen, T. Floor water inrush analysis based on mechanical failure characters and microseismic monitoring. *Tunn. Undergr. Space Technol.* **2021**, *108*, 103698. [CrossRef]
30. Wamriew, D.; Charara, M.; Pissarenko, D. Joint event location and velocity model update in real-time for downhole microseismic monitoring: A deep learning approach. *Comput. Geosci.* **2022**, *158*, 104965. [CrossRef]
31. Zhang, C.; Jin, G.; Liu, C.; Li, S.; Xue, J.; Cheng, R.; Wnag, X.; Zheng, X. Prediction of rockbursts in a typical island working face of a coal mine through microseismic monitoring technology. *Tunn. Undergr. Space Technol.* **2021**, *113*, 103972. [CrossRef]
32. Yi, D.; Xun, L.; Si, G.; Canbulat, I. Seismic source location using the shortest path method based on boundary discretisation scheme for microseismic monitoring in underground mines. *Int. J. Rock Mech. Min. Sci.* **2022**, *149*, 104982.
33. Srinivasan, C.; Arora, S.K.; Benady, S. Precursory monitoring of impending rockbursts in Kolar gold mines from microseismic emissions at deeper levels. *Int. J. Rock Mech. Min. Sci.* **1999**, *36*, 941–948. [CrossRef]
34. Simser, B.P.; Andrieux, P.P.; Peterson, D.A.; Macdonald, T.A.; Alcott, J.M. Advanced monitoring and analysis of microseismic activity as an aid to mining at Brunswick Mines. *Int. J. Rock Mech. Min. Sci.* **1998**, *35*, 389. [CrossRef]
35. Occhiena, C.; Pirulli, M.; Scavia, C. A microseismic-based procedure for the detection of rock slope instabilities. *Int. J. Rock Mech. Min. Sci.* **2014**, *69*, 67–79. [CrossRef]
36. Hassani, H.; Hloušek, F.; Alexandrakis, C.; Buske, S. Migration-based microseismic event location in the Schlema-Alberoda mining area. *Int. J. Rock Mech. Min. Sci.* **2018**, *110*, 161–167. [CrossRef]
37. Mngadi, S.B.; Durrheim, R.J.; Manzi, M.S.D.; Ogasawara, H.; Yabe, Y.; Yilmaz, H.; Wechsler, N.; van Aswegen, G.; Roberts, D.; Ward, A.K.; et al. Integration of underground mapping, petrology, and high-resolution microseismicity analysis to characterise weak geotechnical zones in deep South African gold mines. *Int. J. Rock Mech. Min. Sci.* **2019**, *114*, 79–91. [CrossRef]
38. Fuxing, J. Application of microseismic monitoring technology of strata fracturing in underground coalmine. *Chin. J. Geotech. Eng.* **2002**, *24*, 147–149.
39. Fuxing, J.; Xun, L.; Shuhua, Y. Study on microseismic monitoring for spatial structure of overlying strata and mining pressure field in longwall face. *Chin. J. Geotech. Eng.* **2003**, *25*, 23–25.
40. Cheng, Y.; Jiang, F.; Zhang, X. C-shaped strata spatial structure and stress field in longwall face monitored by microseismic monitoring. *Chin. J. Rock Mech. Eng.* **2007**, *26*, 102–107.
41. Yang, Z.; Yu, R.; Guo, R. Application of microseismic monitoring to deep mines. *Chin. J. Rock Mech. Eng.* **2008**, *27*, 1066–1073.
42. Wang, S.; Mao, D.; Pan, J. Measurement on the whole process of abutment pressure evolution and microseismic activities at the lateral strata of goaf. *J. China Coal Soc.* **2015**, *40*, 2772–2779.
43. Liu, H.; Yang, T.; Zhang, B. Influence factors of overlying coal strata falling and mine pressure behaviours in western coalmines. *J. China Coal Soc.* **2017**, *42*, 460–469.
44. Tang, G.; Zhu, Z.; Han, Y. Relationship between overburden strata movement and strata behavior during dual system seam mining based on microseismic monitoring technology. *J. China Coal Soc.* **2017**, *42*, 212–218.
45. Wang, H.; Jiang, F.; Cheng, Y. Research on microseismic (MS) monitoring of the height of fractured water-conducting zone in overlying rock. *Eng. Geol.* **2006**, *120*, 74–76.
46. Jiang, F.; Ye, G.; Wang, C. Application of high-precision microseismic monitoring technique to water inrush monitoring in coalmine. *Chin. J. Rock Mech. Eng.* **2008**, *27*, 1932–1938.
47. Liu, C.; Li, S.; Xue, J. Identification method of high fractured body for overlying strata in goaf based on microseismic monitoring technology. *J. China Univ. Min. Technol.* **2016**, *45*, 709–716.
48. Han, J.; Zhang, H.; Gao, Z. Failure height of weak overburden by layered fully-mechanized mining in extremely thick coal seam. *J. Min. Saf. Eng.* **2016**, *33*, 226–230.

sustainability

MDPI

Article

Calculation of the Height of the Water-Conducting Fracture Zone Based on the Analysis of Critical Fracturing of Overlying Strata

Yi Tan [1,2,3], Hao Cheng [1,2], Wenyu Lv [3], Weitao Yan [2,*], Wenbing Guo [1,2], Yujiang Zhang [4], Tingye Qi [4], Dawei Yin [5], Sijiang Wei [1,2], Jianji Ren [2] and Yajun Xin [1]

1 School of Energy Science and Engineering, Henan Polytechnic University, Jiaozuo 454000, China; tanyi@hpu.edu.cn (Y.T.); hpuchenghao@163.com (H.C.); guowb@hpu.edu.cn (W.G.); weisj@hpu.edu.cn (S.W.); xinyj2007@163.com (Y.X.)

2 State Collaborative Innovation Center of Coal Work Safety and Clean-Efficiency Utilization, Henan Polytechnic University, Jiaozuo 454000, China; renjianji@hpu.edu.cn

3 State Key Laboratory of Coal Resources in Western China, Xi'an University of Science and Technology, Xi'an 710054, China; lvwenyu@xust.edu.cn

4 College of Mining Engineering, Taiyuan University of Technology, Taiyuan 030024, China; ylczyj@yeah.net (Y.Z.); qty198402@163.com (T.Q.)

5 College of Energy and Mining Engineering, Shandong University of Science and Technology, Qingdao 266000, China; yindawei@sdust.edu.cn

* Correspondence: yanweitao@hpu.edu.cn; Tel.: +86-136-9391-2813

Abstract: Accurate division of the water-conducting fracturing zone (WCFZ) in the mining overburden serves as an important basis to evaluate the stability of coal mining under water bodies. Research on the WCFZ is conducive to controlling surface subsidence and realizing safe coal mining under water. Traditionally, the WCFZ is generally determined by field observation (liquid leakage method, borehole television, etc.) or empirical formula. Although these methods boast high accuracy, they are time-consuming and laborious and have some problems such as weak pertinence and a large value range. In this study, a mechanical model under the critical breakage condition of hard and soft strata was established on the basis of the specific geological and mining information of a mine. Besides, the stability condition for the broken strata to form the "masonry beam" structure and the deflection-based bending deformation formula of hard and soft strata were deduced, and the method of calculating the height of WCFZ based on the analysis of critical fracturing of soft and hard strata (hereafter referred to as the CFSHS-based height calculation method) was proposed. Furthermore, with reference to the results of specific engineering tests, the height of the WCFZ in the working face 15,101 of coal mine XJ was analyzed by means of theoretical analysis, numerical simulation and engineering verification, which verifies the rationality and practicability of the CFSHS-based height calculation method.

Keywords: water-conducting fracture zone; height calculation method; critical fracturing; mechanical model

1. Introduction

Coal mining destroys the equilibrium state of the overburden on the stope, resulting in a series of strata movements such as bending, fracture and caving of overburden. Eventually, the stope is vertically divided into a caving zone, a fracture zone and a bending subsidence zone, collectively called the "three zones". The caving zone and the fracture zone belong to the water-conducting fracture zone (WCFZ) [1–3]. A series of stratum movements induced by coal seam mining can cause certain harm to the environment, which directly leads to surface collapse, soil erosion, and surface building damage [4,5]. With respect to the above problems, Peng et al. put forward the stratum control theory and controlled strata by studying the law of overburden movement induced by coal seam mining [1]. Fan et al. proposed the concept of water conservation coal mining. Afterwards, scholars have carried out a series of theoretical researches accordingly [6]. Accurate division of the WCFZ in the mining overburden serves as an important basis to evaluate the stability of coal mining under water bodies [7]. Therefore, it is of great significance to accurately evaluate the development of the WCFZ in the mining overburden for safe mining.

Researches on the WCFZ in the mining overburden generally adopted the methods of on-site measurement, numerical simulation, physical simulation and micro-seismic monitoring [8–14]. Xu et al. explored the influence of the position of the main key stratum on the development height of the WCFZ; they proposed the conditions to adopt the empirical equation from the perspective of the position of the main key stratum. Moreover, on this basis, they found a method to predict the height of the WCFZ based on the position of the key stratum [15,16]. Peng et al. held that the caving zone and the fracture zone were 2–8 times and 30–50 times as thick as the mining height, respectively [17]. Guo et al. discussed the process of overburden failure in detail and put forward a new method to calculate the height of WCFZ based on the process of overburden failure migration [18]. Fan et al. studied the development characteristics of the WCFZ through on-site measurement and physical simulation; they revealed the dynamic evolution characteristic of "development-development-closure" of the mining-induced fractures and discovered the "mud cap effect" revealed by mining failure under thick mud caprock [6]. Wang et al. proposed a new method and its applicable conditions for the division of the "vertical three zones" of mining overburden based on the key stratum stability and the characteristics of post-fracture movement through physical simulation experiments and the key stratum theory [19,20]. These studies provide good ideas for dividing the WCFZ in the mining overburden under general geological conditions and traditional mining conditions.

With respect to the composite structure of hard and soft strata, this study analyzed the critical condition for the breakage of hard and soft strata, put forward a theoretical judgment method and its applicable conditions for the division of the WCFZ based on strata deflection and bending deformation of the overburden composite structure. Then, this theoretical judgment method, numerical simulation method and engineering analogy method were applied to the division of the WCFZ in the mining overburden, and the rationality of this method was verified by comparing the results with the on-site measurement results.

2. Mining Overburden Structure and Mechanical Analysis

2.1. Determination of the Bearing Strata

In the stope overburden, the thick hard rock plays the bearing role while the thin soft rock plays the loading role. According to the stratum occurrence conditions, the load value and stratum of hard rock are calculated by Equations (1) and (2). From bottom to top, it is named the i-th hard stratum (i = 1, 2, 3...):

$$q_1(x)|_m = \frac{E_1 h_1^3 \sum_{i=1}^{m} \gamma_i h_i}{\sum_{i=1}^{m} E_i h_i^3} \tag{1}$$

$$q_1(x)|_m > q_1(x)|_{m+1} \tag{2}$$

where $q_1(x)|_m$ is the load borne by the first hard stratum for the m-th stratum, MPa; E_i is the elastic modulus of the i-th stratum, MPa; h_i is the thickness of the i-th stratum, m; γ_i is the unit weight of the i-th stratum, MN/m^3.

2.2. Mechanical Analysis on the Hard Rock Strata

2.2.1. Analysis on the Stability of the "Masonry Beam" Structure

With the continuous advancement of the working face, the upper strata exceed the maximum suspending length and breaks, forming neatly arranged rock blocks which squeeze each other to form a masonry beam structure which resembles a beam yet is an arch in essence. Blocks B and C in the "masonry beam" structure are called key blocks whose stability determines the stability of the whole "masonry beam" structure (Figure 1). Based on the instability form of key blocks, the masonry beam structure falls into sliding instability and rotary deformation and instability [21–24].

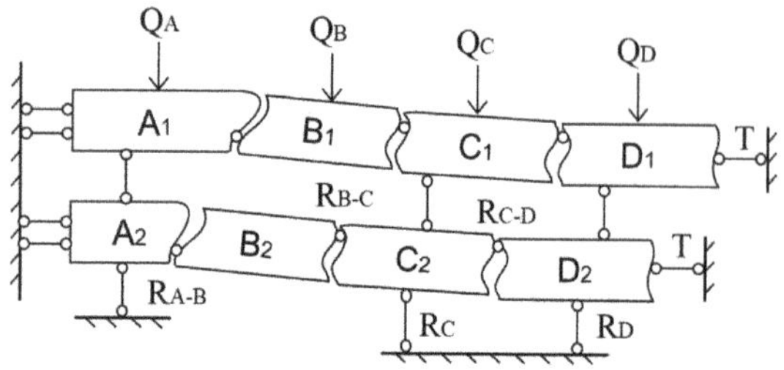

Figure 1. Schematic diagram of the "masonry beam" structure where A_1, B_1, C_1, D_1, A_2, B_2, C_2 and D_2 are articulated rock blocks; R is the jointing force and supporting force between rock blocks; T is the horizontal thrust for the structure; q is the load.

① Sliding instability

The condition where no sliding instability of key blocks in the "masonry beam" occurs is:

$$i \leq \tan\varphi + \frac{3}{4}\sin\theta \tag{3}$$

where $i = h/L_z$; h is the stratum thickness, m; L_z is the length of the periodically broken block, m; φ is the friction angle between broken blocks; θ is the rotary angle between the key blocks, °.

The condition where no sliding instability of this structure occurs is directly related to the block size i and the rotary angle.

② Rotary deformation and instability

The conditions where no rotary deformation and instability of key blocks in the "masonry beam" occurs are:

$$\frac{P}{\left(i - \frac{1}{2}\sin\theta\right)(i - \sin\theta)} \leq 0.15\sigma_c \tag{4}$$

P is the load borne by the key block, MN; σ_c is compressive strength of the bearing stratum, MPa;

$$L_z = h\sqrt{\frac{\sigma_t}{3q}} \tag{5}$$

$$\Delta = M - \sum h(K_p - 1) \tag{6}$$

$$\sin\theta = \frac{\Delta}{L_z} \tag{7}$$

where σ_t is the tensile strength of the stratum, MPa; q is the load borne by the stratum, MPa; Δ is the height of the free space below the stratum, m; M is the thickness of the mining coal seam, m; K_p is the dilatancy coefficient of the stratum.

2.2.2. Analysis on the Bending Subsidence Deformation of the Hard Strata

The maximum stratum subsidence deformation is related to the height of cavity below it, which determines whether the stratum is damaged [25–28]. The stratum without initial breakage is regarded as a fixed-beam model with fixed ends (Figure 2) [29]. The tensile strength served as the stratum failure criterion; then the deflection Equation (10) and the stratum maximum subsidence (11) under the critical failure condition of the stratum are obtained [30,31].

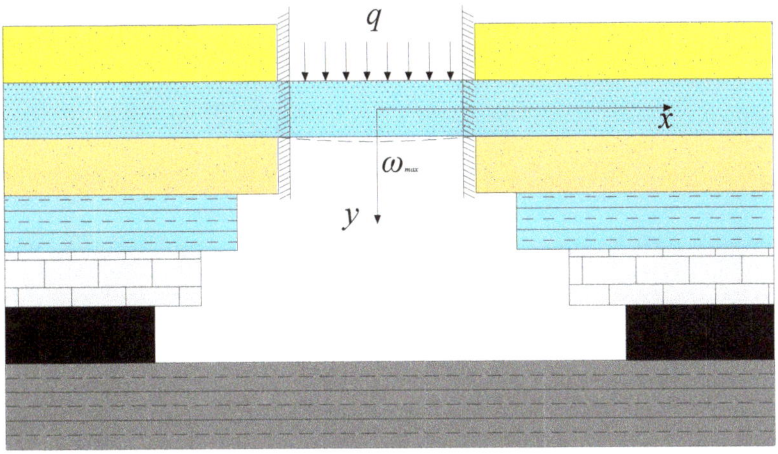

Figure 2. Model of the first breakage of strata.

According to the calculation of the first breakage model, when the normal stress of the stratum reaches the tensile strength, the stratum is cracked.

$$L_c = h\sqrt{\frac{2\sigma_t}{q}} \tag{8}$$

$$I = \frac{bh^3}{12} \tag{9}$$

where L_c is the length of the initial broken block, m; I is the moment of inertia of the rectangular section of the beam, m^4; b is the width of the rectangular section of the beam, m.

When the width of the rectangular section of the stratum is unit width, i.e., $b = 1$:

$$\omega(x) = \frac{q}{24EI}\left(x^2 - \frac{1}{4}L_c^2\right)^2 \tag{10}$$

where ω is the beam deflection, m; E is the elastic modulus of the stratum, MPa.

When $x = 0$, the bending deformation of the fixed beam reaches its maximum value:

$$\omega_{max} = \frac{qL_c^4}{384EI} = \frac{h\sigma_t^2}{8Eq} \tag{11}$$

2.3. Mechanical Analysis on the Soft Strata

When the horizontal tensile strain under load exceeds the tensile strain, the soft stratum breaks:

$$\varepsilon_{max} = \frac{3ql^2}{8Eh^2} > \frac{\sigma_t}{E} \tag{12}$$

where ε_{max} is the horizontal tensile strain.

$$l_S = h\sqrt{\frac{8\sigma_t}{3q}} \tag{13}$$

$$\omega = \frac{5ql^4}{32Eh^3} \tag{14}$$

where l_S is the breaking length of the soft stratum, m; ω is its corresponding deflection of the soft stratum, m.

The breakage of the soft stratum requires the following two conditions: (1) the horizontal tensile strain generated by loading is greater than its tensile strain; (2) the corresponding deflection of the soft stratum is smaller than the height of the free space below.

$$\frac{3ql^2}{8Eh^2} > \frac{\sigma_t}{E} \tag{15}$$

$$\frac{5ql^4}{32Eh^3} < M - \sum h_i\left((K_p)_i - 1\right) \tag{16}$$

where M is the thickness of the mining coal seam, m; h_i is the thickness of the i-th stratum, m; $(K_p)_i$ is the residual dilatancy coefficient of the i-th stratum.

3. CFSHS-Based Height Calculation Method

Based on the above bearing stratum judgment, the specific analysis on the stability of the "masonry beam" structure and the stratum bending and subsidence deformation, the method of calculating the height of WCFZ based on the analysis of critical fracturing of soft and hard strata (hereafter referred to as the CFSHS-based height calculation method) was proposed. The specific calculation procedure is as follows:

Step 1: According to the above bearing stratum judgment method, the load borne by the hard stratum and its stratum position are determined. From bottom to top, it is named the i-th hard stratum (i = 1, 2, 3...). The soft stratum borne by it is named the (i, j) soft stratum from bottom to top (i is number of the corresponding hard stratum; j = 1, 2, 3...).

Step 2: With reference to the thickness of the mining coal seam, the height between the stratum and the coal seam, as well as the dilatancy characteristics of the stratum, the height of the free space Δi below the stratum is calculated according to Equation (12):

$$\Delta_i = M - \sum h_i\left((K_p)_i - 1\right) \tag{17}$$

where Δ_i is the height of the free space below the i-th stratum, m; M is the thickness of the mining coal seam, m; h_i is the thickness of the i-th stratum, m; $(K_p)_i$ is the residual dilatancy coefficient of the i-th stratum.

Step 3: Based on the thickness, tensile strength and bearing load of the hard stratum, the broken block length (Lz)i of the i-th hard stratum is calculated with reference to Equation (5). The rotary angle of the i-th hard stratum θi is calculated with reference to the height of the free space below the stratum and Equation (7).

Step 4: The tensile strength, bearing load and thickness of the hard stratum are substituted into Equation (11) to obtain the maximum bending deformation value of the hard stratum $(\omega_{max})_i$, which is then compared with the height of the free space below it Δ_i:

$$\Delta_i \leq (\omega_{max})_i \tag{18}$$

If Equation (18) is valid, i.e., the height of the free space below the stratum is smaller than its maximum bending deformation value, it is determined that the stratum does not break, only bending deformation occurs, and the stratum is the lower boundary of the bending subsidence zone. Otherwise, if Equation (18) is invalid, i.e., the height of the free space below the stratum is greater than its maximum bending deformation value, the parameters including the rotary angle obtained in Step 3 are substituted into Equations (3) and (4) for judgment. When Equations (3) and (4) are invalid at the same time, the hard stratum belongs to the caving zone. When Equations (3) and (4) are valid at the same time, the hard stratum breaks, and the overburden failure continues to develop upward to the soft stratum which it bears.

Step 5: Determination is made according to the elastic modulus, tensile strength, stratum bearing load and thickness of the (i, j) soft stratum in combination with Equations (15) and (16).

If Equations (15) and (16) are valid at the same time, i.e., when the horizontal tensile strain generated by the loading of the (*i*, *j*) soft stratum is greater than its tensile strain limit and the deflection corresponding to the soft stratum is smaller than the height of the free space below it, the soft stratum breaks and the fracture continues to develop upward. Otherwise, the fracture development stops. Analysis is made stratum by stratum accordingly; then the unfractured stratum is found. The height from the stratum to the coal seam roof is the height of the WCFZ.

The following two conditions are the prerequisites for applying the theoretical judgment method: (1) The working face reaches the overburden failure and full mining in the advancing direction, that is, the height of the WCFZ no longer increases with the advancement of the working face. (2) The lithology, stratum thickness and mechanical parameters of the overburden of the working face are fully grasped.

4. Engineering Case Analysis

4.1. Geological Overview

Coal Mine XJ, which belongs to Changzhi Xinjian Coal Industry Co., Ltd., is located in the west of Nansong Town which is 15 km away in the south of shangdang district in changzhi city. The main coal seam in the mine field is coal seam 15, located at the bottom of Taiyuan Group, 98.86 m above coal seam 3, with an average thickness of 4.20 m. It has a simple structure and its whole area is recoverable. The coal seam roof is limestone; the floor is mudstone and sandy mudstone. The histogram of the overburden is given in Figure 3.

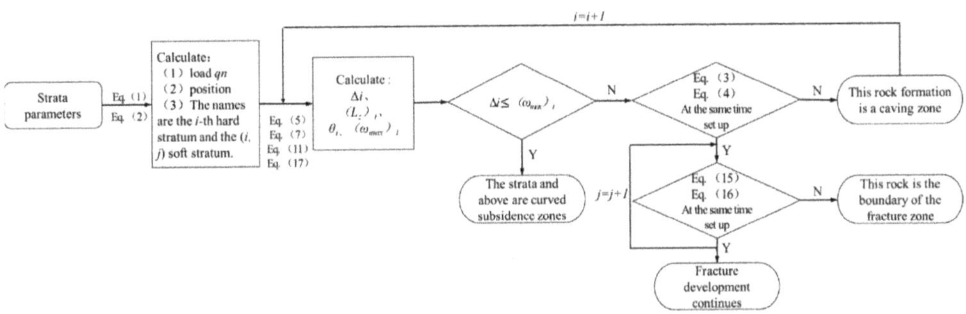

Figure 3. CFSHS-based height calculation method.

This study designs to observe the working face 15,101 which has a strike length of about 1400 m, an inclined length of about 176 m, a horizontal elevation of +849–920 m, a coal seam dip angle of 0–8° and a buried depth of about 276.88 m. The comprehensive mechanized mining is adopted with a mining height of 4.2 m and an advancing speed of 3.6 m/d, and the overburden reaches the fully mining state in the advancing direction.

The normal water inflow of working face 15,101 is 9.19 $m^{3/}$h, and the maximum daily water inflow is 16.9 m^3/h. It is the largest from June to August and the smallest from January to April, which is directly related to rainfall. The hydrogeological type is medium. Most of the surface is exposed bedrock and partially covered by Quaternary loess. The complexity of mine field structure is a simple type. In coal seam 3, there is no large geological structure and large geological events.

According to the on-site test, the dilatancy coefficient K_z of the immediate roof in the goaf of working face 15,101 is 1.15; the residual dilatancy coefficient of the upper rock stratum gradually decreases in the logarithmic form. According to researches [32,33], the relationship between the average residual dilatancy coefficient K_p of the stratum and the mining distance from the coal seam h is:

$$K_p = K_z - 0.017 \ln h (h < 100\ \text{m}) \tag{19}$$

When $h > 100$ m, the upper stratum on the working face is hard stratum, which is broken in large blocks and neatly arranged, with a small residual dilatancy coefficient K_z. Calculation reveals that the average residual dilatancy coefficient of the 16th stratum (mudstone) at a mining distance of 98.86 m from the coal seam is 1.07, so the average residual dilatancy coefficient of the 17th stratum and its upper strata shall be smaller than 1.07; hence both are set as 1.06.

The thickness and mechanical parameters of strata can be obtained from the geological drilling data of the working face (Figure 4).

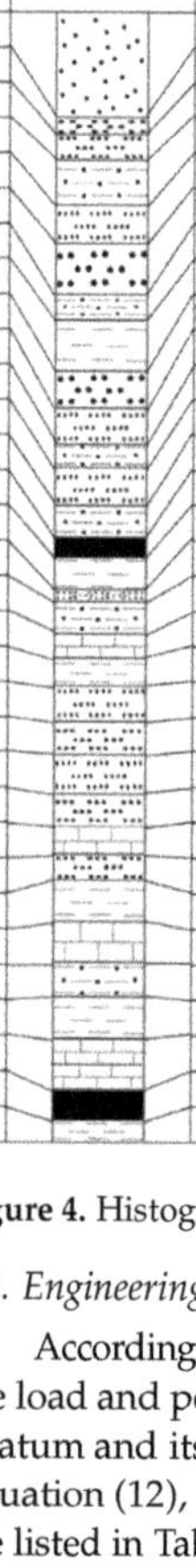

Number	Column	Lithology	Thckness(m)	Buried depth(m)	Bulk density (kg·m⁻³)	Tensile strength(Mpa)	Compressive strength(Mpa)	Elastic modulus(Gpa)	Volume expansion coefficient
30		Quaternary strata	27.85	27.85	1832	0.024	0.192	33.42	1.06
29		Medium sandstone	6.25	34.10	2345	7.01	56.08	19.49	1.06
28		Fine sandstone	8.95	43.05	2319	8.41	67.28	21.13	1.06
27		Sandy mudstone	12.95	56.00	2670	5.43	43.44	18.35	1.06
26		Siltstone	12.20	68.20	2657	4.52	36.16	23.64	1.06
25		Medium sandstone	14.95	83.15	2564	6.14	49.12	29.62	1.06
24		Sandy mudstone	8.80	91.95	2541	5.14	41.12	18.46	1.06
23		Mudstone	26.10	118.05	2432	6.28	50.24	18.66	1.06
22		Medium sandstone	15.85	133.90	2274	6.57	52.56	22.41	1.06
21		Siltstone	13.81	147.71	2560	4.45	35.6	21.54	1.06
20		Sandy mudstone	9.37	157.08	2432	4.78	38.24	19.44	1.06
19		Siltstone	4.35	161.43	2521	4.62	36.96	21.09	1.06
18		Sandy mudstone	12.75	174.18	2630	4.66	37.28	18.56	1.06
17		3# coal	3.84	178.02	1102	1.27	10.16	8.45	1.06
16		Mudstone	12.39	190.41	2345	5.99	47.92	22.07	1.07
15		Siltstone	4.80	195.21	2568	4.61	36.88	23.17	1.07
14		Sandy mudstone	9.48	204.69	2577	4.14	33.12	19.66	1.08
13		Limestone	5.20	209.89	2432	5.17	41.36	27.46	1.08
12		Mudstone	4.13	214.02	2352	6.18	49.44	21.54	1.08
11		Siltstone	7.53	221.55	2671	4.37	34.96	24.88	1.08
10		Fine sandstone	5.77	227.32	2754	7.93	63.44	28.94	1.08
9		Siltstone	8.38	235.70	2530	7.89	56.7	24.77	1.08
8		Fine sandstone	5.98	241.68	2789	8.11	64.88	30.44	1.09
7		Limestone	3.25	244.93	2451	5.63	45.04	25.62	1.09
6		Fine sandstone	2.87	247.80	2741	7.65	61.2	29.51	1.09
5		Mudstone	7.49	255.29	2563	6.21	49.68	20.11	1.09
4		Limestone	4.79	260.08	2511	5.89	47.12	28.17	1.10
3		Sandy mudstone	3.17	263.25	2649	5.84	46.72	18.35	1.10
2		Mudstone	6.13	269.38	2411	5.86	46.88	19.45	1.11
1		Limestone	7.50	276.88	2569	6.33	50.64	27.44	1.15
0		15# coal	4.20	281.08	1330	6.10	7.56	5.04	/
Floor		Mudstone	2.81	283.89	2496	5.66	43.28	18.69	/

Figure 4. Histogram of the overburden.

4.2. Engineering Case of the CFSHS-Based Height Calculation Method

According to the CFSHS-based height calculation method as well as Equations (1) and (2), the load and position of the bearing stratum in the overburden were calculated; the hard stratum and its bearing soft stratum were named from bottom to top. With reference to Equation (12), the free space height Δ_i under the stratum was calculated, and the results are listed in Table 1.

Table 1. Determination of the bearing stratum.

Number of the Stratum	Lithology	Thickness/m	Height from the Coal Seam Roof/m	Δ_i/m
1	Limestone	7.50	7.50	3.08
2	Mudstone	6.13	13.63	2.76
3	Sandy mudstone	3.17	16.80	2.49
4	Limestone	4.79	21.59	2.09
1st hard stratum	Mudstone	7.49	29.08	1.50
(1, 1) soft stratum	Fine sandstone	2.87	31.95	1.29
(1, 2) soft stratum	Limestone	3.25	35.20	1.05
2nd hard stratum	Fine sandstone	5.98	41.18	0.63
3rd hard stratum	Siltstone	8.38	49.56	0.05
(3, 1) soft stratum	Fine sandstone	5.77	55.33	−0.32
(3, 2) soft stratum	Siltstone	7.53	62.86	−0.80
(3, 3) soft stratum	Mudstone	4.13	66.99	−1.06
(3, 4) soft stratum	Limestone	5.20	72.19	−1.38
4th hard stratum	Sandy mudstone	9.48	81.67	−1.94

Based on the mechanical parameters of hard strata and the above Equations (5)–(8) and (11), L_z, θ, L_c and ω_{max} of hard strata can be obtained (Table 2).

Table 2. Calculated parameters of hard strata.

Number of the Stratum	L_z/m	θ/°	L_c/m	ω_{max}/mm
1st hard stratum	24.9	3.5	61.0	9.6
2nd hard stratum	24.5	1.5	60.0	10.0
3rd hard stratum	29.7	0.1	72.6	12.5
4th hard stratum	22.9	/	56.0	4.3

The height of the cavity below the first stratum is 1.50 m, and the bending subsidence deformation before the initial breakage is 9.6 mm. They were substituted in Equation (18) to determine whether that the first hard stratum breaks. Next, the parameters of the first hard stratum were substituted into Equations (3) and (4) to determine whether the first hard stratum can form a stable masonry beam structure. Therefore, the first hard stratum is the boundary stratum under the fracture zone, and the height of the caving zone is the height of the first hard stratum from the coal seam, which is 21.59 m.

The above steps were repeated on the 2nd, 3rd and 4th hard strata. Since the mining coal seam is thick and the 2nd and 3rd hard strata are close to the coal seam, cavities still exist under the 2nd and 3rd hard strata, whose heights are greater than its bending subsidence deformation value. Therefore, the 2nd and 3rd hard strata break, and the overburden failure continues to develop upward. In judging the 4th hard stratum, the cavity under the 4th hard stratum has been filled, which prevents the bending deformation of the 4th hard stratum from occurring, so the 4th hard stratum will not be damaged. Preliminary judgment holds that the height of the overburden failure in the working face 15,101 is between the 3rd and the 4th hard strata, i.e., 49.56–81.67 m.

In summary, there are four soft strata between the 3rd and the 4th hard strata. The four soft strata were judged from bottom to top with reference to Equations (15) and (16). It can be concluded that when the (3, 1) soft stratum is under load, its deflection value is not smaller than the free space height below it, so the soft stratum does not break, and the fracture development stops. Therefore, the (3, 1) soft stratum is the boundary stratum of the bending subsidence zone.

In addition, to explain the relationship between the free space height and the stratum height below the stratum more intuitively, Figure 5 gives the relationship curve between the free space height and the stratum height. The WCFZ develops to the 3rd hard stratum, and the intersection of the curve and the horizontal axis is the maximum development height (49.56 m) of the WCFZ.

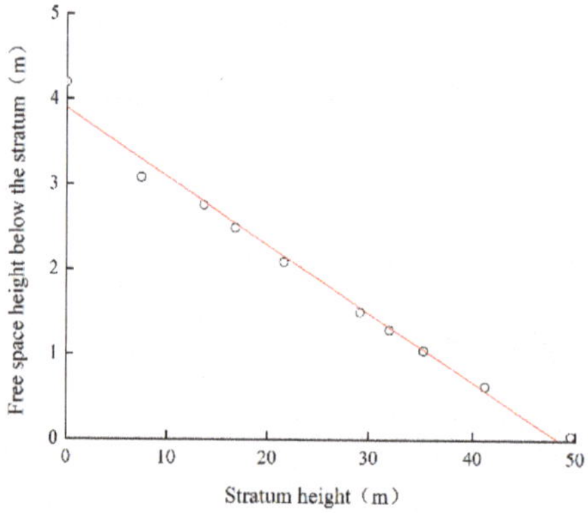

Figure 5. Relationship curve between the free space height below the stratum and the stratum height.

In conclusion, according to the proposed method, the height of the caving zone above the working face 15,101 of coal mine XJ is 21.59 m, and the development height of the WCFZ is 49.56 m. The (3, 1) soft stratum and its above strata belong to the bending subsidence zone (227.32 m thick).

4.3. Engineering Measurement and Analysis

4.3.1. Observation Scheme Design

Abundant aquifers exist in the overburden of most working faces in the mine. Considering this fact, with the working face 15,101 as an engineering case, the height of the WCFZ was detected and analyzed by a borehole television, and the development law of the WCFZ was analyzed by numerical simulation, which provided a basis for safe mining.

According to the theoretical calculation results, the development height of the WCFZ in the working face 15,101 is 49.56 m, so attention should be paid to the observation of fracture development in this range and the development of the WCFZ to the maximum value consumes 1–3 months. Therefore, considering the actual situation of the working face, the observation date is 43 days after recovery, and the observation site is located in the transportation lane of the working face 15,102, which is about 700 m away from the stopping line. The location of the drilling site is shown in Figure 6, and the drilling parameters are listed in Table 3.

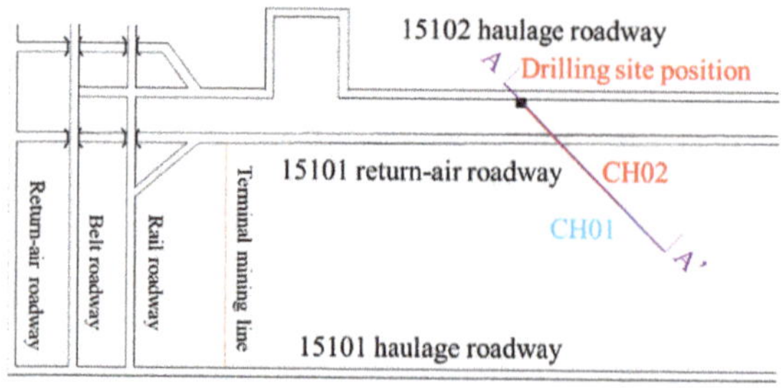

Figure 6. Plan of the drilling site position.

Table 3. Drilling parameters.

Borehole Number	Diameter/mm	Elevation/(°)	Azimuth/(°)	Hole Depth/m	Vertical Depth/m
CH01	Φ89	50	N137	118	90
CH02	Φ89	55	N137	110	90

4.3.2. Observation Method

This observation adopted the borehole television from Tianchen Geophysical Exploration Company (Wuhan, China) to visually measure the development of the WCFZ. The borehole television is composed of a peeping camera, a recorder, an extended drill pipe and an imager. During observation, the camera and the imager are connected through a signal wire; the camera then extends to the depth of the borehole by the extended drill pipe to transmit the images of fracture development in the borehole (Figure 7, an on-site photo).

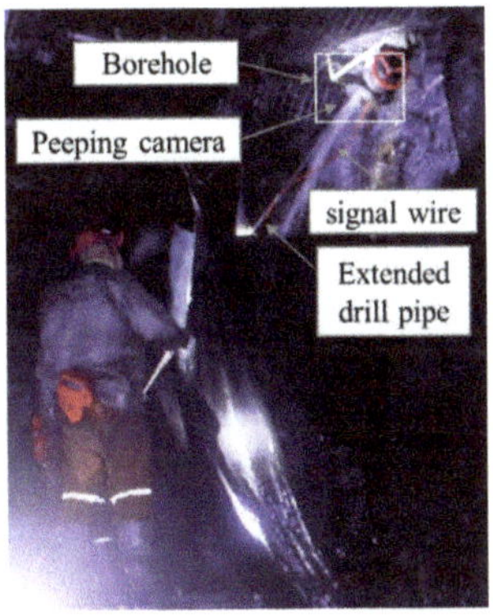

Figure 7. Measured photo of the underground WCFZ.

4.3.3. Analysis on the Observation Results

Figure 8 shows the observation results of mining overburden fractures through the CH02 Borehole. The figures in the right bottom of photos are the distances from the camera to the borehole orifice, and the figures below are the vertical distances between the camera and the coal seam roof.

According to the on-site observation records, Figure 8a,b show the situations of mining overburden within the drilling depth of 12.52–19.78 m, which are 10.25–16.20 m high from coal seam 15 and are above the protective coal pillars between two adjacent working faces. The overburden sees no obvious mining-induced fracture, indicating that the borehole in this range has not entered the fracture zone. Figure 8c,d show the situations of mining overburden with the drilling depth of 30.38–36.30 m, which are 24.88–29.73 m high from coal seam 15. Notable mining-induced fractures begin to emerge in the overburden, and become increasingly obvious as the drilling is approaching the goaf center. The stratum in Figure 8d is seriously damaged; Figure 8e,f illustrate the situations of mining overburden within the drilling depth of 41.33–48.07 m, which are 33.85–39.37 m high from the coal seam 15. Obvious mining-induced fractures can be observed in the strata in this range, suggesting that the strata are significantly affected by mining. However, compared with the overburden of 36.30 borehole depth, the overburden of the drilling depth of 41.33–48.07 m are less damaged and present a certain regularity, and obvious separation occurs at the junction of some adjacent strata. Figure 8g,h show the situations of mining overburden within the drilling depth of 55.37–61.28 m, which are 45.35–50.19 m high from coal seam 15. According to the detection results, as the height from the coal seam increases, fractures become less developed. The distance and the development degree of fractures are negatively correlated, that is, in this range, the influence of mining on the strata weakens as the height from the coal seam increases. When the borehole depth reaches 69.08 m (Figure 8i), that is, when the height from the roof of coal seam 15 is greater than 56.58 m, no obvious mining-induced fracture is found in the rock stratum. With the continuous advancement of the probe, when the borehole depth reaches 74.24 m (Figure 8j), i.e., when the height from the roof of coal seam 15 is beyond 60.08 m, no obvious mining-induced fracture is detected. In summary, the mining overburden of the working face 15,101 develops to 69.08 m borehole depth, and the distance from the roof of coal seam 15 is 56.58 m, that is, the height of the WCFZ detected by the CH02 Borehole is 56.58 m.

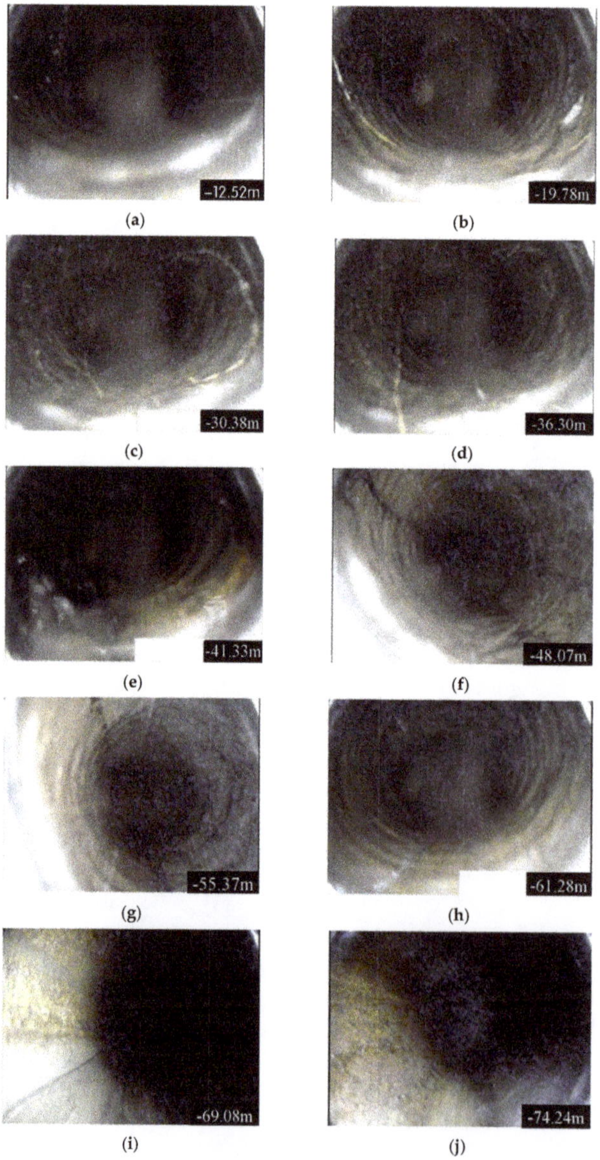

Figure 8. Photos of the overburden fractures after mining. (**a**) 10.25 m (**b**) 16.20 m (**c**) 24.88 m (**d**) 29.73 m (**e**) 33.85 m (**f**) 39.37 m (**g**) 45.35 m (**h**) 50.19 m (**i**) 56.58 m (**j**) 60.08 m.

Based on the measured results, the height of the caving zone in the goaf of the 15,101 working face is 29.73 m; the height of the WCFZ is 56.58 m; and the thickness of the bending subsidence zone is 220.30 m.

Then the calculation results of the above theoretical judgment method of the "vertical three zones" of mining overburden were contrastively analyzed with the on-site measured results. It can be found that the calculation results of the theoretical judgment method (the height of the caving zone 21.59 m, the development height of the WCFZ 49.56 m, and the thickness of the bending subsidence zone 227.32 m) are similar to the measured results, which verifies the rationality and practicability of the theoretical judgment method.

4.4. *Numerical Simulation Analysis*

To verify the rationality of the above analytical results, the geological mining conditions of the fully mechanized mining face 15,101 in coal mine XJ served as the background to simulate the caving and fracture characteristics of the mining overburden, and the migration process of the overburden failure was simulated by 3DEC discrete element simulation software.

4.4.1. Model Design

Based on the actual mining data of the fully mechanized face 15,101 in coal mine XJ, a 3DEC numerical calculation model was established. The advancing length of the working face is 420 m, which is extended outward by 180 m along the working face to prevent the influence of the model boundary on the calculation results. The model size is 600 m × 1 m × 300 m (length × width × height) (Figure 9). The simulated working face was excavated by 30 m each time, a total of 14 excavations.

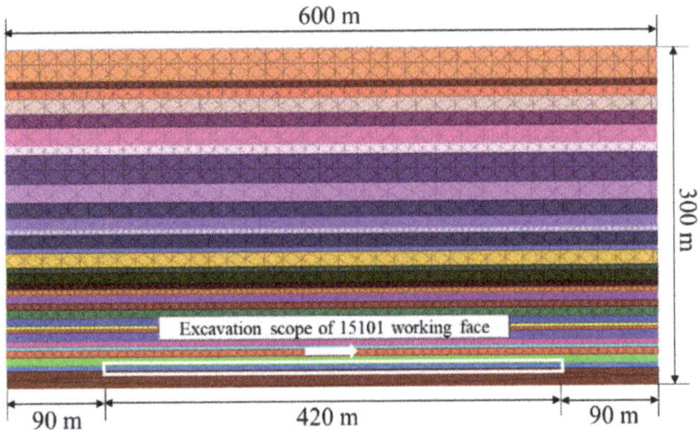

Figure 9. 3DEC numerical calculation model.

In the numerical model, the Mohr-Coulomb model was selected as the block constitutive model, and the Coulomb slip model was selected as the joint constitutive model. The left and right boundaries of the model were fixed in x direction displacement, the front and rear boundaries fixed in y direction displacement and the bottom boundary fixed in z direction displacement. According to the experimental data and model calibration, the physical and mechanical parameters of the overburden were determined (Table 4).

Table 4. Physical and mechanical parameters of the strata.

Lithology	Density /kg·m^{-3}	Bulk Modulus /GPa	Shear Modulus /MPa	Cohesion /MPa	Internal Friction Angle /°	Tensile Strength /MPa	Normal Stiffness /GN*m^{-1}	Tangential Stiffness /GN*m^{-1}
Loess	1832	0.28	0.093	0.85	25	0.35	0.25	0.25
Mudstone	2540	1.23	1.06	2.16	24	1.26	0.18	0.18
Sandy mudstone	2580	3.72	1.62	3.53	25	2.06	0.59	0.59
Medium grained sandstone	2630	2.31	1.26	5.92	26	2.86	0.50	0.50
Siltstone	2580	2.98	1.88	4.28	23	2.55	1.09	1.09
Fine sandstone	2610	2.64	1.68	4.36	22	2.32	0.49	0.49
Coal	1400	1.16	0.73	1.54	22	1.03	0.21	0.21
Limestone	2800	4.45	8.91	4.20	39	3.10	2.60	2.60

4.4.2. Simulation Results and Analysis

With the advancement of the working face, the caving form of the overburden is shown in Figure 10. When the working face advances to 90 m (Figure 11a), the overburden begins to cave in a large area, with 4.39 m maximum vertical displacement; the broken limestone forms a masonry beam structure. At this time, the overburden above the coal seam is separated, with 37.2 m separation height. When the working face advances to 120 m (Figure 11b), the overburden further migrates in a large area, and the separation develops upward to 53.6 m. When the working face advances to 150 m (Figure 11c), the lower separation layer is closed. When the working face is excavated to 420 m (Figure 11d), the separation layer does not develop upward further; according to the closure state of the vertical fractures in Figure 12 (the red and green represent open and closed fractures respectively), the following conclusions can be obtained: (1) when the working face 15,101 reaches the full mining state in the advancing direction, the overburden migrates and the fractures are closed again; in the goaf middle, the number of closed fractures decreases from the middle to the edge, and the fractures display a "V" shape on the edges of both goaf sides. (2) The vertical interpenetrated fractures of the overburden finally develop to the 53.6 m fine sandstone above the coal seam.

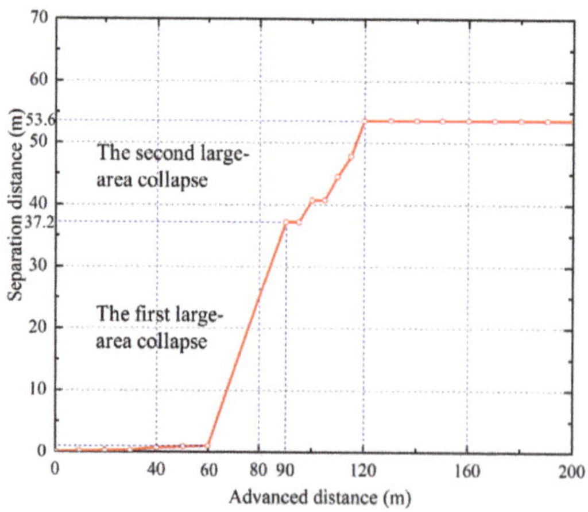

Figure 10. Overburden failure characteristics with different excavation steps.

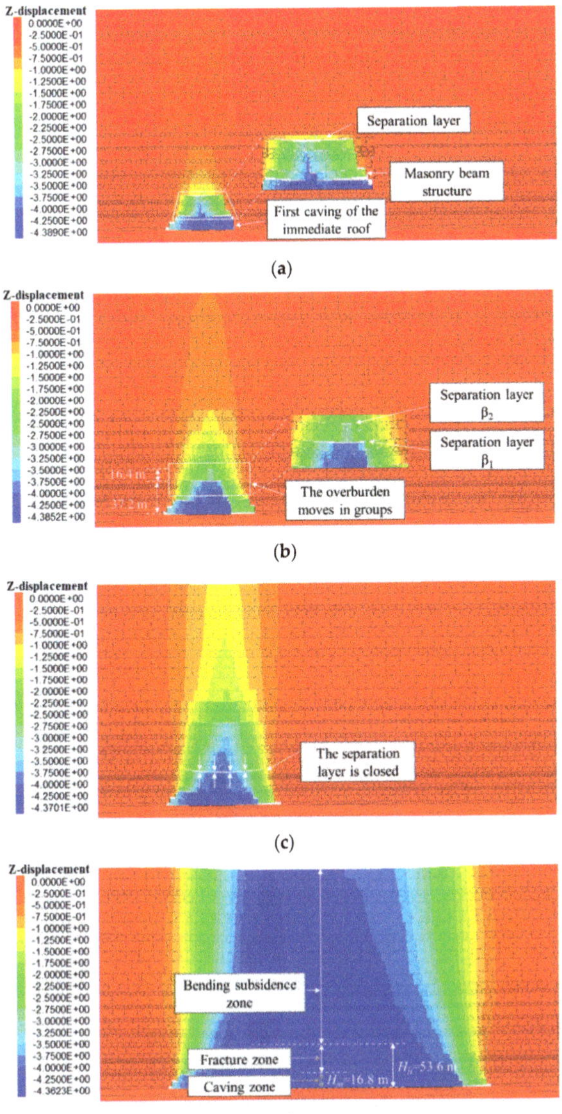

Figure 11. Nephograms of vertical displacement of the overburden during the advancement of working face. (**a**) Excavation to 90 m; (**b**) Excavation to 120 m; (**c**) Excavation to 150 m; (**d**) Excavation to 420 m.

The above numerical simulation experiment leads to the following conclusions: (1) Overburden failure migrates upward in the form of separation layer. When the bending deformation of the stratum below the separation reaches its deformation limit, it interpenetrates the vertical fractures of the overburden, and finally forms a complete overburden water-conducting channel. (2) The upward development of the overburden failure is affected by the hard stratum, so the overburden moves in groups. (3) The numerical simulation reveals that the water-conducting fracture zone is 53.6 m high.

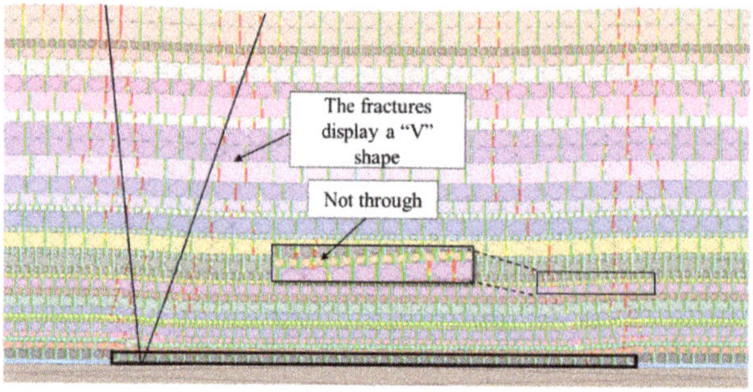

Figure 12. Closure state of the vertical fractures.

4.5. Scope of Application

This model is a theoretical model, and the application of this model is limited. When the movement of strata above gob is only affected by self-weight and overburden pressure, this model is is applied well. But when a place has a wide range of geological structure (e.g., fault, folds) or geological events (e.g., earthquake), this model will be not suitable.

5. Conclusions

(1) The structure of mining overburden failure in the coal mine was analyzed, and the mechanical model of overburden structure conduction was established. The results suggest that the height of overburden failure is controlled by the lithology of soft strata between adjacent hard strata and the height of free space below.

(2) The bearing stratum breakage distance, maximum deflection and free space of overburden were studied with reference to the overburden histogram data. On this basis, the CFSHS-based height calculation method was proposed, and its applicable conditions were given.

(3) Through the engineering field test, the height of overburden failure in the working face was investigated by means of field measurement and numerical simulation. Moreover, the results obtained through the CFSHS-based height calculation method, numerical simulation and field measurement were comparatively analyzed. The comparison verified accuracy of the CFSHS-based height calculation method.

Author Contributions: Conceptualization, Y.T. and H.C.; methodology, W.Y.; software, W.L. and W.Y.; validation, S.W. and Y.X.; formal analysis, W.G.; investigation, D.Y. and W.Y.; resources, T.Q.; data curation, Y.T. and J.R.; writing—original draft preparation, Y.T.; writing—review and editing, W.Y.; visualization, W.G.; supervision, W.Y.; project administration, Y.Z.; funding acquisition, Y.T. All authors have read and agreed to the published version of the manuscript.

Funding: This work was financially supported by National Natural Science Foundation of China (U21A20108, 52174108, 51974105 and 52104127), Support Plan for Science & Technology Innovation Talents in Universities of Henan Province (21HASTIT024), State Key Laboratory of Coal Resources in Western China (SKLCRKF1912), Scientific and technological innovation research team of Henan Polytechnic University (T2021-5), Henan Excellent Youth Science Foundation (222300420045), Henan Science and Technology Research Project (212102310399, 222102320058), Key Scientific Research Projects of Colleges and Universities in Henan Province (21A440003).

Institutional Review Board Statement: Not applicable.

Informed Consent Statement: Not applicable.

Data Availability Statement: Not applicable.

Acknowledgments: The authors gratefully acknowledge the editor and reviewers for their helpful comments.

Conflicts of Interest: The authors declare no conflict of interest.

References

1. Peng, S.S. *Coal Mine Ground Control*; Wiley: New York, NY, USA, 1978.
2. Kratzsch, H. *Mining Subsidence Engineering*; Springer: Berlin/Heidelberg, Germany, 1983.
3. Gray, R.E. Mining subsidence- past, present, future. *Int. J. Min. Geol. Eng.* **1990**, *8*, 400–408. [CrossRef]
4. Kang, Y.; Jin, R. Actuality and developing trend of long wall top coal caving mining under water. *Coal Min. Technol.* **2003**, *1*, 15–18.
5. Yan, W.; Chen, J.; Yang, W.; Liu, X.; Wang, W.; Zhang, W. On-Site Measurement on Surface Disturbance Law of Repeated Mining with High Relief Terrain. *Sustainability* **2022**, *14*, 3166. [CrossRef]
6. Fan, G.; Zhang, D.; Ma, L. Overburden movement and fracture distribution induced by longwall mining of the shallow coal seam in the Shendong coalfield. *J. China Univ. Min. Technol.* **2011**, *40*, 196–201.
7. State Administration of Safety; State Administration of Coal Mine Safety; State Energy Administration; State Railway Administration. *Buildings Water, Railways and Main Well Lane of Coal Pillar and Mining Regulations*; China Coal Industry Publishing House: Beijing, China, 2017.
8. Sun, Y.; Xu, Z.; Dong, Q. Monitoring and simulation research on development of water flowing fractures for coal mining under Xiaolangdi reservoir. *Chin. J. Rock Mech. Eng.* **2011**, *31*, 3444–3451.
9. Cao, D.; Li, W. Estimation method for height of fratured zone with in coal mining area. *Chin. J. Geol. Hazard Control* **2014**, *1*, 63–69.
10. Vervoor, A. Various phases in surface movements linked to deep coal longwall mining: From start-up till the period after closure. *Int. J. Coal Sci. Technol.* **2021**, *8*, 412–426. [CrossRef]
11. Guo, C.; Yang, Z.; Li, S.; Lou, J. Predicting the Water-Conducting Fracture Zone (WCFZ) Height Using an MPGA-SVR Approach. *Sustainability* **2020**, *12*, 1809. [CrossRef]
12. Wanner, C.; Bucher, K.; von Strandmann, P.A.P.; Waber, H.N.; Pettke, T. On the use of Li isotopes as a proxy for water–rock interaction in fractured crystalline rocks: A case study from the Gotthard rail base tunnel. *Geochim. Cosmochim. Acta* **2017**, *198*, 396–418. [CrossRef]
13. Khanzode, V.V.; Maiti, J.; Ray, P.K. A methodology for evaluation and monitoring of recurring hazards in underground coal mining. *Saf. Sci.* **2011**, *49*, 1172–1179. [CrossRef]
14. Rezaei, M.; Hossaini, M.F.; Majdi, A. A time-independent energy model to determine the height of destressed zone above the mined panel in longwall coal mining. *Tunn. Undergr. Space Technol.* **2015**, *47*, 81–92. [CrossRef]
15. Xu, Y.; Li, J.; Liu, S.; Zhou, L. Calculation formula of "two-zone" height overlying strata and its adaptability analysis. *Coal Min. Technol.* **2011**, *2*, 4–7.
16. Xu, Y.; Liu, S. Study on method to set safety coal and rock pillar for full mechanized top coal caving mining under water body. *Coal Sci. Technol.* **2011**, *11*, 1–4.
17. Qian, M.; Miao, X. Theoretical analysis of the structural form and stability of overlying strata in Long wang mining. *Chin. J. Rock Mech. Eng.* **1995**, *2*, 97–106.
18. Guo, H.; Sun, Z.; Ji, M.; Wu, Y.; Nian, L. An Investigation on the Impact of Unloading Rate on Coal Mechanical Properties and Energy Evolution Law. *Int. J. Environ. Res. Public Health* **2022**, *19*, 4546. [CrossRef] [PubMed]
19. Wang, Y.; Wang, X.; Zhang, J.; Chen, X.; Zhu, W.; Zhang, Y. Roof Subsidence and Movement Law of Composite Strata Mining: Insights from Physical and Numerical Modeling. *Minerals* **2022**, *12*, 3. [CrossRef]
20. Zhai, W.; Li, W.; Huang, Y.; Ouyang, S.; Ma, K.; Li, J.; Gao, H.; Zhang, P. A Case Study of the Water Abundance Evaluation of Roof Aquifer Based on the Development Height of Water-Conducting Fracture Zone. *Energies* **2020**, *13*, 4095. [CrossRef]
21. Sun, Q.; Mu, Y.; Yang, X. Study on "two zones" height of overlying of fully-mechanized technology with high mining height at Hongliu Coal Mine. *J. China Coal Soc.* **2013**, *38*, 283–286.
22. Xun, Y. Research on movement and evolution law of breaking of overlying strata in shallow coal seam with a thin bedrock. *Rock Soil Mech.* **2008**, *2*, 512–516.
23. Yan, W.; Chen, J.; Tan, Y.; Zhang, W.; Cai, L. Theoretical Analysis of Mining Induced Overburden Subsidence Boundary with the Horizontal Coal Seam Mining. *Adv. Civ. Eng.* **2021**, *2021*, 6657738. [CrossRef]
24. Sun, Y.; Zuo, J.; Karakus, M.; Liu, L.; Zhou, H.; Yu, M. A New Theoretical Method to Predict Strata Movement and Surface Subsidence due to Inclined Coal Seam Mining. *Rock Mech. Rock Eng.* **2021**, *54*, 2723–2740. [CrossRef]
25. Thongprapha, T.; Fuenkajorn, K.; Daemen, J.K. Study of surface subsidence above an underground opening using a trap door apparatus. *Tunn. Undergr. Space Technol.* **2015**, *46*, 94–103. [CrossRef]
26. Vervoort, A. Surface movement above an underground coal longwall mine after closure. *Nat. Hazards Earth Syst. Sci.* **2016**, *9*, 2107–2121. [CrossRef]
27. Sasaoka, T.; Takamoto, H.; Shimada, H.; Oya, J.; Hamanaka, A.; Matsui, K. Surface subsidence due to underground mining operation under weak geological condition in Indonesia. *J. Rock Mech. Geotech. Eng.* **2015**, *3*, 337–344. [CrossRef]
28. Mokhov, A.V. A rock mass permeability model within the subsidence zone in workings of coal fields. *Dokl. Earth Sci.* **2017**, *473*, 390–393. [CrossRef]
29. Yang, W.; Guo, W.; Zhao, G.; Ma, Z.; Yang, D. Theoretical judgement method of overburden "Three-zone" based on rock strata deflection deformation and its engineering application. *Coal Sci. Technol.* **2021**, 1–9. Available online: http://kns.cnki.net/kcms/detail/11.2402.TD.20211206.2350.003.html (accessed on 17 March 2022).

30. Hou, E.; Wen, Q.; Ye, Z.; Chen, W.; Wei, J. Height prediction of water-flowing fracture zone with a genetic-algorithm support-vector-machine method. *Int. J. Coal Sci. Technol.* **2020**, *7*, 740–751. [CrossRef]
31. Cao, Z.; Ju, J.; Xu, J. Distribution model of water-conducted fracture main channel and its flow characteristics. *J. China Coal Soc.* **2019**, *12*, 3719–3728.
32. Gao, B.; Wang, X.; Zhu, M.; Zhou, J. Dynamic development characteristics of two zones of overburden strata under conditions of compound roof highly gassy and thick coal seam in full-mechanized top cole caving faces. *Chin. J. Rock Mech. Eng.* **2013**, *31*, 3444–3451.
33. Cao, J.; Ma, Q.; Wang, Y. Study of Wide Strip Mining Based on Spatial Structure Principle of Overlying Strata. *Coal Sci. Technol.* **2008**, *1*, 68–72.

MDPI AG
Grosspeteranlage 5
4052 Basel
Switzerland
Tel.: +41 61 683 77 34

Sustainability Editorial Office
E-mail: sustainability@mdpi.com
www.mdpi.com/journal/sustainability

www.ingramcontent.com/pod-product-compliance
Lightning Source LLC
Chambersburg PA
CBHW061452210926
51850CB00008B/1690